Witold Pedrycz and Shyi-Ming Chen (Eds.)

Granular Computing and Intelligent Systems

Intelligent Systems Reference Library, Volume 13

Editors-in-Chief

Prof. Janusz Kacprzyk
Systems Research Institute
Polish Academy of Sciences
ul. Newelska 6
01-447 Warsaw
Poland
E-mail: kacprzyk@ibspan.waw.pl

Prof. Lakhmi C. Jain
University of South Australia
Adelaide
Mawson Lakes Campus
South Australia 5095
Australia
E-mail: Lakhmi.jain@unisa.edu.au

Further volumes of this series can be found on our homepage: springer.com

Vol. 1. Christine L. Mumford and Lakhmi C. Jain (Eds.)
Computational Intelligence: Collaboration, Fusion and Emergence, 2009
ISBN 978-3-642-01798-8

Vol. 2. Yuehui Chen and Ajith Abraham
Tree-Structure Based Hybrid Computational Intelligence, 2009
ISBN 978-3-642-04738-1

Vol. 3. Anthony Finn and Steve Scheding
Developments and Challenges for Autonomous Unmanned Vehicles, 2010
ISBN 978-3-642-10703-0

Vol. 4. Lakhmi C. Jain and Chee Peng Lim (Eds.)
Handbook on Decision Making: Techniques and Applications, 2010
ISBN 978-3-642-13638-2

Vol. 5. George A. Anastassiou
Intelligent Mathematics: Computational Analysis, 2010
ISBN 978-3-642-17097-3

Vol. 6. Ludmila Dymowa
Soft Computing in Economics and Finance, 2011
ISBN 978-3-642-17718-7

Vol. 7. Gerasimos G. Rigatos
Modelling and Control for Intelligent Industrial Systems, 2011
ISBN 978-3-642-17874-0

Vol. 8. Edward H.Y. Lim, James N.K. Liu, and Raymond S.T. Lee
Knowledge Seeker – Ontology Modelling for Information Search and Management, 2011
ISBN 978-3-642-17915-0

Vol. 9. Menahem Friedman and Abraham Kandel
Calculus Light, 2011
ISBN 978-3-642-17847-4

Vol. 10. Andreas Tolk and Lakhmi C. Jain
Intelligence-Based Systems Engineering, 2011
ISBN 978-3-642-17930-3

Vol. 11. Samuli Niiranen and Andre Ribeiro (Eds.)
Information Processing and Biological Systems, 2011
ISBN 978-3-642-19620-1

Vol. 12. Florin Gorunescu
Data Mining, 2011
ISBN 978-3-642-19720-8

Vol. 13. Witold Pedrycz and Shyi-Ming Chen (Eds.)
Granular Computing and Intelligent Systems, 2011
ISBN 978-3-642-19819-9

Witold Pedrycz and Shyi-Ming Chen (Eds.)

Granular Computing and Intelligent Systems

Design with Information Granules of Higher Order and Higher Type

Springer

Prof. Witold Pedrycz
University of Alberta
Department of Electrical & Computer Engineering Edmonton, AB T6G 2G6,
Canada
E-mail: pedrycz@ee.ualberta.ca

Prof. Shyi-Ming Chen
Department of Computer Science and Information Engineering, National Taiwan
University of Science and Technology,
43, Section 4, Keelung Road
Taipei 106, Taiwan
E-mail: smchen@mail.ntust.edu.tw

ISBN 978-3-642-26800-7 ISBN 978-3-642-19820-5 (eBook)

DOI 10.1007/978-3-642-19820-5

Intelligent Systems Reference Library ISSN 1868-4394

Softcover reprint of the hardcover 1st edition 2011

Typeset & Cover Design: Scientific Publishing Services Pvt. Ltd., Chennai, India.

Printed on acid-free paper

9 8 7 6 5 4 3 2 1

springer.com

Preface

Information granulation has emerged as one of the fundamental concepts of information processing giving rise to the discipline of Granular Computing. The concept itself permeates through a large variety of information systems. The underlying idea is intuitive and appeals to our commonsense reasoning. We perceive the world by structuring our knowledge, perceptions, and acquired evidence in terms of information granules-entities, which are abstractions of the complex word and phenomena. By being abstract constructs, information granules and their ensuing processing done under the umbrella of Granular Computing, provides a conceptual and algorithmic framework to deal with an array of decision-making, control, and prediction problems. Granular Computing supports human-centric processing, which becomes an inherent feature of intelligent systems. The required level of detail becomes conveniently controlled by making suitable adjustments to the size of information granules and their distribution in the space in which the problem at hand is being positioned and handled. In this sense, an inherent flexibility, which comes hand in hand with Granular Computing can be effectively exploited in various ways. It is not surprising at all that there have been a number of formal frameworks in which information granules are described and processed to meet the specification of the problem. Interestingly, we are witnessing here a growing diversity of formalisms stemming from the realm of fuzzy sets, interval analysis, probability, rough sets, and shadowed sets.

Three general tendencies encountered in Granular Computing can be identified: (a) a design of information granules of *higher order*, (b) a development of information granules of *higher type*, and (c) a formation of *hybrid* information granules bringing together various formalisms of information granulation.

The essence of these three directions is quite distinct and there are quite different agendas behind each of them. The higher order information granularity is concerned with an effective formation of information granules over the space already constructed by information granules of lower order and a design of models utilizing such information granules. This construct is directly tied with the concept of hierarchy of systems where we form successive layers characterized by the increasing levels of abstraction. This idea of layered, hierarchical realization of models of complex systems has gained a significant level of visibility in fuzzy modeling with the well-established concept of hierarchical fuzzy models. In particular, there has been an interesting realization of this concept in rule-based systems where by forming successive nested layers of the model, especially the associated rule bases, one strives to achieve a sound tradeoff between accuracy and a level of detail captured by the model and its level of interpretability.

Higher type information granules arise in situations where the information granules themselves cannot be fully characterized in a purely numerical fashion. Instead in their realization (definition) it becomes necessary or at least convenient to confine to their realization in the form of other types of information granules. For instance, we could envision membership grades of fuzzy sets being characterized (described) by fuzzy sets defined in the unit interval (in which case we refer to such constructs as type-2 fuzzy sets) or intervals (interval-valued fuzzy sets), probability density functions expressed in [0,1] (where we allude to probabilistic fuzzy sets).

The hybridization of information granules brings forward concepts that harmoniously capture various facets of information granules acknowledging that various formalisms are orthogonal. For instance, we can envision an interesting hybrid structure of information granules in which probability is of non-numeric nature but emerges in the form of interval probabilities or linguistically quantified probabilities.

The ultimate objective of this volume is to offer a comprehensive, fully updated and systematic exposure to the subject matter, carefully addressing concepts, methodology, algorithms, and case studies/applications. All the three general directions of Granular Computing are authoritatively covered in the chapters of the volume. The individual contributions of the volume are reflective of the diversity of the conceptual developments, underlying methodologies, algorithms and a wealth of applications.

We would like to take this opportunity to express our sincere thanks to the authors for reporting on their innovative research and sharing insights into the area. The reviewers deserve our thanks for their constructive input. We highly appreciate a continuous support and encouragement from the Editor-in-Chief, Professor Janusz Kacprzyk whose leadership and vision makes this book series a unique vehicle to disseminate the most significant accomplishments in Computational Intelligence and Granular Computing.

We hope that the readers will find this publication of genuine interest and help in research, educational, and practical endeavors.

Contents

From Interval (Set) and Probabilistic Granules to Set-and-Probabilistic Granules of Higher Order

Vladik Kreinovich

Abstract. In this chapter, we provide a natural motivation for granules of higher order, and we show that these granules provide a unified description of different uncertainty formalisms such as random sets, Dempster-Shafer approach, fuzzy sets, imprecise probabilities, and Bayesian statistics. We also prove that for fuzzy uncertainty, granules of second order are sufficient.

Keywords: Granules of Higher Order, Random Sets, Dempster-Shafer Approach, Fuzzy Sets, Imprecise Probabilities, Bayesian Statistics.

1 Introduction

Techniques for representing uncertainty: current situation. Many techniques have been proposed (and successfully used) to describe and process uncertainty:

- sets, in particular, intervals (Rabinovich 2005);
- probability distributions (Jaynes 2003), (Rabinovich 2005);
- imprecise probabilities (Walley 1989), (Ferson 2002), (Ferson et al. 2003);
- Dempster-Shafer approach;
- fuzzy sets;
- interval-valued (and, more generally, type-2) fuzzy sets (Mendel 2001), (Mendel and Wu 2010);
- Bayesian statistical techniques (Jaynes 2003), (Gelman et al. 2004);
- and many other different approaches.

Natural questions. From this variety, come natural questions:

- Which of the techniques should we apply in different practical situations?
- Are there yet-to-be-discovered better techniques for processing uncertainty?

Vladik Kreinovich
University of Texas at El Paso, El Paso, TX 79968, USA
e-mail: vladik@utep.edu

W. Pedrycz and S.-M. Chen (Eds.): Granular Computing and Intell. Sys., ISRL 13, pp. 1–16.
springerlink.com

- How can we combine uncertainty described by different formalisms?

How to answer these questions: a need for classification. To answer all these questions, it is necessary to come up with a reasonable classification of different uncertainty techniques. Such classification would enable us:

- to meaningfully compare different techniques, for the purpose of deciding which is better in different situations,
- to check whether all possible techniques in this classification have been invented, or new (possible better) techniques are still possible, and
- to meaningfully combine these techniques – since they appear as particular cases of a general approach to uncertainty.

What we do in this paper. In this paper, we start with the question of where our data comes from, what are the corresponding uncertainties. Based on the corresponding from-scratch analysis, we explain how different uncertainty techniques appear, and thus, come up with a natural classification.

In the process of this classification, we provide a natural motivation for granules of higher order. We also show that these granules provide a unified description of different uncertainty formalisms.

2 From an Ideal Exact Description to Interval (Set) and Probabilistic Granules

Idealized objects. To describe the physical world, we identify *objects*: elementary particles, atoms, molecules, solid bodied, stars, etc.

An ideal object should be well-defined. For example, in a geographic description, when we define forests, lakes, rivers, etc., we should be able to determine the exact boundary between a lake and a river that flows into this lake, the exact boundary between a forest and a nearby grassy area, etc.

An ideal object should also be reasonably stable with time. From this viewpoint,

- a river is a reasonable geographic object, because its path does not change much for a long time, while
- a puddle – which is often easily visible too – is not a reasonable geographic object, because it can disappear in the course of hours.

Group objects. The above description applied to *individual* objects: we can talk about the height of an individual man, the speed with which the individual river flows, etc.

In physics, we are also interested in "group" objects: the mass of an electron, the magnetic moment of an ion of a specific type. Similarly, in manufacturing, we are interested in the speed and/or fuel efficiency of a certain type of a car, in the frequency of a certain type of a laser, etc. In all these cases, instead of dealing with an *individual* object, we have a *collection* of similar objects.

In the ideal case, we assume that all these objects are *identical*, so whatever we can observe based on one of these objects can be applied to others as well. For example, if we apply safety tests to several cars of the same make and model, and these tests are successful, we conclude that all the cars of this make and model are safe.

Idealized description of idealized objects. Ideally, we should have a full description of each (idealized) object. In other words, for each possible quantity (such as location, mass, height, etc.), we should know the exact value of this quantity for the given object.

For many physical quantities, there are several different ways to measure the value of this quantity; ideally, all these ways should lead to the same numerical result. For example, a GPS location of the car can be measured based on different parts of this car; all these measurements should lead to the same result. Similarly, a width of a wooden plank should be the same no matter where we measure it.

In mathematical terms, an (idealized) exact description of an (idealized) object means that we know, for this object, the exact values $s_1,\ldots,s_n$ of all possible quantities of interest.

All such tuples $s=(s_1,\ldots,s_n)$ of real numbers form an n-dimensional space $\mathbb{R}^n$.

Usually, not all tuples are physically possible. For example, mass is always non-negative, velocity is always limited by the speed of light, etc. The actual state must therefore belong to the set $S \subseteq \mathbb{R}^n$ of all physically possible tuples.

Comment. In this paper, for simplicity, we assume that this set S is known. To get a more realistic description, we must take into account that this set is not exactly known – e.g., the speed of light is only approximately known.

Objects are not ideal: aleatoric uncertainty. As we have mentioned, ideally, objects should be well-defined and stable (not changing with time).

In practice, objects are often not well defined. For example, when a river flows into the lake, it is often not clear where is the boundary between the river and the lake.

Objects are also not perfectly stable: they change, slightly but change. For example, the weight of a person slightly changes – when she breathes in and breathes out, when she sweats, etc.

As a result, for the same object, even an ideally accurate measuring instrument can measure different values:

- the measurement results differ with time because the object changes,
- these results differ because we may select different boundaries for the object, etc.

This "objective" difference is known as *aleatoric* uncertainty.

Objects are not perfectly identical: another case of aleatoric uncertainty. In the ideal case, we assumed that all the "instance" of a group object are identical. In reality, different objects are slightly different. For example, different cars of the same make and model may have slightly different fuel economy characteristics.

Different measuring procedures can lead to different values: yet another case of objective (aleatoric) uncertainty. In the ideal case, all possible procedures for measuring a given quantity should lead to the exact same result. In practice, different procedure may lead to slightly different results.

For example, for a non-ideal wooden plank, its width may differ slightly from one location to another.

Measurements are not 100% accurate: epistemic uncertainty. In the ideal case, we assumed that all the measurements lead to the exact values of the measured quantity. In practice, measurements are never 100% accurate; the result $\widetilde{x}$ of each measurement is, in general, slightly different form the actual (unknown) value x of the desired quantity: $\Delta x \stackrel{\text{def}}{=} \widetilde{x} - x \neq 0$; see, e.g., (Rabinovich 2005).

Thus, even in the ideal case, when the object is well-defined and stable – i.e., where there is no aleatoric ("objective") uncertainty, measurement results can be different. This uncertainty is "subjective" in the sense that it characterizes not the object itself, but rather our *knowledge* about this object. Such uncertainty is called *epistemic*.

How to describe aleatoric uncertainty: set (interval) granules. In the ideal case, each object is characterized by a single state $s \in S$.

Due to aleatoric uncertainty, for the same object, we may get *different* states $s \in S$, depending on:

- at what time we measure the corresponding quantities;
- how we define the boundaries of the object;
- which object from the group of (almost) identical objects we take; and
- which of the possible measuring procedures we use to measure the corresponding quantities.

Thus, to fully characterize the situation, we must know which states s are possible for this object and which are not. In other words, in view of the aleatoric uncertainty, to describe an object, we need to know the *set* $\mathbf{s} \subseteq S$ of all possible states characterizing this object. Different states $s \in \mathbf{s}$ are "equally" possible, so this set forms a single set *granule* characterizing the object.

This set $\mathbf{s}$ is usually *connected* – in the sense that there is a continuous transition between every two possible states $s, s' \in \mathbf{s}$, In the 1-D case, this connectivity implies that with every two possible states s and s', all intermediate states are also possible – i.e., that the set $\mathbf{s}$ of all possible states is an *interval*.

How to describe aleatoric uncertainty: probabilistic granules. In addition to knowing which states are possible and which are not, it is also desirable to know how frequent are different possible states.

For example, within a population of cars of the same make and model, it is desirable not only to know that the fuel efficiency of an individual car may be lower than on average, it is also desirable to know how frequent are such low-efficiency situation:

- if they are rare, it is acceptable, but
- if such situations are frequent, this is a strong argument against buying this particular model.

Thus, it is desirable to know, for each possible value $s \in \mathbf{s}$, a frequency (= probability) that this value occurs when measuring this object. If we know this information, then, instead of a set granule, we have a *probabilistic* granule.

Description of epistemic uncertainty: set (interval) and probabilistic granules. Similarly, we can describe *epistemic* uncertainty.

Indeed, once we know the results $\widetilde{s} = (\widetilde{s}_1, \ldots, \widetilde{s}_n)$ of measuring the desired quantities, we would like to know which states s are compatible with these measurement results. In other words, we need to know the set $\mathbf{s}$ of all such states – i.e., a set (interval) granule.

In addition to knowing which states $s \in S$ are possible and which are not possible, it is also desirable to know which values $s \in \mathbf{s}$ are more frequent and which are less frequent – i.e., it is desirable to know the *frequency* (*probability*) of different value s. Thus, we also arrive at the need to consider a probability distribution on the set $\mathbf{s}$ of possible states – i.e., a probabilistic granule.

Yet another reason for granulation: need to speed up decision making. Even when we know the actual element s from S with a good accuracy, it is still often reasonable, when making a decision, to ignore this difficult-to-process accurate information and to base our decision on the *type* of an object – i.e., on the granule to which this object belongs.

For example, when an animal attacks a person, it makes sense to ignore the animal's eye color and other details of the animal and concentrate of the type of the animal: e.g., is it a small (mostly harmless) dog or a dangerous tiger.

3 Need for Granules of Granules – i.e., for Granules of Higher Order

Need for granules of higher order. Ideally, to characterize the uncertainty, we describe

- either the set of possible values
- or the probability distribution on the set of possible values.

This description assumes that we know *exactly* which values are possible and which values are not possible – and we know the *exact* values of the corresponding probabilities.

In practice, we are not always sure which values are possible and which are not, and do not know the exact values of the corresponding probabilities. In other words, our knowledge of the corresponding uncertainty is also uncertain.

Thus, instead of a single set-valued granule, we have a *granule* of possible set-valued granules – a construction which can be naturally described as a granule of higher order. Similarly, instead of a single probabilistic granule, we have a *granule* of possible probabilistic granules – a construction which can also be naturally described as a granule of higher order.

Motivation for restricting ourselves to granules of second order. For example, instead of a single set of possible states, we can have a class of possible sets – a second order construction. This class may also not be exactly known – so we should consider class of possible classes, a third order construction.

However, from the computational viewpoint, a set is already difficult to process, a class of sets is even more complex, and a class of classes is practically impossible to analyze.

Comment. For fuzzy uncertainty, a deeper argument in favor of second-order granules is given in the Appendix; this argument was first outlined in (Nguyen and Kreinovich 1998) and (Kreinovich and Nguyen 2001).

4 Second-Order Granules: Natural Classification

Classification: general idea. An ideal description of uncertainty is to describe as a granule – i.e.,

- either a set of possible values
- or a probability distribution on the set of possible values.

In reality, we do not have the exact knowledge of the corresponding granule (i.e., of the set or of the probability distribution).

When we did not know the exact state, we considered either the set of all possible states or a probability distribution on the set of possible states. Similarly, when we do not know the exact granule, we have to consider:

- either a set of possible granules,
- or a probability distribution on the set of possible granules.

In each of these two cases, we have two subcases, depending on whether we consider set-valued (interval) or probabilistic granules. Thus, we arrive at the four possible situations:

1. a set of possible sets;
2. a set of possible probability distributions;
3. a probability distribution on the class of possible sets; and
4. a probability distribution on the class of possible probability distributions.

What we plan to show. At first glance, we have a very mathematical classification. However, as we will show, these four types of second order granules correspond to well-known and well-used types of uncertainty – such as random sets, Dempster-Shafer approach, fuzzy sets, imprecise probabilities, and Bayesian statistics.

Thus, the idea of second order granules provides a natural unified description of different formalisms for describing uncertainty.

First case: set of possible sets. In the first case, instead of selecting a single set of possible values, we select a *class* of possible sets. This idea is actively used in representation of uncertainty. For example, in the *rough set* approach, each set S is represented by a set $\underline{S}$ that is contained in S and a set $\overline{S}$ that contains S. In this case, the only information that we have about the actual (unknown) set S is that $\underline{S} \subseteq S \subseteq \overline{S}$, i.e., that S belongs to the *set interval* $[\underline{S}, \overline{S}] \stackrel{\text{def}}{=} \{S : \underline{S} \subseteq S \subseteq \overline{S}\}$. General set intervals – not necessarily generated by rough sets – are also actively used; see, e.g., (Yao and Li 1996), (Yao et al. 2008).

Second case: a set of probability distributions. In this case, instead of selecting a single probability distribution, we select a set of possible probability distributions. This description of uncertainty is known as *imprecise probability*; see, e.g., (Walley 1991).

An important particular case of imprecise probability is the case of p-boxes (probability boxes), i.e., interval bounds on the values of the cumulative distribution function $F(x) \stackrel{\text{def}}{=} \text{Prob}(\xi \leq x)$; see, e.g., (Ferson 2002), (Ferson et al. 2003).

Third case: a probability distribution on the class of possible sets. In this case, instead of selecting a single set, we select a probability distribution on the class of all possible sets. In other words, we assign, to each set, a probability that this is indeed the right set.

When we assign a probability to each number, we get a *random number*. When we assign a probability to each vector, we get a *random vector*. Similarly, when we assign a probability to each set, we get a *random set*. Random sets have indeed been actively applied in description and processing of uncertainty; see, e.g., (Nguyen 2008) and references therein.

In the discrete case, a random set means that we assign, to different sets A, values $m(A) \geq 0$ such that $\sum\limits_{A} m(A) = 1$. This is exactly the widely used *Dempster-Shafer* approach to describing uncertainty.

Relation to fuzzy. One of the reason why random sets are important in describing uncertainty is that they provide a reasonable alternative description of *fuzzy sets*. Specifically, one way to assign a membership degree to a statement like "25 is young" is to take N experts, ask these experts whether a person who is 25 years old is young, and take the portion $N(25)/N$ of experts who agree with this statement as the desired degree $\mu_{\text{young}}(25)$. This procedure assumes

- that the experts are equally important – because we give equal weight to opinions of different experts, and
- that each expert is capable of giving a precise (crisp) answer to each such question.

For each of N experts i ($1 \leq i \leq N$), there us denote, by S_i, the set of all ages which for which, for this expert, the person is young. In general, different experts have different sets S_i, but it is possible that two or more different experts have the same set.

As we have mentioned, we assume that the experts are equally important, i.e., that each expert gets assigned the same probability $1/N$. So, if a set S_i occurs as an opinion of only one expert, we assign it the probability $1/N$; if it occurs as the common opinion of several (k) experts, we assign it the probability k/N. Thus, we have probabilities assigned to different sets, i.e., we have a random set. In terms of this random set, the degree to which, say, 25 is young, is simply equal to the sum of the probabilities of all the sets S_i that include 25, i.e., to the probability that 25 belongs to the random set.

In general, for the corresponding random set S, for every value x, the membership degree $\mu(x)$ is equal to the probability that $x \in S$: $\mu(x) = \text{Prob}(x \in S)$.

Fourth case: a probability distribution on the class of probability distributions. In this case, instead of selecting a single probability distribution, we select a probability distribution on the set of these distributions.

Theoretically, we can have an infinite-dimensional class of probability distributions, i.e., a class in which we need to know the values of infinitely many parameters to uniquely determine the distribution. In practice, in a computer, we can only store finitely many values of the parameters. Thus, from the practical viewpoint, each class of probability distributions is characterized by the values of finitely many parameters. There may be only two parameters – like in 1-D Gaussian distributions, there can be many more parameters as in more sophisticated classes, but there are always finitely many parameters.

In this case, selecting a probability distribution from the class means selecting the values of these parameters. Thus, the above situation can be described as follows:

- instead of selecting a unique set of parameters characterizing a probability distribution,
- we select a probability distribution on the set of these parameters.

This idea describes the *Bayesian* statistical approach, whose main idea is indeed to select a (prior) distribution on the set of all possible values of different parameters; see, e.g., (Jaynes 2003), (Gelman et al. 2004).

Summary. Second order granules approach covers many known uncertainty formalisms as particular cases:

	set of ...	probability distributions on ...
... sets	set intervals; rough sets	random sets; Dempster-Shafer approach; fuzzy approach
... probability distributions	imprecise probability; p-box	Bayesian statistics

What if we also consider fuzzy. In the above text, we only consider set-valued (interval) and probabilistic granules. These granules correspond to "objective" (aleatoric) and measurement uncertainty. Expert estimates lead to fuzzy uncertainty, where we have fuzzy granules – i.e., fuzzy sets. If we add the possibility of fuzzy granules, then we get new possibilities of second-order granules:

- a set of fuzzy granules, i.e., a set of fuzzy sets – e.g., an interval-valued fuzzy set;
- a probability distribution on the set of fuzzy granules, i.e., a *random fuzzy set*; see, e.g., (Li et al. 2002);
- a fuzzy set of fuzzy sets – i.e., a general type-2 fuzzy set;
- a fuzzy class of sets or probability distributions – something which was tried in imprecise probability research.

Third order granules? In contrast to the second order granules which have many practical applications, third order ones are rarely used. A few practical examples include interval-valued fuzzy sets; see, e.g., (Nguyen and Kreinovich 1995), (Nguyen et al. 1997), (Mendel 2001), (Mendel and Wu 2010):

- since a fuzzy set can be interpreted as a random set,
- an interval-valued fuzzy set – i.e., a set of possible fuzzy sets – can be interpreted as a set of possible random sets, i.e., as a third order granule.

5 Conclusion

In this paper, we provided a natural motivation for granules of higher order, and showed that these granules provide a unified description of different uncertainty formalisms such as random sets, Dempster-Shafer approach, fuzzy sets, imprecise probabilities, and Bayesian statistics.

We also prove that within this general description, most reasonable uncertainty formalisms have already been discovered: for example, we prove that for fuzzy uncertainty, granules of second order are sufficient.

Acknowledgments

This work was supported in part by the National Science Foundation grants HRD-0734825 and DUE-0926721, by Grant 1 T36 GM078000-01 from the National Institutes of Health, by Grant MSM 6198898701 from MŠMT of Czech Republic, and by Grant 5015 "Application of fuzzy logic with operators in the knowledge based systems" from the Science and Technology Centre in Ukraine (STCU), funded by European Union.

The author is greatly thankful to the volume editors and to the anonymous referees for valuable suggestions.

References

Ferson, S.: RAMAS Risk Calc 4.0: Risk assessment with uncertain numbers. CRC Press, Boca Raton (2002)

Ferson, S., Kreinovich, V., Ginzburg, L., Myers, D.S., Sentz, K.: Constructing probability boxes and Dempster-Shafer structures. Sandia National Laboratories, Report SAND2002-4015 (January 2003)

Gelman, A., Carlin, J.B., Stern, H.S., Rubin, D.B.: Bayesian data analysis. Chapman & Hall/CRC, Boca Raton (2004)

Jaynes, E.T.: Probability Theory: The logic of science. Cambridge University Press, Cambridge (2003)

Kreinovich, V., Nguyen, H.T.: 1st order, 2nd order, what next? Do we really need third-order descriptions: a view from a realistic (granular) viewpoint. In: Proceedings of the Joint 9th World Congress of the International Fuzzy Systems Association and 20th International Conference of the North American Fuzzy Information Processing Society IFSA/NAFIPS 2001, Vancouver, Canada, July 25-28, pp. 1908–1913 (2001)

Li, S., Ogura, Y., Kreinovich, V.: Limit theorems and applications of set valued and fuzzy valued random variables. Kluwer Academic Publishers, Dordrecht (2002)

Mendel, J.: Uncertain Rule-Based Fuzzy Logic Systems: Introduction and New Directions. Prentice-Hall, Upper Saddle River (2001)

Mendel, J.M., Wu, D.: Perceptual computing: aiding people in making subjective judgments. IEEE Press and Wiley, Los Alamitos (2010)

Nguyen: An introduction to random sets. Chapman & Hall/CRC, Boca Raton (2008)

Nguyen, H.T., Kreinovich, V.: Towards theoretical foundations of soft computing applications. International Journal of Uncertainty, Fuzziness, and Knowledge-Based Systems (IJUFKS) 3(3), 341–373 (1995)

Nguyen, H.T., Kreinovich, V.: Possible new directions in mathematical foundations of fuzzy technology: a contribution to the mathematics of fuzzy theory. In: Phuong, N.P., Ohsato, A. (eds.) Proceedings of the Vietnam-Japan Bilateral Symposium on Fuzzy Systems and Applications VJFUZZY 1998, HaLong Bay,Vietnam, September 30–October 2, pp. 9–32 (1998)

Nguyen, H.T., Kreinovich, V., Zuo, Q.: Interval-valued degrees of belief: applications of interval computations to expert systems and intelligent control. International Journal of Uncertainty, Fuzziness, and Knowledge-Based Systems (IJUFKS) 5(3), 317–358 (1997)

Rabinovich, S.: Measurement errors and uncertainties: theory and practice. Springer, New York (2005)

Walley, P.: Statistical reasoning with imprecise probabilities. Chapman&Hall, New York (1991)

Yao, J.T., Yao, Y.Y., Kreinovich, V., Pinheiro da Silva, P., Starks, S.A., Xiang, G., Nguyen, H.T.: Towards more adequate representation of uncertainty: from intervals to set intervals, with the possible addition of probabilities and certainty degrees. In: Proceedings of the IEEE World Congress on Computational Intelligence WCCI 2008,Hong Kong,chaina,June 1-6, pp. 983–990 (2008)

Yao, Y.Y., Li, X.: Comparison of rough-set and interval-set models for uncertain reasoning. Fundamenta Informaticae 27, 289–298 (1996)

Appendix: 2nd Order Is Sufficient for Fuzzy Uncertainty

Second order descriptions: the main idea. Experts are often not 100% certain in the statements they make; therefore, in the design of knowledge-based systems, it is desirable to take this uncertainty into consideration. Usually, this uncertainty is described by a number from the interval $[0,1]$; this number is called *subjective probability*, *degree of certainty*, etc.

One of the main problems with this approach is that we must use *exact* numbers from the interval $[0,1]$ to represent experts' degrees of certainty; an expert may be able to tell whether his degree of certainty is closer to 0.9 or to 0.5, but it is hardly possible that an expert would be able to meaningfully distinguish between degrees of certainty, say, 0.7 and 0.701. If you ask the expert whether his degree of certainty about a certain statement A can be described by a certain number d (e.g., $d = 0.701$), the expert will, sometimes, not be able to give a definite answer, she will be uncertain about it. This uncertainty can be, in its turn, described by a number from the interval $[0,1]$. It is, therefore, natural to represent our degree of certainty in a statement A not by a *single* (crisp) number $d(A) \in [0,1]$ (as in the $[0,1]$-based description), but rather by a *function* $\mu_{\mathbf{d}(A)}$ which assigns, to each possible real number $d \in [0,1]$, a *degree* $\mu_{\mathbf{d}(A)}(d)$ with which this number d can be the (desired) degree of certainty of A. This is called a *second-order* description of uncertainty.

Third and higher order descriptions. In second-order description, to describe a degree with which a given number $d \in [0,1]$ can be a degree of certainty of a statement A, we use a *real number* $\mu_{\mathbf{d}(A)}(d)$. As we have already mentioned, it is difficult to describe our degree of certainty by a single number. Therefore, to make this description even more realistic, we can represent each degree of certainty $d(P(x))$ not by a (more traditional) $[0,1]$-based description, but by a *second order* description. As a result, we get the *third order* description.

Similarly, to make our description even more realistic, we can use the third order descriptions to describe degrees of certainty; then, we get *fourth order* uncertainty, etc.

Third order descriptions are not used: why? Theoretically, we can define third, fourth order, etc., descriptions, but in practical applications, only second order descriptions were used so far; see, e.g., (Nguyen and Kreinovich 1995), (Nguyen et al. 1997), (Mendel 2001), (Mendel and Wu 2010). Based on this empirical fact, it is natural to conclude that third and higher order descriptions are not really necessary. We will show that this conclusion can be theoretically justified.

First step in describing uncertainty: set of uncertainty-describing words. Let us first describe the problem formally. An expert uses words from a natural language to describe his degrees of certainty. In every language, there are only finitely many words, so we have a finite set of words that needs to be interpreted. We will denote this set of words by W.

Second step: a fuzzy property described by a word-valued "membership function". If we have any property P on a universe of discourse U, an expert can describe, for each element $x \in U$, his degree of certainty $d(x) \in W$ that the element x has the property P.

Traditional fuzzy logic as a first approximation: numbers assigned to words describing uncertainty. Our ultimate goal is to provide a computer representation for each word $w \in W$. In the traditional $[0,1]$-based description, this computer representation assigns, to every word, a *real number* from the interval $[0,1]$; in general, we may have some other computer representations (examples will be given later). Let us denote the set of all possible computer representations by S.

In the first approximation, i.e., in the first order description, we represent each word $w \in W$, which describes a degree of uncertainty, by an element $s \in S$ (e.g., by a real number from the interval $[0,1]$). In this section, we will denote this first-approximation computer representation of a word w by $s = \|w\|$.

If the set S is too small, then it may not contain enough elements to distinguish between different expert's degree of belief: this was exactly the problem with classical $\{0,1\}$-based description, in which we only have two possible computer representations – "true" and "false" – that are not enough to adequately describe the different degrees of certainty. We will therefore assume that the set S is rich enough to represent different degrees of certainty.

In particular, the set $[0,1]$ contains infinitely many points, so it should be sufficient; even if we only consider computer-representable real numbers, there are still much more of them (millions and billions) than words in a language (which is usually in hundreds of thousands at most), so we can safely make this "richness" assumption. In mathematical terms, it means that two different degrees of belief are represented by different computer terms, i.e., that if $w_1 \neq w_2$, then $\|w_1\| \neq \|w_2\|$.

First approximation is not absolutely adequate. The problem with the first-order representation is that the relation between words $w \in W$ and computer representation $s \in S$ is, in reality, also imprecise. Typically, when we have a word $w \in W$, we cannot pick a single corresponding representative $s \in S$; instead, we may have *several* possible representatives, with different degrees of adequacy.

Actual description of expert uncertainty: word-valued degree to which a word describes uncertainty. In other words, instead of a *single* value $s = \|w\|$ assigned to a word w, we have *several* values $s \in S$, each with its own degree of adequacy; this degree of adequacy can also be described by an expert, who uses an appropriate word $w \in W$ from the natural language.

In other words, for every word $w \in W$ and for ever representation $s \in S$, we have a degree $w' \in W$ describing to what extent s is adequate in representing w. Let us represent this degree of adequacy by $a(w,s)$; the symbol a represents a function $a : W \times S \to W$, i.e., a function that maps every pair (w,s) into a new word $a(w,s)$.

Second-order description of uncertainty as a second approximation to actual uncertainty. So, the meaning of a word $w \in W$ is represented by a *function* a which assigns, to every element $s \in S$, a degree of adequacy $a(w,s) \in W$. We want to represent this degree of adequacy in a computer; therefore, instead of using the word $a(w,s)$ itself, we will use the computer representation $\|a(w,s)\|$ of this word. Hence, we get a *second-order* representation, in which a degree of certainty corresponding to a word $w \in W$ is represented not by a *single* element $\|w\| \in S$, but by a *function* $\mu_w : S \to S$, a function which is defined as $\mu_w(s) = \|a(w,s)\|$.

Second-order description is not 100% adequate either; third-, fourth-order descriptions, etc. The second-order representation is also not absolutely adequate, because, to represent the degree $a(w,s)$, we used a single number $\|a(w,s)\|$. To get a more adequate representation, instead of this single value, we can use, for each element $s' \in S$, a degree of adequacy with which the element s' represents the word $a(w,s)$. This degree of adequacy is also a word $a(a(w,s),s')$, so we can represent it by an appropriate element $\|a(a(w,s),s')\|$. Thus, we get a *third-order* representation, in which to every element s, we assign a second-order representation. To get an even more adequate representation, we can use fourth- and higher order representations.

Let us express this scheme formally.

Definition 1

- *Let W be a finite set; element of this set will be called* words.
- *Let U be set called a* universe of discourse. *By a* fuzzy property P, *we mean a mapping which maps each element $x \in U$ into a word $P(x) \in W$; we say that this word* described the degree of certainty *that x satisfies the property P.*
- *By a* first-approximation uncertainty representation, *we mean a pair $\langle S, \|.\|\rangle$, where:*
  - *S is a set; elements of this set will be called computer representations; and*
  - *$\|.\|$ is a function from W to S; we say that an element $\|w\| \in S$* represents *the word w.*
- *We say that an uncertainty representation is* sufficiently rich *if for every two words $w_1, w_2 \in W$, $w_1 \neq w_2$ implies $\|w_1\| \neq \|w_2\|$.*

Definition 2. *Let W be a set of words, and let S be a set of computer representations. By an* adequacy *function, we mean a function $a : W \times S \to W$; for each word $w \in W$, and for each representation $s \in S$, we say that $a(w,s)$ describes the degree to which the element s* adequately describes *the word w.*

Definition 3. *Let U be a universe of discourse, and let S be a set of computer representations. For each $n = 1,2,\ldots$, we define the notions of n*-th order degree of certainty *and of a n-th order fuzzy set, by the following induction over n:*

- *By a* first-order degree of certainty, *we mean an element $s \in S$ (i.e., the set S_1 of all first-order degrees of certainty is exactly S).*
- *For every n, by a n*-th order fuzzy set, *we mean a function $\mu : U \to S_n$ from the universe of discourse U to the set S_n of all n-th order degrees of certainty.*
- *For every $n > 1$, by a n*-th order degree of certainty, *we mean a function s_n which maps every value $s \in S$ into an $(n-1)$-th order degree of certainty (i.e., a function $s_n : S \to S_{n-1}$).*

Definition 4. *Let W be a set of words, let $\langle S, \|.\|\rangle$ be an uncertainty representation, and let a be an adequacy function. For every $n > 1$, and for every word $w \in W$, we* *define the* n-th order degree of uncertainty $\|w\|_{a,n} \in S_n$ corresponding to the word w *as follows:*

- *As a first order degree of uncertainty* $\|w\|_{a,1}$ *corresponding to the word* w*, we simply take* $\|w\|_{a,1} = \|w\|$.
- *If we have already defined degrees of orders* $1,\ldots,n-1$*, then, as an n-th order degree of uncertainty* $\|w\|_{a,n} \in S_n$ *corresponding to the word* w*, we take a function* s_n *which maps every value* $s \in S$ *into a* $(n-1)$*-th order degree* $\|a(w,s)\|_{a,n-1}$.

Definition 5. *Let* W *be a set of words, let* $\langle S, \|.\|\rangle$ *be an uncertainty representation, let* a *be an adequacy function, and let* P *be a fuzzy property on a universe of discourse* P*. Then, by a n-th order fuzzy set (or a* n-th order membership function*)* $\mu^{(n)}_{P,a}(x)$ *corresponding to* P*, we mean a function which maps every value* $x \in U$ *into an n-th order degree of certainty* $\|P(x)\|_{a,n}$ *which corresponds to the word* $P(x) \in W$.

We will prove that for properties which are *non-degenerate* in some reasonable sense, it is sufficient to know the *first* and *second* order membership functions, and then the others can be uniquely reconstructed. Moreover, if we know the membership functions of first two orders for a non-degenerate *class* of fuzzy properties, then we will be able to reconstruct the higher order membership functions for *all* fuzzy properties from this class.

Definition 6

- *We say that a fuzzy property* P *on a universe of discourse* U *is* non-degenerate *if for every* $w \in W$*, there exists an element* $x \in U$ *for which* $P(x) = w$.
- *We say that a class* $\mathscr{P}$ *of fuzzy properties* P *on a universe of discourse* U *is* non-degenerate *if for every* $w \in W$*, there exists a property* $P \in \mathscr{P}$ *and an element* $x \in U$ *for which* $P(x) = w$.

Comment. For example, if $W \neq \{0,1\}$, then every crisp property, i.e., every property for which $P(x) \in \{0,1\}$ for all x, is *not* non-degenerate (i.e., degenerate).

Proposition 1. *Let* W *be a set of words, let* $\langle S, \|.\|\rangle$ *be a sufficiently rich uncertainty representation, let* U *be a universe of discourse. Let* P *and* P' *be fuzzy properties, so that* P *is non-degenerate, and let* a *and* a' *be adequacy functions. Then, from* $\mu^{(1)}_{P,a} = \mu^{(1)}_{P',a'}$ *and* $\mu^{(2)}_{P,a} = \mu^{(2)}_{P',a'}$*, we can conclude that* $\mu^{(n)}_{P,a} = \mu^{(n)}_{P',a'}$ *for all* n.

Comments

- In other words, under reasonable assumptions, for each property, the information contained in the first and second order fuzzy sets is sufficient to reconstruct all higher order fuzzy sets as well; therefore, in a computer representation, it is sufficient to keep only first and second order fuzzy sets.
- This result is somewhat similar to the well-known result that a Gaussian distribution can be uniquely determined by its moments of first and second orders, and all higher order moments can be uniquely reconstructed from the moments of the first two orders.

- It is possible to show that the non-degeneracy condition is needed, because if a property P is not non-degenerate, then there exist adequacy functions $a \neq a'$ for which $\mu^{(1)}_{P,a} = \mu^{(1)}_{P,a'}$ and $\mu^{(2)}_{P,a} = \mu^{(2)}_{P,a'}$, but $\mu^{(3)}_{P,a} \neq \mu^{(3)}_{P,a'}$ already for $n = 3$.

Proposition 2. *Let W be a set of words, let $\langle S, \|.\|\rangle$ be a sufficiently rich uncertainty representation, let U be a universe of discourse. Let $\mathscr{P}$ and $\mathscr{P}'$ be classes of fuzzy properties, so that the class $\mathscr{P}$ is non-degenerate, and let $\varphi : \mathscr{P} \to \mathscr{P}'$ be a 1-1-transformation, and let a and a' be adequacy functions. Then, if for every $P \in \mathscr{P}$, we have $\mu^{(1)}_{P,a} = \mu^{(1)}_{\varphi(P),a'}$ and $\mu^{(2)}_{P,a} = \mu^{(2)}_{\varphi(P),a'}$, we can conclude that $\mu^{(n)}_{P,a} = \mu^{(n)}_{\varphi(P),a'}$ for all n.*

Comment. So, even if we do not know the adequacy function (and we do not know the corresponding fuzzy properties $P \in \mathscr{P}$), we can still uniquely reconstruct fuzzy sets of all orders which correspond to all fuzzy properties P.

Proof of Propositions 1 and 2. Proposition 1 can be viewed as a particular case of Proposition 2, when $\mathscr{P} = \{P\}$, $\mathscr{P}' = \{P'\}$, and φ maps P onto P'. Therefore, to prove both Propositions 1 and 2, it is sufficient to prove Proposition 2.

We will show that under the conditions of Proposition 2, from $\mu^{(1)}_{P,a} = \mu^{(1)}_{\varphi(P),a'}$ and $\mu^{(2)}_{P,a} = \mu^{(2)}_{\varphi(P),a'}$, we will be able to conclude that $\varphi(P) = P$ for all $P \in \mathscr{P}$, and that $a = a'$; therefore, we will easily conclude that $\mu^{(n)}_{P,a} = \mu^{(n)}_{\varphi(P),a'}$ for all n.

Indeed, by definition of the first membership function, for every $x \in U$, we have $\mu^{(1)}_{P,a}(x) = \|P(x)\|$. Thus, from the equality $\mu^{(1)}_{P,a} = \mu^{(1)}_{\varphi(P),a'}$, we conclude that for every $P \in \mathscr{P}$, we have $\|P(x)\| = \|\varphi(P)(x)\|$ for all $x \in U$. Since the uncertainty representation is assumed to be sufficiently rich, we can conclude that $\varphi(P)(x) = P(x)$ for all $x \in U$, i.e., that $\varphi(P) = P$ for every $P \in \mathscr{P}$.

Let us now show that $a = a'$, i.e., that for every $w \in W$ and for every $s \in S$, we have $a(w,s) = a'(w,s)$. Indeed, since $\mathscr{P}$ is a non-degenerate class, there exists a value $x \in U$ and a property $P \in \mathscr{P}$ for which $P(x) = w$. Let us consider the equality of the second order membership functions for this very P. Since $\varphi(P) = P$, the given equality $\mu^{(2)}_{P,a} = \mu^{(2)}_{\varphi(P),a'}$ can be simplified into the following form: $\mu^{(2)}_{P,a} = \mu^{(2)}_{P,a'}$. Let us consider this equality for the above-chosen value x (for which $P(x) = w$). For this x, by definition of the second-order membership function, $\mu^{(2)}_{P,a}(x) = \|P(x)\|_{a,2} = \|w\|_{a,2}$; and similarly, $\mu^{(2)}_{P,a'}(x) = \|P(x)\|_{a,2} = \|w\|_{a',2}$; thus, $\|w\|_{a,2} = \|w\|_{a',2}$.

By definition, $\|w\|_{a,2}$ is a function which maps every value $s \in S$ into a 1-st order degree $\|a(w,s)\|_{a,1} = \|a(w,s)\|$. Thus, from the equality of the functions $\|w\|_{a,2}$ and $\|w\|_{a',2}$, we can conclude that their values at a given s are also equal, i.e., that $\|a(w,s)\| = \|a'(w,s)\|$. Since the uncertainty structure is sufficiently rich, we conclude that $a(w,s) = a'(w,s)$. The proposition is proven.

Proof of a comment after Proposition 1. Since P is *not* non-degenerate, there exists a value $w_0 \in W$ which cannot be represented as $P(x)$ for any $x \in U$. Let us pick arbitrary elements $x_0 \in U$ and $s_0 \in S$, and define $a(w,s)$ and $a'(w,s)$ as follows:

- first, we define $a(w,s) = a'(w,s)$ for all words w of the type $w = P(x)$: namely, we take $a(P(x_0),s_0) = a'(P(x_0),s_0) = w_0$ and take arbitrary other values for different pairs (w,s) with $w = P(x)$;
- then, we define $a(w,s)$ and $a'(w,s)$ for the remaining pairs (w,s): namely, we take $a(w_0,s_0) = w_0$, $a'(w_0,s_0) = P(x_0) \neq w_0$, and we define a and a' arbitrarily for all other pairs (w,s).

Let us show that for thus chosen adequacy functions, the membership functions of first and second order coincide, but the membership functions of the third order differ. Indeed:

For the *first* order, we have, for every x, $\mu^{(1)}_{P,a}(x) = \|P(x)\|$ and similarly, $\mu^{(1)}_{P,a'}(x) = \|P(x)\|$; therefore, $\mu^{(1)}_{P,a}(x) = \mu^{(1)}_{P,a'}(x)$ for all x. Hence, $\mu^{(1)}_{P,a} = \mu^{(1)}_{P,a'}$.

For the *second* order, for every x, $\mu^{(2)}_{P,a}(x)$ is a function which maps $s \in S$ into a value $\|a(P(x),s)\|_{a,1} = \|a(P(x),s)\|$. Similarly, $\mu^{(2)}_{P,a'}(x)$ is a function which maps $s \in S$ into a value $\|a'(P(x),s)\|_{a',1} = \|a'(P(x),s)\|$. For words w of the type $P(x)$, we have defined a and a' in such a way that $a(w,s) = a'(w,s)$; therefore, $\|a(P(x),s)\| = \|a'(P(x),s)\|$ for all x and s. Thus, $\mu^{(2)}_{P,a} = \mu^{(2)}_{P,a'}$.

Finally, let us show that the *third* order membership functions differ. We will show that the values of the functions $\mu^{(3)}_{P,a}$ and $\mu^{(3)}_{P,a'}$ differ for $x = x_0$. Indeed, by definition of the third order membership function,

- $\mu^{(3)}_{P,a}(x_0)$ is a function which maps every s into the value $\|a(P(x_0),s)\|_{a,2}$, and
- $\mu^{(3)}_{P,a'}(x_0)$ is a function which maps every s into the value $\|a'(P(x_0),s)\|_{a',2}$.

To prove that these function are different, it is sufficient to show that their values differ for *some* values s; we will show that they differ for $s = s_0$, i.e., that $\|a(P(x_0),s_0)\|_{a,2} \neq \|a'(P(x_0),s_0)\|_{a',2}$. By our construction of a, we have $a(P(x_0),s_0) = a'(P(x_0),s_0) = w_0$, so the inequality that we need to prove takes the form $\|w_0\|_{a,2} \neq \|w_0\|_{a',2}$.

By definition, $\|w_0\|_{a,2}$ is a function which maps every value $s \in S$ into $\|a(w_0,s)\|_{a,1} = \|a(w_0,s)\|$. Similarly, $\|w_0\|_{a',2}$ is a function which maps every value $s \in S$ into $\|a'(w_0,s)\|_{a,1} = \|a'(w_0,s)\|$. For s_0, according to our construction of a and a', we have $a(w_0,s_0) = w_0 \neq P(x_0) = a'(w_0,s_0)$. Thus, since the uncertainty representation is sufficiently rich, we conclude that $\|a(w_0,s_0)\| \neq \|a'(w_0,s_0)\|$, and therefore, that $\|w_0\|_{a,2} \neq \|w_0\|_{a',2}$ and $\mu^{(3)}_{P,a} = \mu^{(3)}_{P,a'}$.

The statement is proven.

Artificial Intelligence Perspectives on Granular Computing

Yiyu Yao

Abstract. Granular computing concerns a particular human-centric paradigm of problem solving by means of multiple levels of granularity and its applications in machines. It is closely related to Artificial Intelligence (AI) that aims at understanding human intelligence and its implementations in machines. Basic ideas of granular computing have appeared in AI under various names, including abstraction and reformulation, granularity, rough set theory, quotient space theory of problem solving, hierarchical problem solving, hierarchical planning, learning, etc. However, artificial intelligence perspectives on granular computing have not been fully explored. This chapter will serve the purpose of filling in such a gap. The results will have bidirectional benefits. A synthesis of results from artificial intelligence will enrich granular computing; granular computing philosophy, methodology, and tools may help in facing the grand challenge of reverse-engineering the brain, which has significant implications to artificial machine intelligence.

1 Introduction

There are many views, interpretations and models of granular computing, encompassing various theories and techniques used in, for example, Artificial Intelligence (AI), computer programming, human and machine problem solving, information granulation, temporal granulation, spatial granulation, discretization, rough sets, quotient space theory, interval computing, fuzzy sets, qualitative reasoning, computing with words, and many others (Bargiela and Pedrycz 2009; Inuiguchi et al. 2003; Keet 2008; Lin et al. 2002; Pedrycz 2001; Pedrycz et al. 2008; Yao JT 2007, 2010; Yao 2000; Zadeh 1979, 1997). On the other hand, there does not exist a precise and satisfactory definition of "granular computing" that is general enough to cover a wide range of theories and techniques and, at the same

Yiyu Yao
Department of Computer Science
University of Regina
Regina, Saskatchewan, Canada S4S 0A2
e-mail: yyao@cs.uregina.ca

W. Pedrycz and S.-M. Chen (Eds.): Granular Computing and Intell. Sys., ISRL 13, pp. 17–34.
springerlink.com

time, specific enough to differentiate it from other paradigms of computing. Nevertheless, it is generally agreed that granular computing is a new paradigm of computation inspired by human problem solving (Bargiela and Pedrycz 2002; Hobbs 1985; Yao 2010; Zadeh 1997; Zhang and Zhang 1992). Although the basic ideas of granular computing are not entirely new and have been implicitly used under different names in many disciplines, the explicit formulation of granular computing, as a separate field of study, by re-interpreting and integrating existing ideas, is instrumental to human and machine problem solving (Yao 2007a).

An understanding of the intuitive powerful notion of granular computing is given as follows (Yao 2000, 2007a, 2008b). Roughly speaking, granular computing explores multiple levels of granularity in problem solving. The levels of granularity may be interpreted as the levels of abstraction, detail, complexity, and control in specific contexts (Yao 2009b). The objects of granular computing are families of granules that represent a problem at multiple levels. Studies on granular computing attempt to provide a new framework and a new discipline-independent and domain-independent language for describing ideas, principles, strategies, methodologies, techniques that make use of granules and have been studied in isolation across many disciplines. By examining and synthesizing results from existing studies in the light of the unified framework of granular computing and extracting their commonalities, it may be possible to develop a general theory for problem solving (Yao 2000, 2007a, 2008a, 2008b). The triarchic theory of granular computing (Yao 2005, 2006, 2008a, 2008b) offers a conceptual framework, in which granular computing is approached from the philosophical, methodological and computational perspectives based on granular structures.

Continuing this line of study, the present chapter examines artificial intelligence perspectives on granular computing. There are bidirectional benefits from such a study. On the one hand, results from artificial intelligence, such as concept formation, categorization, and learning, abstraction and reformulation, hierarchical planning and hierarchical problem solving, etc., are reviewed and woven together to enrich granular computing. On the other hand, human-inspired granular computing offers suggestions for future development of artificial intelligence.

Instead of covering exhaustively topics from artificial intelligence, we restrict ourselves to the information processing aspects. The information-processing theory, framework or metaphor, has been widely used for modeling the human mind, cognition, the human brain, computers, human and machine problem solving, and many more (Lindsay and Norman 1977; Nakashima 1999; Newell and Simon 1972; Pinker 1997; Simon 1978, 1979). Results from psychological analysis of human problem and artificial intelligence based machine problem solving suggest that problem solving may be viewed as a form of information processing (Gilhooly 1989; Lindsay and Norman 1977; Newell et al. 1958; Newell and Simon 1972), involving, for example, manipulation of symbols or change in representations. In the context of granular computing, Bargiela and Pedrycz (2002, 2008, 2009; Pedrycz 2008), through their book and several recent publications, promote granular computing as a paradigm of, and computational methods of, human-centric information processing. This view is shared by many other authors (Jankowski and Skowron 2009; Pedrycz et al. 2008; Yao JT 2007, 2008a, 2008b,

2010; Yao 2008a). Granular computing contributes in its unique way to the information-processing framework by focusing on granularity, an essential notion of problem solving that has been used in many studies (Cermak and Craik 1979; Hobbs 1985; Wiederhold 1992) but has not received its due attention.

In this chapter, we only briefly cover various notions and topics in artificial intelligence that are pertinent to granular computing, by focusing on the essence of ideas rather than their mathematical formulations. A reader can consult the extensive list of references for more detailed discussions.

2 A Triarchic Theory of Granular Computing

The triarchic theory of granular computing has three components: the philosophy of structured thinking, the methodology of structured problem solving, and the computation of structured information processing. Fig. 1 illustrates the triarchic theory by the granular computing triangle, in which each node represents a particular perspective and each edge represents the mutual support of any two perspectives. The complementary three perspectives focus on a central notion known as granular structures characterized by multilevel and multiview. While a multilevel hierarchy represents a particular view, a collection of many hierarchies represents a multiview description.

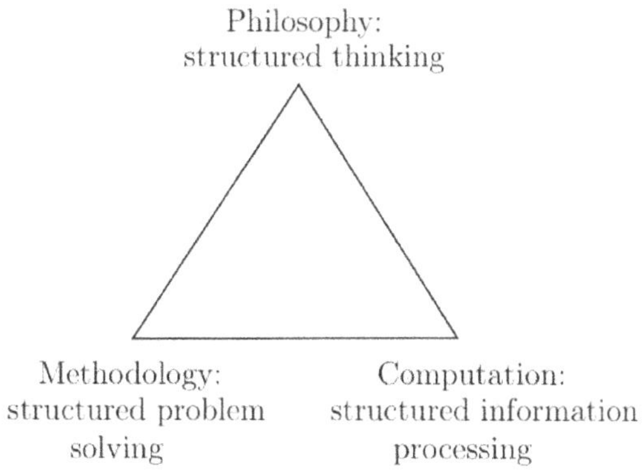

Fig. 1 Granular Computing Triangle

The theory is developed in series of papers (Yao 2000, 2005, 2006, 2007a, 2008a, 2008b, 2010). The main ingredients of the triarchic theory are granular structures and three perspectives. They are summarized as follows.

Granular Structures: Multilevel and Multiview. A granular structure consists of inter-connected and inter-acting granules. A granule represents a focal point of our observation or a unit of our discussion in the process of problem solving. In the information-processing framework, a granule may be interpreted as a particular representation or summarization of some information. A family of granules of similar size or nature provides one level of representation. Many families of granules provides hierarchically ordered many levels of representation. A structure of partially ordered multiple levels is known as a hierarchy. It is the results of a structured understanding, interpretation, representation, and description of a real-world problem or system. A hierarchy represents a problem from one particular angle or point-of-view with multiple levels of granularity. It is inevitable that a particular view given by one hierarchy is of limited power. A complete understanding of the problem requires the use and comparison of multiple hierarchies, and hence a multiview approach. Granular structures should reflect multiview and multilevel in each view.

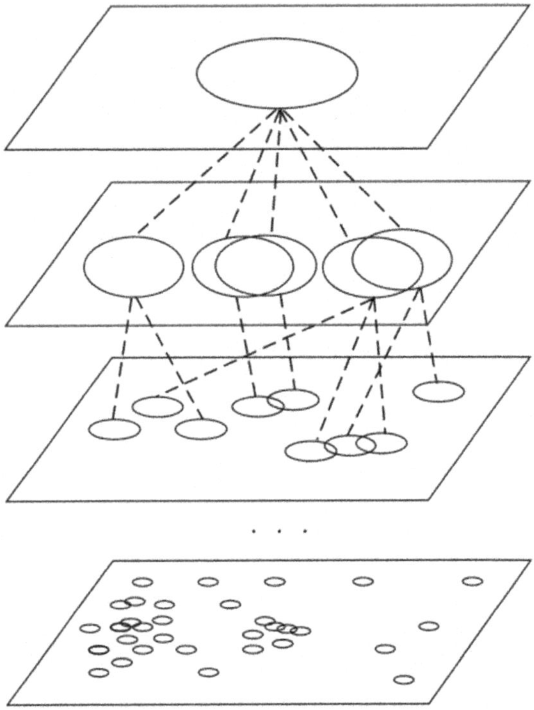

Fig. 2 A multilevel, hierarchical granular structure

Fig. 2 is an illustration of a multilevel, hierarchical granular structure. A few features of a granular structure may be commented. First, a granule at a higher level may be related to, or expressed by, several smaller granules at a lower level. We typically have more granules at a lower level. Second, granules at the same level may be of the same nature. At the different levels, one may use different representation schemes to describe granules, and use different languages to explain the granules involved and consider different strategies of granular computing. In other words, at a higher level we may have higher order and higher type of information granules. Consider an example of granular data analysis. At the very lower level, granules may be raw, unprocessed data. At the next level, granules may be coded and processed data that provide useful information. At yet another higher level, granules may be represented as rules that summary knowledge derived from data. Moving up the hierarchy, granules may be further interpreted as some models built based on the data. Third, granules at the same level may be connected to each other. In the figure, we only illustrate a simple connection in terms of overlapping and non-overlapping granules. In general, one may consider more complex relationships between granules. The inter-connection of granules at a particular provides useful structural information. Fourth, the multilevel, hierarchical structure may be interpreted as an information-processing pyramid proposed by Bargiela and Pedrycz (2002), in which more semantics information is processed when moving bottom up in the pyramid. Alternatively, the hierarchy may be interpreted as sequential transformations of different types of information, for example, from quantitative information to qualitative information. The commonly used data-information-knowledge-wisdom hierarchy is a good example to demonstrate such sequential transformations. Thus, a multilevel granular structure provides a paradigm of level-wise information processing (Yao 2009b).

Philosophy: Structured Thinking. As a way of structured thinking, granular computing draws results from two complementary philosophical views about the complexity of real-world problems, i.e., the traditional reductionist thinking and the more recent systems thinking. It combines analytical thinking for decomposing a whole into parts and synthetic thinking for integrating parts into a whole. Granular computing stresses the importance of the conscious effects in thinking with hierarchical structures that model a complex system or problem in terms of the whole and parts.

Methodology: Structured Problem Solving. As a general method of structured problem solving, granular computing promotes systematic approaches, effective principles, and practical heuristics and strategies that have been used effectively by humans for solving real-world problems. A central issue is the exploration of granular structures. This involves three basic tasks: constructing granular structures, working within a particular level of the structure, and switching between levels. The methodology of granular computing is inspired by human problem solving.

Computation: Structured Information Processing. As a paradigm of structured information processing, granular computing focuses on implementing knowledge-intensive systems based on granular structures. Two related basic issues are representations and processes. Representation covers the formal and precise description of granules and granular structures. Processes may be broadly divided into two classes: granulation and computation with granules. Granulation processes involve the construction of the building blocks and structures, namely, granules, levels, and hierarchies. Computation processes explore the granular structures. This involves two-way communications up and down in a hierarchy, as well as switching between levels and between hierarchies.

Granular computing emphasizes on structured approaches suggested by granular structures. The three perspectives of granular computing are interwoven together and mutually support each other; there may not exist dividing lines that separate them. Our simplest division only serves the purpose to emphasize that granular computing needs to be investigated from many perspectives.

3 Granular Computing for Artificial Intelligence

Granular computing is a particular way of human and machine problem solving. The study of granular computing serves two purposes, one for humans and the other for machines.

3.1 Scope and Goals of Granular Computing

Humans tend to organize, and categorization is essential to mental life (Pinker 1997). Results of such organizations are some types of structures. Humans tend to form multiple versions of the same world and to have several kinds of data presentations in the brain (Minsky 2007; Pinker 1997). For a particular problem, we normally have several versions of descriptions and understanding. Humans consider a problem at multiple levels of granularity. This allows us to focus on solving a problem at the most appropriate level of granularity by ignoring unimportant and irrelevant details (Hobbs 1985). We can readily switch between levels of granularity at different stages of problem solving (Hobbs 1985). We can also easily switch from one description to another. Granular computing research aims to formalize some or all of them.

It becomes evident that granular computing focuses on a special class of approaches to problem solving. The class is characterized by multiple levels of granularity. Regarding human intelligence, Minsky (2007) points out that humans have many "Ways to Think." We can easily switch among them and create new "Ways to Think," if none of existing ones works. The use of multiple levels of granularity is a fundamental one in the repository of ways of human problem solving. It may be more realistic for the study of granular computing not to cover the whole spectrum of ways of human problem solving. Therefore, we restrict the study of granular computing to human-inspired and multiple-granularity-based ways of problem solving.

The study of granular computing has two goals (Yao 2010). One is to understand the nature, the underlying principles and mechanisms of a particular way of human problem solving; the other is to apply them in the design and implementation of human-inspired machines and systems. They in turn lead to two classes of research on granular computing, namely, human-oriented studies and machine-oriented studies. For human-oriented studies, granular computing can be viewed as the fourth R, in addition to the classical three Rs (reading, writing and arithmetic). Once we articulate and master the principles of granular computing, we become a better problem solver. Granular computing is for everyone and can be used to solve a wide spectrum of problems. For machine-oriented studies, an understanding of human problem solving is a prerequisite for building machines having the similar power. Results from human-oriented studies may serve as a solid basis for machine-oriented studies. If we have a full understanding of human problem solving, we can design machines and systems based on the same principles. The two types of studies are relatively independent and mutually support each other. The former focuses on human problem solving and the latter on machine problem solving.

Accordingly, the study of granular computing serves two purposes. One purpose aims at an understanding of the underlying principles of human problem solving so that more people can consciously apply these principles. We use the phrase "granular computing for humans" to denote this aspect. The other purpose is to design machines and systems based on the same principles. We use the phrase "granular computing for machines" to denote the second aspect. In summary, granular computing is for both humans and machines.

3.2 Implications of Granular Computing to AI

Artificial Intelligence (AI), to a large extent, studies the principles of human problem solving and the construction of intelligent systems for problem solving based on the same principles, but (maybe) with different implementations. Holte and Choueiry (2003) state two reasons for paying attention to human reasoning and problem solving in early AI research. One is to develop computational models of human cognition, and the other is to obtain suggestions about how to program a computer to perform a particular cognitive task. Thus, as pointed out by Simon (Stewart 1994), "AI can have two purposes. One is to use the power of computers to augment human thinking, just as we use motors to augment human or horse power. ... The other is to use a computer's artificial intelligence to understand how humans think." These two purposes are closely related to the goals of granular computing in terms of machine-oriented and human-oriented approaches (Yao 2010). We study granular computing in order to understand a particular way of human problem solving and to create intelligent systems.

A major challenge faced by artificial intelligence researchers may be described by the Moravec's paradox (Moravec 1988) observed more than 20 years ago and is still applicable today. An insightful finding by artificial intelligence researchers is the dichotomy between machines and humans regarding the complexity or easiness in solving different problems. Contrary to what researchers first assumed and believed, machines can do well things that humans find hard (i.e., theorem proving, playing chess, etc.), and perform poorly on what is seemingly effortless for humans (i.e., perception, image processing, natural language understanding, etc.). A plausible explanation for the Moravec's paradox is based on the theory of evolution (Moravec 1988). The human brain and all human skills are results of Darwinian evolution. This natural selection process gradually and continually improves and optimizes the biological designs and implementations of the brain. Older skills, such as recognizing faces, recognizing voices, moving around in space, etc., are fully evolved and mastered by humans. We can perform these perception-based tasks almost unconsciously and they therefore appear to us to be effortless. In contrast, abstract thought is a new trick developed more recently in human evolutionary history and we have not fully mastered it yet. It therefore seems intrinsically difficult when we do it, as we must follow some precisely defined and constructed procedures.

Newer skills are not difficult to reengineer and thus machines may easily duplicate them. On the other hand, reengineering the working principles underlying the human brain and older human skills are much more difficult. Consequently, we do not have satisfactory success in duplicating them in computers yet. The human brain is perhaps the only device that represents the highest level of intelligence for problem solving. Unlocking the mechanisms of human brain and human problem solving may provide the necessary hints on designing intelligent machines.

The National Academy of Engineering (2008) points out the same difficulty with artificial intelligence research and lists "reverse-engineer the brain" as one of the fourteen grand challenges for engineering in the 21st century. More specifically, it states,

> *While some of thinking machines have mastered specific narrow skills -playing chess, for instance - general-purpose artificial intelligence (AI) has remained elusive.*
>
> *Part of the problem, some experts now believe, is that artificial brains have been designed without much attention to real ones. Pioneers of artificial intelligence approached thinking the way that aeronautical engineers approached flying without much learning from birds. It has turned out, though, that the secrets about how living brains work may offer the best guide to engineering the artificial variety. Discovering those secrets by reverse-engineering the brain promises enormous opportunities for reproducing intelligence the way assembly lines spit out cars or computers.*
>
> *Figuring out how the brain works will offer rewards beyond building smarter computers. Advances gained from studying the brain may in return pay dividends for the brain itself. Understanding its methods will enable engineers to simulate its activities, leading to deeper insights about how and why the brain works and fails.*

In fact, several research initiatives have been proposed along the same line. A brief summary of some of these initiatives is given in another paper (Yao 2008c).

Granular computing is relevant to the task of reverse-engineering the mechanisms of human problem solving and, hence, may have a significant impact on artificial intelligence. One reason for believing this is that granular computing focuses on a particular way of problem solving, instead of the whole spectrum of ways. At the same time, the granular way of problem solving is general and flexible enough to cover an important class of methods in problem solving. There is, therefore, a better chance to produce fruitful results. The second reason is that granular computing introduces a new view, and new associated vocabulary, to understand and study problems of artificial intelligence. The insights obtained from multiple levels of granularity may lead to new findings in understanding human intelligence and human problem solving, and the results can be eventually applied to artificial intelligence.

4 Ideas of Granular Computing in Artificial Intelligence

The basic ideas and strategies of granular computing, namely, the use of granules and multiple levels of granularity, have appeared in many places in artificial intelligence. A few key notions and issues of artificial intelligence are reviewed and re-interpreted in the light of granular computing.

4.1 Categorization, Concept Formation and Learning

Concepts are the basic units of thought that underlie human intelligence and communication. Concept formation and learning play an essential role in making sense of our perception of the world, understanding our experiences, and organizing our knowledge. Categorization, concept formation and learning are some of the fundamental human cognitive activities. Using the terminology of granular computing, a concept may be interpreted as a granule, categorization, concept formation and learning as granule construction (i.e., granulation) and granule characterization (Yao 2009a). These connections can be made explicit by considering, for example, the classical view of concepts.

In the classical view, a concept is understood as a unit of thought that consists of two parts, the intension and the extension of the concept (Smith 1989; Sowa 1984; Van Mechelen et al. 1993). The intension of a concept consists of all properties or attributes that are valid for all those objects to which the concept applies. The extension of a concept is the set of objects or entities which are instances of the concept. While the intension of a concept is an abstract description of the elements in the extension, the extension is a set of concrete examples of the concept. Typically, one associates a natural language word with a concept and calls it the name of the concept. The triplet of the intension, extension, and name of a concept forms the so-called the meaning triangle, in which one corner represents the concept, intension, thought, idea, or sense, another corner represents symbol or word, and the third corner represents the referent, object, or extension (Ogden and Richards 1946; Sowa 1984). Extensional objects are mapped into intensional concepts through perception, and concepts are coded by words in speech. The two mappings of perception and speech define an indirect mapping between words and objects (Ogden and Richards 1946; Sowa 1984).

The classical view of concepts can be immediately applied to construct a particular model of granular computing. Specifically, the extension of a concept is viewed as the set of objects forming a granule, the intension of a concept as a description or representation of the granule, and the name of the concept as the name of the granule. In this way, we obtain a meaning triangle to interpret a particular type of concept-based granules, as shown in Fig. 3. We can study granular computing in a logic setting in terms of intensions of concepts and in a set-theoretic setting in terms of extensions of concepts. Reasoning about intensions is based on logic. Inductive inference and learning attempt to derive relationships between the intensions of concepts based on the relations between the extensions of concepts.

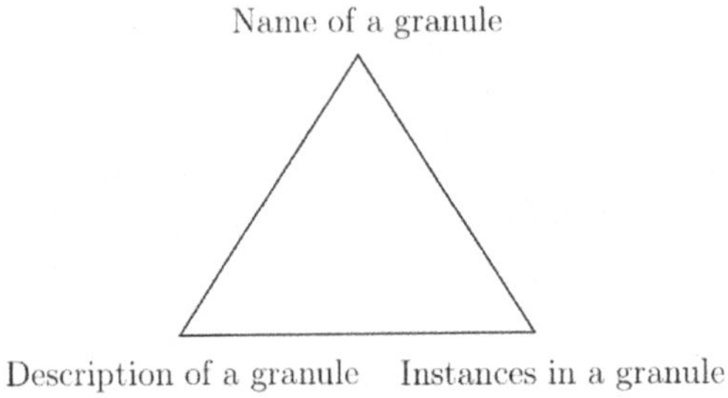

Fig. 3 Meaning triangle of a granule

There are broadly two classes of approaches for categorization, concept formation and learning, namely, unsupervised and supervised methods. In unsupervised categorization and learning, say cluster analysis (Anderberg 1973), one starts with a set of unlabeled objects. The goal is to first group similar objects together to form clusters and then to find a meaningful description or to give a name for each cluster. Clustering algorithms can be divided into non-hierarchical or hierarchical methods (Anderberg 1973). Typically, a clustering algorithm uses a distance function or a similarity function that represents a certain semantics relationship between objects. The results of a clustering algorithm may be interpreted as follows: a cluster represents the extension of a concept and its description represents the intension of the concept. They are meaningful if objects within the same cluster are closer to each other and, at the same time, objects in different clusters are further away from each other. For hierarchical clustering, objects within a cluster at a lower level in the hierarchy show a much stronger similarity than objects within a cluster at a higher level. In addition, it is possible to find a meaningful description of each cluster. To some extent, finding a meaningful description of a cluster can be interpreted as supervised concept learning, where instances and non-instances of a concept are given and the goal is to finding a description of the concept or rules defining the concept.

According to the correspondence between concepts and granules, theories, methodologies, strategies, algorithms and tools of unsupervised and supervised categorization, concept formation and learning can be easily adapted for granular computing. In fact, the former may be treated as concrete approaches to granular computing.

The theory of rough sets proposed by Pawlak (1982, 1991, 1998) can be used as an example to illustrate the ideas of granular computing with respect to categorization, concept formation and learning. A key notion of the rough set theory is an indiscernibility relation defined by a set of attributes. Assume that a finite universe of objects is described by a finite set of attributes. Given a subset of attributes, one can define an indiscernibility relation based on their values. More specifically, two objects are indiscernible if and only if they have the same values on the given subset of attributes. The indiscernibility relation is an equivalence relation and induces a partition of the set of objects, namely, a family of nonempty subsets of the universe whose union is the universe. Each equivalence class can be precisely defined by the conjunction of attribute values of an element in the class and is referred to as an elementary definable category or granule (Pawlak 1991; Yao 2007b). The union of a family of elementary granules produces another definable granule whose intension is given by a disjunction of the conjunctions of component elementary granules. A key problem of the rough set theory is the approximation of a subset of the universe, representing the extension of a concept, by using definable granules. This can be viewed as supervised concept learning. Thus, rough set theory not only provides a method to construct definable granules but also gives a method for supervised concept learning by using definable granules.

4.2 Abstraction and Granulation

Abstraction is a central concept in computer science and artificial intelligence (Colburn and Shute 2007; Frorer et al. 1997; Giunchglia and Walsh 1992; Holte and Choueiry 2003; Kramer 2007; Wing 2006; Zucker 2003). Abstraction and granulation share many common features. Results from studies of abstraction are equally applicable to granulation. Granulation may be viewed as a particular way of abstraction, in which granularity is specifically emphasized.

Although abstractions are pervasive in human reasoning and problem solving and have been used extensively in many subareas of artificial intelligence (Holte and Choueiry 2003; Nayak and Levy 1995; Zucker 2003), there still does not exist a precise, satisfactory, universal definition of abstraction (Holte and Choueiry 2003). This is not surprising as the notion of abstraction has many meanings and is interpreted differently by different people, in different disciplines and in different contexts (Frorer et al. 1997; Saitta 2003). Instead of giving a precise definition of what is abstraction, it might be more constructive to examine the main features of abstraction or the act of abstracting.

In the context of cognitive science, Barsalou (2003) considers six senses of abstraction, namely, abstraction can be interpreted in terms of categorical knowledge, the behavioral ability to generalize across category members, summary representation, schematic representation, flexible representation, and abstract concepts. In the context of artificial intelligence, Hotle and Choueiry (2003) treat abstraction as a kind of reformulation. Reformulation refers to the general idea of changing a statement or a representation of a given problem. Reformulations enable us to look at a given problem from many different points of view and are essential to creative human problem solving. Abstractions serve the same purpose.

Human problem solving depends crucially on multiple levels of abstraction. With respect to levels, Giunchglia and Walsh (1992) propose a theory of abstraction by considering an abstraction as a mapping between two formal systems at different levels; the mapping must preserve certain properties. Nayak and Levy (1995) suggest a semantic theory of abstractions by viewing abstractions as model level mappings. Zucker (2003) points out that in AI abstraction is related to the use of various levels of details and the change between levels while preserving certain useful properties. From an information quantity point of view, Zucker proposes a grounded theory of abstraction, where abstraction is represented by using abstraction operators.

Kramer (2007) considers two essential aspects of abstraction. Abstraction is a process of removing detail that allows us to simplify and to focus attention. Abstraction is a process of generalization that allows us to identify the common core or essence. Colburn and Shute (2007) compare two objectives of abstractions in mathematics and computer science, namely, information neglect and information hiding. They argue that abstraction in mathematics is information neglect, i.e., abstraction abstracts away from the details, and abstraction in computer science is information hiding, i.e., abstraction hides the underlying details.

Parnas (2007) emphasizes the meaningfulness and usefulness of abstraction, and quotes Dijkstra's definition, "An abstraction is one thing that represents several real things equally well." Parnas also identifies two distinct human skills related to abstractions: one is the ability to work with a given abstraction and the other is ability to develop a useful abstraction. In artificial intelligence, finding a useful abstraction or an adequate representation is a difficult problem when designing intelligent systems (Zucker 2003).

Based on the review of these examples of studies, although a non-exhaustive one, a few essential aspects of abstractions can be identified for granular computing:

- ignoring/eliminating detail,
- hiding detail,
- simplifying and reducing complexity,
- preserving certain properties,
- extracting core and essence,
- generalizing and summarizing,
- focusing on specific aspects,
- abstracting in multiple levels,
- truthfully representing the real thing.

They reflect the task, the process, the functionality and the meaningfulness of abstractions.

Granulation as an abstraction has been considered by some authors in artificial intelligence (Euzenat 2001; Hobbs 1985; Ye and Tsotsos 1998; Zhang and Zhang 1992), and in particular in temporal granulation (Allen 1983; Bettini and Montanari 2000; Euzenat 2001; Hornsby 2001) and spatial granulation (Bettini and Montanari 2000; Stell and Worboys 1998). Compared with abstraction, granulation focuses on the notion of granularity in the process. Hobbs (1985) outlines the general idea of granularity. A theory of granularity is motivated by the fact that humans view the world under various grain sizes and abstract only those things relevant to the present interests (Hobbs 1985). Human intelligence and flexibility, to a large degree, depend on the ability to conceptualize the world at different granularity and to switch granularity. With the theory of granularity, we can map the complexities of the real world around us into simpler theories that are computationally tractable to reason in.

The notion of granularity can be applied in many ways. For example, Poria and Garigliano (1997) consider granularity in natural language and show that granularity plays an important role in constructing explanations. Ye and Tsotsos (1998) use the notion of knowledge granularity to study an agent's action selection process. Mani (1998) applies granularity to solve the problem of underspecification of meaning in natural languages. McCalla et al. (1992) introduce the concept of granularity hierarchies and apply it in intelligent tutoring systems.

By interpreting granulation as abstraction, it is possible to adopt many features of abstraction to granular computing. At the same time, granulation offers a new view of abstraction, which has been less considered in the study of abstraction. More specifically, in some abstraction tasks, we must consider granularity introduced by abstraction.

4.3 Hierarchical Problem Solving

Abstraction or granulation usually produces multiple representations of the same problem. At different levels, different vocabularies may be used. In addition, these representations can be partially ordered to produce a hierarchical structure, namely, a granular structure. This immediately suggests a way of problem solving known as hierarchical problem solving (Knoblock 1993; Zhang and Zhang 1992).

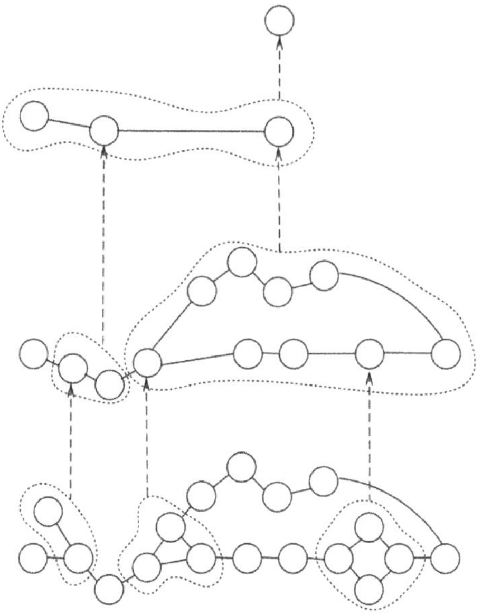

Fig. 4 Hierarchical abstractions of problem space

An effective problem solving method used in artificial intelligence is graph-based state (problem) space search, where each node of a graph represents a partial problem solution state and each arc represents a step in the problem-solving process. A solution is obtained by finding a path from a start state to a goal state in the graph. Hierarchical problem solving explores multiple levels of state spaces. Fig. 4 is an example of multilevel state space, in which each state may be viewed as a granule. A set of states at a lower level is combined into a macro state at a higher level, and thus the number of states is reduced. It is crucial that macro states are properly constructed and the macro and micro states at the different level are meaningfully linked together, so that hierarchical problem solving is possible and meaningful (Bacchus and Yang 1994; Holte et al. 1996; Knoblock 1993; Zhang and Zhang 1992).

A main reason for hierarchical problem solving is that it leads to an efficient solution under certain conditions (Bacchus and Yang 1994; Holte et al. 1996; Knoblock 1993; Zhang and Zhang 1992). Another important reason, although less considered, is the effectiveness of abstraction. A lower level with a large number of states may contain too much detail that would interfere with a right understanding of the problem. A higher level may in fact bring out more insight without the distraction from less-relevant details. The use of the right level of abstraction or granulation may produce easier to interpret and simpler results. As pointed by Kluger (2008), the trick for studying complexity and simplicity is a question of focal point. In other words, it is a question of finding a useful abstraction and knowing when to look at details and when not.

There are basically three modes of hierarchical problem solving. They are top-dwon, bottom-up and middle-out approaches (Allen and Fulton 2010; Bacchus and Yang 1994; Ginat 2001; Kintsch 2005; Knuth 1984; Lindsay and Norman 1977; Shiu and Sin 2006; Sun and Zhang 2004; Wolfe et al. 2003). As their names suggest, top-down approaches move from more abstract (i.e., coarser granulation) to more concrete (i.e., finer granulation) and in the process more detailed information is added, bottom-up approaches work the other direction and in the process some information is ignored, and middle-out approaches start from a particular level and move either up or down, depending which way is more fruitful. In general, one may use a mixture of the three modes, which is particularly useful at the exploration stage of problem solving.

Hierarchical problem solving reveals some important aspects of granular computing, for example, communicating up and down the different levels of granularity and switching between differing granularity. Methodology of hierarchical problem solving in artificial intelligence and other fields can be adapted for granular computing.

5 Conclusion

Granular computing is inspired by humans and aims at serving both humans and machines. A grand challenge for granular computing is to revere-engineer a particular way of human problem solving based on multiple levels of granularity. Once we understand the underlying principles, we can empower everyone to be a better problem solver on one hand and implement machine problem solving on the other. Studies of granular computing and artificial intelligence support each other. Existing results from artificial intelligence enrich granular computing and granular computing research sheds new lights on methods in artificial intelligences.

Acknowledgments. This work is partially supported by a Discovery Grant from NSERC Canada. This chapter is based on an unpublished statement at the RSCTC 2008 Panel on RS and SC Challenges in AI and KDD organized by Professors Andrzej Skowron and Chien-Chung Chan; the author is grateful to their kind support and encouragement. The author thanks the editors of the volume, Professors Witold Pedrycz and Shyi-Ming Chen, and the reviewers for their constructive comments.

References

Allen, J.F.: Maintaining knowledge about temporal intervals. Communications of the ACM 26, 832–843 (1983)

Allen, J.I., Fulton, E.A.: Top-down, bottom-up or middle-out? avoiding extraneous detail and over-generality in marine ecosystem models. Progress in Oceanography 84, 129–133 (2010)

Anderberg, M.R.: Cluster Analysis for Applications. Academic Press, New York (1973)

Bacchus, F., Yang, Q.: Downward refinement and the efficiency of hierarchical problem solving. Artificial Intelligence 71, 43–100 (1994)

Bargiela, A., Pedrycz, W.: Granular Computing: An Introduction. Kluwer Academic Publishers, Boston (2002)

Bargiela, A., Pedrycz, W.: Toward a theory of granular computing for human-centered information processing. IEEE Transactions on Fuzzy Systems 16, 320–330 (2008)

Bargiela, A., Pedrycz, W. (eds.): Human-Centric Information Processing Through Granular Modeling. Springer, Berlin (2009)

Barsalou, L.W.: Abstraction in perceptual symbol systems. Philosophical Transactions of the Royal Society B: Biological Sciences 358, 1177–1187 (2003)

Bettini, C., Montanari, A. (eds.): Spatial and Temporal Granularity: Papers from the AAAI Workshop, Technical Report WS-00-08. The AAAI Press, Menlo Park, CA (2000)

Cermak, L.S., Craik, F.I.M. (eds.): Levels of Processing in Human Memory. Lawrence Erlbraum Associates, Hillsdale (1979)

Colburn, T., Shute, G.: Abstraction in computer science. Minds & Machines 17, 169–184 (2007)

Euzenat, J.: Granularity in relational formalisms - with application to time and space representation. Computational Intelligence 17, 703–737 (2001)

Frorer, P., Hazzan, O., Manes, M.: Revealing the faces of abstraction. International Journal of Computers for Mathematical Learning 2, 217–228 (1997)

Gilhooly, K.J. (ed.): Human and Machine Problem Solving. Plenum Press, New York (1989)

Ginat, D.: Starting top-down, refining bottom-up, sharpening by zoom-in. ACM SIGCSE Bulletin 33, 28–31 (2001)

Giunchglia, F., Walsh, T.: A theory of abstraction. Artificial Intelligence 56, 323–390 (1992)

Hobbs, J.R.: Granularity. In: Proceedings of the Ninth International Joint Conference on Artificial Intelligence, Los Angeles, California, USA, pp. 432–435 (1985)

Holte, R.C., Choueiry, B.Y.: Abstraction and reformulation in artificial intelligence. Philosophical Transactions of the Royal Society B: Biological Sciences 358, 1197–1204 (2003)

Holte, R.C., Mkadmi, T., Zimmer, R.M., MacDonald, A.J.: Speeding up problem-solving by abstraction: a graph oriented approach. Artificial Intelligence 85, 321–361 (1996)

Hornsby, K.: Temporal zooming. Transactions in GIS 5, 255–272 (2001)

Inuiguchi, M., Hirano, S., Tsumoto, S. (eds.): Rough Set Theory and Granular Computing. Springer, Berlin (2003)

Jankowski, A., Skowron, A.: Rough-granular computing in human-centric information processing. In: Cyran, K.A., Kozielski, S., Peters, J.F., Stanczyk, U., Wakulicz-Deja, A. (eds.) Man-Machine Interactions. AISC, vol. 59, pp. 23–42. Springer, Berlin (2009)

Keet, C.M.: A Formal Theory of Granularity.Dissertation, KRDB Research Centre for Knowledge and Data, Faculty of Computer Science, Free University of Bozen-Bolzano, Italy (2008), http://www.meteck.org/files/AFormalTheoryOfGranularity_CMK08.pdf

Kintsch, W.: An overview of top-down and bottom-up effects in comprehension: the CI perspective. Discourse Processes 39, 125–128 (2005)

Kluger, J.: Simplicity, Why Simple Things Become Complex and Complex Thing Can be Made Simple. Hyperion, New York (2008)

Knoblock, C.A.: Generating Abstraction Hierarchies: An Automated Approach to Reducing Search in Planning. Kluwer Academic Publishers, Boston (1993)

Knuth, D.E.: Literate programming. The Computer Journal 27, 97–111 (1984)

Kramer, J.: Is abstraction the key to computing? Communications of the ACM 50, 36–42 (2007)

Lin, T.Y., Yao, Y.Y., Zadeh, L.A. (eds.): Data Mining, Rough Sets and Granular Computing. Physica-Verlag, Heidelberg (2002)

Lindsay, P.H., Norman, D.A.: An Introduction to Psychology, 2nd edn. Academic Press, New York (1977)

Mani, I.: A theory of granularity and its application to problems of polysemy and underspecification of meaning. In: Cohn, A., Schubert, L., Shapiro, S. (eds.) Proceedings of the 6th International Conference on Principles of Knowledge Representation and Reasoning (KR 1998), San Francisco, CA, pp. 245–255 (1998)

McCalla, G., Greer, J., Barrie, B., Pospisil, P.: Granularity hierarchies. Computers & Mathematics with Applications 23, 363–375 (1992)

Minsky, M.: The Emotion Machine: Commonsense Thinking, Artificial Intelligence, and the Future of the Human Mind. Simon & Schuster Paperbacks, New York (2007)

Moravec, H.: Mind Children, the Future of Robot and Human Intelligence. Harvard University Press, Cambridge (1988)

Nakashima, H.: AI as complex information processing. Minds and Machines 9, 57–80 (1999)

The National Academy of Engineering ,Grand Challenges for Engineering (2008) , http://www.engineeringchallenges.org (accessed September 19, 2008)

Nayak, P.P., Levy, A.Y.: A semantic theory of abstraction. In: Proceedings of the 14th International Joint Conference on Artificial Intelligence (JICAI), Montreal, Quebec, Canada, pp. 196–202 (1995)

Newell, A., Shaw, J.C., Simon, H.A.: Elements of a theory of human problem solving. Psychological Review 65, 151–166 (1958)

Newell, A., Simon, H.A.: Human Problem Solving. Prentice-Hall, Englewood Cliffs (1972)

Ogden, C.K., Richards, I.A.: The Meaning of Meaning: a Study of the Influence of Language upon Thought and of the Science of Symbolism, 8th edn. Harcourt Brace, New York (1946)

Parnas, D.L.: Use the simplest model, but not too simple. Communications of the ACM 50, 7 (2007)
Pawlak, Z.: Rough sets. International Journal of Computer and Information Sciences 11, 341–356 (1982)
Pawlak, Z.: Rough Sets. Theoretical Aspects of Reasoning about Data (1991)
Pawlak, Z.: Granularity of knowledge, indiscernibility and rough sets. In: Proceedings of the IEEE International Conference on Fuzzy Systems, Anchorage, AK, USA, pp. 106–110 (1998)
Pedrycz, W. (ed.): Granular Computing: An Emerging Paradigm. Physica-Verlag, Heidelberg (2001)
Pedrycz, W.: Fuzzy sets as a user-centric processing framework of granular computing. In: Pedrycz, W., Skowron, A., Kreinovich, V. (eds.) Handbook of Granular Computing, pp. 97–139. Wiley Interscience, New York (2008)
Pedrycz, W., Skowron, A., Kreinovich, V. (eds.): Handbook of Granular Computing. Wiley Interscience, New York (2008)
Pinker, S.: How the Mind Works. WW Norton & Company, New York (1997)
Poria, S., Garigliano, R.: Granularity for explanation. In: Costa, E. (ed.) EPIA 1997. LNCS, vol. 1323, pp. 331–336. Springer, Heidelberg (1997)
Saitta, L.: A Theme Issue "The abstraction paths: from experience to concept". Philosophical Transactions of the Royal Society B: Biological Sciences 358, 1173–1307 (2003)
Shiu, L.P., Sin, C.Y.: Top-down, middle-out, and bottom-up processes: a cognitive perspective of teaching and learning economics. International Review of Economics Education 5, 60–72 (2006)
Simon, H.A.: Information-processing theory of human problem solving. In: Estes, W.K. (ed.) Handbook of Learning and Cognitive Processes, Human Information Processing, Hillsdale, Erlbaum, NJ, vol. 5, pp. 271–295 (1978)
Simon, H.A.: Information processing models of cognition. Annual Review of Psychology 30, 363–396 (1979)
Smith, E.E.: Concepts and induction. In: Posner, M.I. (ed.) Foundations of Cognitive Science, pp. 501–526. The MIT Press, Cambridge (1989)
Sowa, J.F.: Conceptual Structures: Information Processing in Mind and Machine, Reading. Addison-Wesley, Massachusetts (1984)
Stell, J.G., Worboys, M.F.: Stratified map spaces: a formal basis for multi-resolution spatial databases. In: Proceedings of the 8th International Symposium on Spatial Data Handling, Vancouver, BC, Canada, pp. 180–189 (1998)
Stewart, D.: Interview with Herbert Simon. OMNI Magazine (1994)
Sun, R., Zhang, X.: Top-down versus bottom-up learning in cognitive skill acquisition. Cognitive Systems Research 5, 63–89 (2004)
Van Mechelen, I., Hampton, J., Michalski, R.S., Theuns, P. (eds.): Categories and Concepts, Theoretical Views and Inductive Data Analysis. Academic Press, New York (1993)
Wiederhold, G.: Mediators in the architecture of future information systems. IEEE Computer 25, 38–49 (1992)
Wing, J.: Computational thinking. Communications of the ACM 49, 33–35 (2006)
Wolfe, J.M., Butcher, S.J., Lee, C., Hyle, M.: Changing your mind: on the contributions of top-down and bottom-up guidance in visual search for feature singletons. Journal of Experimental Psychology 29, 483–502 (2003)
Yao, J.T.: A ten-year review of granular computing. In: Proceedings of 2007 IEEE International Conference on Granular Computing, Silicon Valley, CA, USA, pp. 734–739 (2007)

Yao, J.T.: Granular computing: a new paradigm in information processing. In: Proceedings of 2008 North American Simulation Technology Conference, Montreal, Canada, pp. 5–6 (2008a)
Yao, J.T.: Recent developments in granular computing: a bibliometrics study. In: Proceedings of 2008 IEEE International Conference on Granular Computing, Hangzhou, China, pp. 74–79 (2008b)
Yao, J.T. (ed.): Novel Developments in Granular Computing: Applications for Advanced Human Reasoning and Soft Computation. IGI Global, Hershey (2010)
Yao, Y.Y.: Granular computing: basic issues and possible solutions. In: Proceedings of the 5th Joint Conference on Information Sciences, Atlantic, NJ, USA, pp. 186–189 (2000)
Yao, Y.Y.: Perspectives of granular computing. In: Proceedings of 2005 IEEE International Conference on granular computing, Beijing, China, vol. 1, pp. 85–90 (2005)
Yao, Y.Y.: Three perspectives of granular computing. Journal of Nanchang Institute of Technology 25, 16–21 (2006)
Yao, Y.: The art of granular computing. In: Kryszkiewicz, M., Peters, J.F., Rybiński, H., Skowron, A. (eds.) RSEISP 2007. LNCS (LNAI), vol. 4585, pp. 101–112. Springer, Heidelberg (2007a)
Yao, Y.: A note on definability and approximations. In: Peters, J.F., Skowron, A., Marek, V.W., Orłowska, E., Słowiński, R., Ziarko, W.P. (eds.) Transactions on Rough Sets VII. LNCS, vol. 4400, pp. 274–282. Springer, Heidelberg (2007b)
Yao, Y.Y.: Granular computing: past, present and future. In: Proceedings of 2008 IEEE International Conference on Granular Computing, Hangzhou, China, pp. 80–85 (2008a)
Yao, Y.Y.: A unified framework of granular computing. In: Pedrycz, W., Skowron, A., Kreinovich, V. (eds.) Handbook of Granular Computing, pp. 401–410. Wiley Interscience, New York (2008b)
Yao, Y.Y.: The rise of granular computing. Journal of Chongqing University of Posts and Telecommunications (Natural Science Edition) 20, 299–308 (2008c)
Yao, Y.Y.: Interpreting concept learning in cognitive informatics and granular computing. IEEE Transactions on System, Man and Cybernetics, B 39, 855–866 (2009a)
Yao, Y.Y.: Integrative levels of granularity. In: Bargiela, A., Pedrycz, W. (eds.) Human-Centric Information Processing. SCI, vol. 182, pp. 31–47. Springer, Berlin (2009b)
Yao, Y.Y.: Human-inspired granular computing. In: Yao, J.T. (ed.) Novel Developments in Granular Computing: Applications for Advanced Human Reasoning and Soft Computation, pp. 1–15. IGI Global, Hershey (2010)
Ye, Y., Tsotsos, J.K.: Knowledge granularity and action selection. In: Giunchiglia, F. (ed.) AIMSA 1998. LNCS (LNAI), vol. 1480, pp. 475–488. Springer, Heidelberg (1998)
Zadeh, L.A.: Fuzzy sets and information granularity. In: Gupta, N., Ragade, R., Yager, R. (eds.) Advances in Fuzzy Set Theory and Applications, pp. 3–18. North-Holland, Amsterdam (1979)
Zadeh, L.A.: Towards a theory of fuzzy information granulation and its centrality in human reasoning and fuzzy logic. Fuzzy Sets and Systems 90, 111–127 (1997)
Zhang, B., Zhang, L.: Theory and Applications of Problem Solving. North-Holland, Amsterdam (1992)
Zucker, J.D.: A grounded theory of abstraction in artificial intelligence. Philosophical Transactions of the Royal Society B: Biological Sciences 358, 1293–1309 (2003)

Calculi of Approximation Spaces in Intelligent Systems

Andrzej Skowron, Jarosław Stepaniuk, and Roman Swiniarski

Abstract. Solving complex real-life problems requires new approximate reasoning methods based on new computing paradigms. One such recently emerging computing paradigm is Rough–Granular Computing (Pedrycz et al. 2008, Stepaniuk 2008) (RGC, in short). The RGC methods have been successfully applied for solving complex problems in areas such as identification of behavioral patterns by autonomous systems, web mining, and sensor fusion. In RGC, an important role play special information granules (Zadeh 1979, Zadeh 2006) called as approximation spaces. These higher order granules are used for approximation of concepts or, in a more general sense, complex granules. We discuss some generalizations of the approximation space definition introduced in 1994 (Skowron and Stepaniuk 1994, Skowron and Stepaniuk 1996, Stepaniuk 2008). The generalizations are motivated by real-life applications of intelligent systems and are related to inductive extensions of approximation spaces.

Keywords: Rough Sets, Approximation Space, Approximation Space Extension, Granular Computing, Complex Granule Approximation, Intelligent Systems.

Andrzej Skowron
Institute of Mathematics, The University of Warsaw, Banacha 2, 02-097 Warsaw, Poland
e-mail: skowron@mimuw.edu.pl

Jarosław Stepaniuk
Department of Computer Science, Białystok University of Technology,
Wiejska 45A, 15-351 Białystok, Poland
e-mail: j.stepaniuk@pb.edu.pl

Roman Swiniarski
Department of Mathematical and Computer Sciences,
5500 Campanile Drive San Diego, CA 92182, USA and
Institute of Computer Science Polish Academy of Sciences,
Ordona 21, 01-237 Warsaw, Poland
e-mail: rswiniar@sciences.sdsu.edu

W. Pedrycz and S.-M. Chen (Eds.): Granular Computing and Intell. Sys., ISRL 13, pp. 35–55.
springerlink.com

1 Introduction

Rough sets, due to Zdzisław Pawlak, can be represented by pairs of sets which give the lower and the upper approximation of the original sets. In the standard version of rough set theory, an approximation space is based on the indiscernibility equivalence relation. Approximation spaces belong to the broad spectrum of basic issues investigated in rough set theory (see, *e.g.*, (Bazan et al. 2006, Greco et al. 2008, Jankowski and Skowron 2008, Skowron and Stepaniuk 1994, Skowron and Stepaniuk 1996, Słowiński and Vanderpooten 2000, Skowron et al. 2006, Stepaniuk 2008, Zhu 2009). Over the years different aspects of approximation spaces were investigated and many generalizations of the approach based on indiscernibility equivalence relation (Pawlak and Skowron 2007) were proposed. In this chapter, we discuss some aspects of generalizations of approximation spaces investigated in (Skowron and Stepaniuk 1994, Skowron and Stepaniuk 1996, Stepaniuk 2008) that are important for real-life applications, e.g., in searching for approximation of complex concepts (see, e.g., (Bazan et al. 2006, Bazan 2008). Rough set based strategies for extension of such approximation spaces from samples of objects onto their extensions are discussed. The extensions of approximation spaces can be treated as operations for inductive reasoning. The investigated approach enables us to present the uniform foundations for inducing approximations of different kinds of higher order granules such as concepts, classifications, or functions. In particular, we emphasize the fundamental role of approximation spaces for inducing diverse kinds of classifiers used in machine learning or data mining. The searching problem for relevant approximation spaces and their extensions is of high computational complexity. Hence, efficient heuristics should be used in searching for approximate solutions of this problem. Moreover, in hierarchical learning of complex concepts many different approximation spaces should be discovered. Learning of such concepts can be supported by domain knowledge and ontology approximation (see, e.g., (Bazan 2008, Bazan et al. 2006)).

The chapter is organized as follows. In Section 2 we discuss basic notions for our approach. In Section 3 we present a generalization of the approximation space definition from (Skowron and Stepaniuk 1994, Skowron and Stepaniuk 1996, Stepaniuk 2008). In particular, in Subsection 3.2 we present new rough set approach to function approximation.

2 Basic Notions

2.1 Attributes, Signatures of Objects and Two Semantics

In (Pawlak and Skowron 2007) any attribute a is defined as a function from the universe of objects U into the set of attribute values V_a. However, in applications we expect that the value of attribute should be also defined for objects from extensions of U, *i.e.*, for new objects which can be perceived in the future[1]. The universe U

[1] Objects from U are treated as labels of real perceived objects.

is only a sample of possible objects. This requires some modification of the basic definitions of attribute and signature of objects (Pawlak 1991, Pawlak and Skowron 2007).

One can give interpretation of attributes using a concept of interaction. In this case, we treat attributes as granules and we consider their interactions with environments. If a is a given attribute and e denotes a state of environment then the result of interaction between a and e is equal to a pair (l_e, v), where l_e is a label of e and $v \in V_a$. Analogously, if $IS = (U, A)$ is a given information system and e denotes a state of environment then by interaction of IS and e we obtain the information system $IS' = (U \cup \{l_e\}, A^*)$, where $A^* = \{a^* : a \in A\}$ and $a^*(u) = a(u)$ for $u \in U$ and $a^*(l_e) = v$ for some $v \in V_a$. Hence, information systems are dynamic objects created through interaction of already existing information systems with environments. Note that the initial information system can be empty, i.e. the set of objects of this information system is empty. Moreover, let us observe that elements of U are labels of the environment states rather than states.

One can represent any attribute by a family of formulas and interpret the attribute as the result of interaction of this set with the environment. In this case, we assume that for any attribute a under consideration there is given a relational structure R_a. Together with the simple structure $(V_a, =)$ (Pawlak and Skowron 2007), some other relational structures R_a with the carrier V_a for $a \in A$ and a signature τ are considered. We also assume that with any attribute a is identified a set of some generic formulas $\{\alpha_i\}_{i \in J}$ (where J is a set of indexes) interpreted over R_a as a subsets of V_a, i.e., $\|\alpha_i\|_{R_a} = \{v \in V_a : R_a, v \models \alpha_i\}$. Moreover, it is assumed that the set $\{\|\alpha_i\|_{R_a}\}_{i \in J}$ is a partition of V_a. Perception of an object u by a given attribute a is represented by selection of a formula α_i and a value $v \in V_a$ such that $v \in \|\alpha_i\|_{R_a}$. Using an intuitive interpretation one can say that such a pair (α_i, v) is selected from $\{\alpha_i\}_{i \in J}$ and V_a, respectively, as the result of sensory measurement. We assume that for a given set of attributes A and any object u the signature of u relative to A is given by $Inf_A(u) = \{(a, \alpha_u^a, v) : a \in A\}$, where (α_u^a, v) is the result of sensory measurement by a on u.

Let us observe that a triple (a, α_u^a, v) can be encoded by the atomic formula $a = v$ with interpretation

$$\|a = v\|_{U^*} = \{u \in U^* : (a, \alpha_u^a, v) \in Inf_a(u) \text{ for some } \alpha_u^a\}. \quad (1)$$

We also write (a, v) instead of (a, α_u^a, v) if this not lead to confusion.

One can also consider a soft version of the attribute definition. In this case, we assume that the semantics of the family $\{\alpha_i\}_{i \in J}$ is given by fuzzy membership functions for α_i and the set of these functions define a fuzzy partition (Klir 2007).

We construct granular formulas from atomic formulas corresponding to the considered attributes. In the consequence, the satisfiability of such formulas is defined if the satisfiability of atomic formulas is given as the result of sensor measurement. Hence, one can consider for any constructed formula α over atomic formulas its semantics $\|\alpha\|_U \subseteq U$ over U as well as the semantics $\|\alpha\|_U^* \subseteq U^*$ over U^*, where $U \subseteq U^*$ (see Figure 1). The difference between these two cases is the following. In the case of U, one can compute $\|\alpha\|_U \subseteq U$ but in the case $\|\alpha\|_{U^*} \subseteq U^*$, for objects from $U^* \setminus U$, there is no information about their membership relative to $\|\alpha\|_{U^* \setminus \|\alpha\|_U}$. One can estimate the satisfiability of α for objects $u \in U^* \setminus U$ only after the relevant sensory measurements on u are performed. In particular, one can use some methods for estimation of relationships among semantics of formulas over U^* using the relationships among semantics of these formulas over U. For example, one can apply statistical methods. This step is crucial in investigation of extensions of approximation spaces relevant for inducing classifiers from data (see, e.g., (Bazan et al 2006, Skowron et al. 2006, Pedrycz et al. 2008).

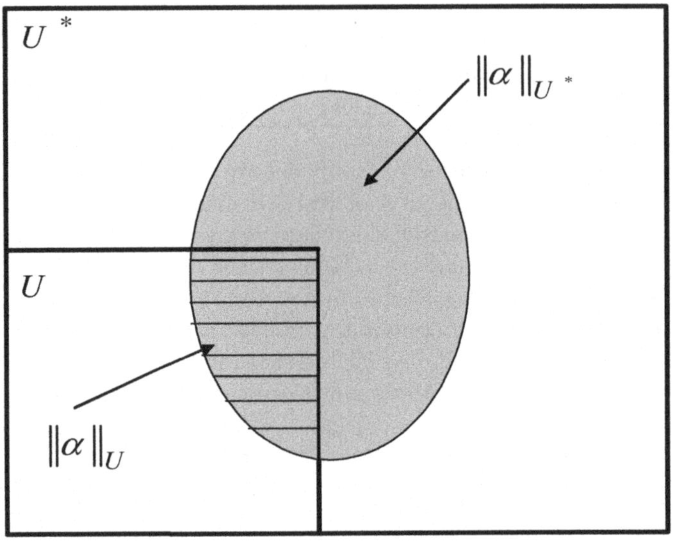

Fig. 1 Two semantics of α over U and U^*, respectively

2.2 Uncertainty Function

In (Skowron and Stepaniuk 1994, Skowron and Stepaniuk 1996, Stepaniuk 2008) the uncertainty function I defines for every object x from a given sample U of objects, a set of objects described similarly to x. The set $I(x)$ is called the neighborhood of x.

Let $P_\omega(U^*) = \bigcup_{i \geq 1} P^i(U^*)$, where $P^1(U^*) = P(U^*)$ and $P^{i+1}(U^*) = P(P^i(U^*)))$ for $i \geq 1$. For example, if $card(U^*) = 2$ and $U^* = \{x_1, x_2\}$, then we obtain $P^1(U^*) = \{\emptyset, \{x_1\}, \{x_2\}, \{x_1, x_2\}\}$, $P^2(U^*) = \{\emptyset, \{\emptyset\}, \{\{x_1\}\}, \{\{x_2\}\}, \{\{x_1, x_2\}\}, \{\emptyset, \{x_1\}\}, \{\emptyset, \{x_2\}\}, \{\emptyset, \{x_1, x_2\}\}, \ldots\}, \ldots$. If $card(U^*) = n$, where n is a positive natural number, then $card(P^1(U^*)) = 2^n$ and $card(P^{n+1}(U^*)) = 2^{card(P^n(U^*))}$, for $n \geq 1$. For example, $card(P^3(U^*)) = 2^{2^{2^n}}$.

In this chapter, we consider uncertainty functions of the form $I : U^* \longrightarrow P_\omega(U^*)$. The values of uncertainty functions are called granular neighborhoods. These granular neighborhoods are defined by the so called granular formulas. The values of such uncertainty functions are not necessarily from $P(U^*)$ but from $P_\omega(U^*)$. In the following sections, we will present more details on granular neighborhoods and granular formulas. Figure 2 presents an illustrative example of the uncertainty function with values in $P^2(U^*)$ rather than in $P(U^*)$. The discussed here generalization of neighborhoods are also motivated by the necessity of modeling or discovery of complex structural objects in solving problems of pattern recognition, machine learning, or data mining. These structural objects (granules) can be defined as sets on higher levels of the powerset hierarchy. Among examples of such granules are indiscernibility or similarity classes of patterns or relational structures discovered in images, clusters of time widows, indiscernibility or similarity classes of sequences of time windows representing processes, behavioral graphs (for more details see, e.g., (Skowron and Szczuka 2010, Bazan 2008)).

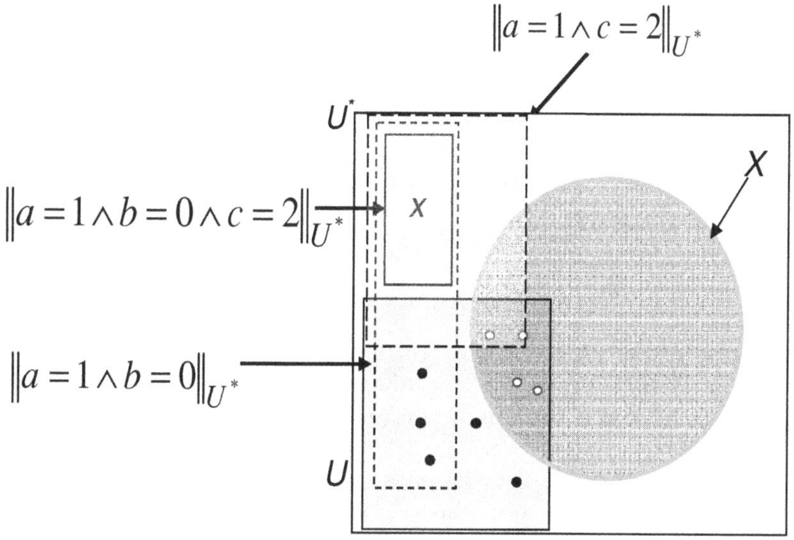

Fig. 2 Uncertainty function $I : U^* \to P^2(U^*)$. The neighborhood of $x \in U^* \setminus U$, where $Inf_A(x) = \{(a,1),(b,0),(c,2)\}$, does not contain training cases from U. The generalizations of this neighborhood described by formulas $\|a=1 \wedge c=2\|_{U^*}$ and $\|a=1 \wedge b=2\|_{U^*}$ have non empty intersections with U.

If $X \in P_\omega(U^*)$ and $U \subseteq U^*$ then by $X \upharpoonright U$ we denote the set defined as follows (i) if $X \in P(U^*)$ then $X \upharpoonright U = X \cap U$ and (ii) for any $i \geq 1$ if $X \in P^{i+1}(U^*)$ then $X \upharpoonright U = \{Y \upharpoonright U : Y \in X\}$. For example, if $U = \{x_1\}$, $U^* = \{x_1, x_2\}$ and $X = \{\{x_2\}, \{x_1, x_2\}\}$ ($X \in P^2(U^*)$), then $X \upharpoonright U = \{Y \upharpoonright U : Y \in X\} = \{Y \cap U : Y \in X\} = \{\emptyset, \{x_1\}\}$.

2.3 Rough Inclusion Function

The second component of any approximation space is the rough inclusion function (Skowron and Stepaniuk 1996, Stepaniuk 2008).

One can consider general constraints which the rough inclusion functions should satisfy. In this section, we present only some examples of rough inclusion functions.

The rough inclusion function $\nu : P(U) \times P(U) \to [0,1]$ defines the degree of inclusion of X in Y, where $X,Y \subseteq U$.[2]

In the simplest case the standard rough inclusion function can be defined by (see, e.g., (Skowron and Stepaniuk 1996, Pawlak and Skowron 2007)):

$$\nu_{SRI}(X,Y) = \begin{cases} \frac{card(X \cap Y)}{card(X)} & \text{if } X \neq \emptyset \\ 1 & \text{if } X = \emptyset. \end{cases} \quad (2)$$

Some illustrative example is given in Table 1.

Table 1 Illustration of Standard Rough Inclusion Function

X	Y	$\nu_{SRI}(X,Y)$
$\{x_1,x_3,x_7,x_8\}$	$\{x_2,x_4,x_5,x_6,x_9\}$	0
$\{x_1,x_3,x_7,x_8\}$	$\{x_1,x_2,x_4,x_5,x_6,x_9\}$	0.25
$\{x_1,x_3,x_7,x_8\}$	$\{x_1,x_2,x_3,x_7,x_8\}$	1

It is important to note that an inclusion measure expressed in terms of the confidence measure, widely used in data mining, was considered by Łukasiewicz (Łukasiewicz 1913) long time ago in studies on assigning fractional truth values to logical formulas.

The rough inclusion function was generalized in rough mereology (Polkowski and Skowron 1996). For definition of inclusion function for more general granules, e.g., partitions of objects one can use measure based on positive region (Pawlak and Skowron 2007), entropy (Hastie et al. 2008) or rough entropy (Pal et al. 2005, Małyszko and Stepaniuk 2010). Inclusion measures for more general granules were also investigated (Skowron 2001, Bianucci and Cattaneo 2009). However, more work in this direction should be done, especially on inclusion of granules with complex structures, in particular for granular neighborhoods.

3 Approximation Spaces

In this section, we present a generalization of the approximation space definition from (Skowron and Stepaniuk 1994, Skowron and Stepaniuk 1996, Stepaniuk 2008).

In applications, approximation spaces are constructed for a given concept or a family of concepts creating a partition (in the case of classification) rather than for the class of all concepts. Then the searching for components of approximation space relevant for the concept approximation becomes feasible. The concept is given only

[2] We assume that U is a finite sample of objects.

an a sample U of objects. We often restrict the definition of components of approximation space to objects from U^* and/or some patterns from $P_\omega(U^*)$ or from $P_\omega(U^\bullet)$, where $U^* \subseteq U^\bullet$ necessary for approximation of a given concept X only. The definitions of the uncertainty function and the inclusion function can be restricted to some subsets of U^* and $P_\omega(U^*)$ (or $U^* \subseteq U^\bullet$), respectively, which are relevant for approximated concept(s).

Moreover, the optimization requirements for the lower approximation and upper approximation are then also restricted to the given concept X. These requirements are expressing closeness of the induced concept approximations to the approximation of the concept(s) on the given sample of objects and are combined with the description length of constructed approximations for obtaining the relevant quality measures. Usually, the uncertainty functions and the rough inclusion functions are parameterized. Then searching (in the family of the these functions defined by possible values of parameters) for the (semi)optimal uncertainty function and the rough inclusion function relative to the selected quality measure becomes feasible.

Definition 1. An approximation space over a set of attributes A for a concept $X \subseteq U^*$ given on a sample $U \subseteq U^*$ of objects is a system

$$AS = (U, U^*, I, \nu, L) \tag{3}$$

where

- U is a sample of objects with known signatures relative to a given set of attributes A,
- L is a language of granular formulas defined over atomic formulas corresponding to generic formulas from signatures (see Section 2.2),
- the set U^* is such that for any object $u \in U^*$ the signature $Inf_A(u)$ of u relative to A can be obtained as the result of sensory measurements on u,
- $I: U^* \to P_\omega(U^\bullet)$ is an uncertainty function, where $U^* \subseteq U^\bullet$; we assume that the granular neighborhood $I(u)$ is computable from $Inf_A(u)$, i.e., from $Inf_A(u)$ it is possible to compute a formula $\alpha_{Inf_A(u)} \in L$ such that $I(x) = \|\alpha_{Inf_A(u)}\|_{U^*}$,
- $\nu : P_\omega(U^\bullet) \times P_\omega(U^\bullet) \to [0,1]$ is a partial rough inclusion function, such that form any $x \in U^*$ the value $\nu(I(x), X)$ is defined for the considered concept X.

In Section 3.2, we consider an uncertainty function with values in $P(P(U^*) \times P(R_+))$, where R_+ is the set of reals. Hence, we assume that the values of the uncertainty function I may belong to the space of possible patterns from $P_\omega(U^\bullet)$, where $U^* \subseteq U^\bullet = U^* \cup R_+$.

The partiality of the rough inclusion makes it possible to define the values of this functions on relevant patterns for approximation only.

We assume that the lower approximation operation $LOW(AS, X)$ and the upper approximation operation $UPP(AS, X)$ of the concept X in the approximation space AS satisfy the following condition:

$$\nu(LOW(AS,X), UPP(AS,X)) = 1. \tag{4}$$

Usually the uncertainty function and the rough inclusion function are parameterized. In this parameterized family of approximation spaces, one can search for an approximation space enabling us to approximate the concept X restricted to a given sample U with the satisfactory quality. The quality of approximation can be expressed by some quality measures. For example, one can use the following criterion:

1. $LOW(AS,X) \upharpoonright U$ is included in $X \upharpoonright U$ to a degree at least deg, i.e., $\nu(LOW(AS,X) \upharpoonright U, X \upharpoonright U)) \geq deg$,
2. $X \upharpoonright U$ is included in $UPP(AS,X) \upharpoonright U$ to a degree at least deg, this means that, $\nu(UPP(AS,X \upharpoonright U), X \upharpoonright U) \geq deg$,

where deg is a given threshold from the interval $[0,1]$.

The above condition expresses the degree to which at least the induced approximations in AS are close to the concept X on the sample U. One can also introduce the description length of the induced approximations. A combination of these two measures can be used as the quality measure for the induced approximation space. Then the searching problem for the relevant approximation space can be considered as the optimization problem relative to this quality measure. This approach may be interpreted as a form of the minimal description length principle (Rissanen 1985). The result of optimization can be checked against a testing sample. This enables us to estimate the quality of approximation. Note that further optimization can be performed relative to parameters of the selected quality measure.

3.1 Approximations and Decision Rules

In this section, we discuss generation of approximations on extensions of samples of objects.

In the example we illustrate how the approximations of sets (concepts) can be estimated using only partial information about these sets. Moreover, the example introduces uncertainty functions with values in $P^2(U)$ and rough inclusion functions defined for sets from $P^2(U)$.

Let us assume that $DT = (U, A \cup \{d\})$ is a decision table, where $U = \{x_1, \ldots, x_9\}$ is a set of objects and $A = \{a,b,c\}$ is a set of condition attributes (see Table 2).

In DT we compute two decision reducts: $\{a,b\}$ and $\{b,c\}$. We obtain the set $Rule\_set = \{r_1, \ldots, r_{12}\}$ of minimal (reduct based) decision rules (Pawlak and Skowron 2007).

From x_1 we obtain two rules:
r_1 : **if** $a = 1$ **and** $b = 1$ **then** $d = 1$, r_2 : **if** $b = 1$ **and** $c = 0$ **then** $d = 1$.
From x_2 and x_4 we obtain two rules:
r_3 : **if** $a = 0$ **and** $b = 2$ **then** $d = 1$, r_4 : **if** $b = 2$ **and** $c = 0$ **then** $d = 1$.
From x_5 we obtain one new rule:
r_5 : **if** $a = 0$ **and** $b = 1$ **then** $d = 1$.
From x_3 we obtain two rules:
r_6 : **if** $a = 1$ **and** $b = 0$ **then** $d = 0$, r_7 : **if** $b = 0$ **and** $c = 1$ **then** $d = 0$.
From x_6 we obtain two rules:
r_8 : **if** $a = 0$ **and** $b = 0$ **then** $d = 0$, r_9 : **if** $b = 0$ **and** $c = 0$ **then** $d = 0$.

Table 2 Decision table over the set of objects U

	a	b	c	d
x_1	1	1	0	1
x_2	0	2	0	1
x_3	1	0	1	0
x_4	0	2	0	1
x_5	0	1	0	1
x_6	0	0	0	0
x_7	1	0	2	0
x_8	1	2	1	0
x_9	0	0	1	0

From x_7 we obtain one new rule:
r_{10} : **if** $b = 0$ **and** $c = 2$ **then** $d = 0$.
From x_6 we obtain two rules:
r_{11} : **if** $a = 1$ **and** $b = 2$ **then** $d = 0$, r_{12} : **if** $b = 2$ **and** $c = 1$ **then** $d = 0$.
Let $U^* = U \cup \{x_{10}, x_{11}, x_{12}, x_{13}, x_{14}\}$ (see Table 3).

Table 3 Decision table over the set of objects $U^* - U$

	a	b	c	d	d_{class}
x_{10}	0	2	1	1	1 from r_3 or 0 from r_{12}
x_{11}	1	2	0	0	1 from r_4 or 0 from r_{11}
x_{12}	1	2	0	0	1 from r_4 or 0 from r_{11}
x_{13}	0	1	2	1	1 from r_5
x_{14}	1	1	2	1	1 from r_1

Let $h : [0,1] \to \{0, 1/2, 1\}$ be a function defined by

$$h(t) = \begin{cases} 1 & \text{if } t > 1/2 \\ 1/2 & \text{if } t = 1/2 \\ 0 & \text{if } t < 1/2. \end{cases} \tag{5}$$

Below we present an example of the uncertainty and rough inclusion functions:

$$I(x) = \{\|lh(r)\|_{U^*} : x \in \|lh(r)\|_{U^*} \text{ and } r \in Rule\_set\}, \tag{6}$$

where $x \in U^*$ and $lh(r)$ denotes the formula on the left hand side of the rule r, and

$$\nu_U(X,Z) = \begin{cases} h\left(\frac{card(\{Y \in X : Y \cap U \subseteq Z\})}{card(\{Y \in X : Y \cap U \subseteq Z\}) + card(\{Y \in X : Y \cap U \subseteq U^* \setminus Z\})}\right) & \text{if } X \neq \emptyset \\ 0 & \text{if } X = \emptyset, \end{cases} \tag{7}$$

where $X \subseteq P(U^*)$ and $Z \subseteq U^*$.

The uncertainty and rough inclusion functions can now be used to define the lower approximation $LOW(AS^*,Z)$, the upper approximation $UPP(AS^*,Z)$, and the boundary region $BN(AS^*,Z)$ of $Z \subseteq P(U^*)$ by:

$$LOW(AS^*,Z) = \{x \in U^* : \nu_U(I(x),Z) = 1\}, \tag{8}$$

and

$$UPP(AS^*,Z) = \{x \in U^* : \nu_U(I(x),Z) > 0\}, \tag{9}$$

$$BN(AS^*,Z) = UPP(AS^*,Z) \setminus LOW(AS^*,Z). \tag{10}$$

In the example, we classify objects from U^* to the lower approximation of Z if majority of rules matching this object are voting for Z and to the upper approximation of Z if at least half of the rules matching x are voting for Z. Certainly, one can follow many other voting schemes developed in machine learning or by introducing less crisp conditions in the boundary region definition. The defined approximations can be treated as estimations of the exact approximations of subsets of U^* because they are induced on the basis of samples of such sets restricted to U only. One can use some standard quality measures developed in machine learning to calculate the quality of such approximations assuming that after estimation of approximations on U^* full information about membership for objects relative to the approximated subsets of U^* is uncovered analogously to the testing sets in machine learning.

Let $C_1^* = \{x \in U^* : d(x) = 1\} = \{x_1,x_2,x_4,x_5,x_{10},x_{13},x_{14}\}$. We obtain the set $U^* \setminus C_1^* = C_0^* = \{x_3,x_6,x_7,x_8,x_9,x_{11},x_{12}\}$. The uncertainty function and rough inclusion are presented in Table 4.

Table 4 Uncertainty function and rough inclusion over the set of objects U^*

	$I(\cdot)$	$\nu_U(I(\cdot),C_1^*)$
x_1	$\{\{x_1,x_{14}\},\{x_1,x_5\}\}$	$h(2/2) = 1$
x_2	$\{\{x_2,x_4,x_{10}\},\{x_2,x_4,x_{11},x_{12}\}\}$	$h(2/2) = 1$
x_3	$\{\{x_3,x_7\},\{x_3,x_9\}\}$	$h(0/2) = 0$
x_4	$\{\{x_2,x_4,x_{10}\},\{x_2,x_4,x_{11},x_{12}\}\}$	$h(2/2) = 1$
x_5	$\{\{x_5,x_{13}\},\{x_1,x_5\}\}$	$h(2/2) = 1$
x_6	$\{\{x_6,x_9\},\{x_6\}\}$	$h(0/2) = 0$
x_7	$\{\{x_3,x_7\},\{x_7\}\}$	$h(0/2) = 0$
x_8	$\{\{x_8,x_{11},x_{12}\},\{x_8,x_{10}\}\}$	$h(0/2) = 0$
x_9	$\{\{x_6,x_9\},\{x_3,x_9\}\}$	$h(0/2) = 0$
x_{10}	$\{\{x_2,x_4,x_{10}\},\{x_8,x_{10}\}\}$	$h(1/2) = 1/2$
x_{11}	$\{\{x_8,x_{11},x_{12}\},\{x_2,x_4,x_{11},x_{12}\}\}$	$h(1/2) = 1/2$
x_{12}	$\{\{x_8,x_{11},x_{12}\},\{x_2,x_4,x_{11},x_{12}\}\}$	$h(1/2) = 1/2$
x_{13}	$\{\{x_5,x_{13}\}\}$	$h(1/1) = 1$
x_{14}	$\{\{x_1,x_{14}\}\}$	$h(1/1) = 1$

Thus, in our example from Table 4 we obtain

$$LOW(AS^*, C_1^*) = \{x \in U^* : \nu_U(I(x), C_1^*) = 1\} = \{x_1, x_2, x_4, x_5, x_{13}, x_{14}\}, \tag{11}$$

$$UPP(AS^*, C_1^*) = \{x \in U^* : \nu_U(I(x), C_1^*) > 0\} =$$

$$\{x_1, x_2, x_4, x_5, x_{10}, x_{11}, x_{12}, x_{13}, x_{14}\}, \tag{12}$$

$$BN(AS^*, C_1^*) = UPP(AS^*, C_1^*) \setminus LOW(AS^*, C_1^*) = \{x_{10}, x_{11}, x_{12}\}. \tag{13}$$

3.2 Function Approximations

In this subsection, we discuss the rough set approach to function approximation from available incomplete data. Our approach can be treated as a kind of rough clustering of functional data (Ramsay 2002).

Let us consider an example of function approximation. We assume that a partial information is only available about a function. This means that, some points from the graph of the function are known.

Before presenting a more formal description of function approximation we introduce some notation.

A function $f : U \longrightarrow R_+$ will be called a sample of a function $f^* : U^* \rightarrow R_+$, where R_+ is the set of non-negative reals and $U \subseteq U^*$ is a finite subset of U^*, if f^* is an extension of f.

By Gf (Gf^*) we denote the graph of f (f^*), respectively, i.e., the set $\{(x, f(x)) : x \in U\}$ ($\{(x, f^*(x)) : x \in U^*\}$). For any $Z \subseteq U^* \times R_+$ by $\pi_1(Z)$ and $\pi_2(Z)$ we denote the set $\{x \in U^* : \exists y \in R_+ \ (x,y) \in Z\}$ and $\{y \in R_+ : \exists x \in U^* \ (x,y) \in Z\}$, respectively.

First we define approximations of Gf given on a sample U of objects and next we show how to induce approximations of Gf^* over U^*, i.e., on extension of U.

Let Δ will be a partition of $f(U)$ into sets of reals of diameter less than $\delta > 0$, where δ is a given threshold. We also assume that $IS = (U, A)$ is a given information system. Let us also assume that for any object signature $Inf_A(x) = \{(a, a(x)) : a \in A\}$ (Pawlak and Skowron 2007) there is assigned an interval of non-negative reals with diameter less than δ. We denote this interval by $\Delta_{Inf_A(x)}$. Hence, $\Delta = \{\Delta_{Inf_A(x)} : x \in U\}$.

We consider an approximation space $AS_{IS,\Delta} = (U, I, \nu^*)$ (relative to given IS and Δ), where

$$I(x) = [x]_{IND(A)} \times \Delta_{Inf_A(x)}, \tag{14}$$

and

$$\nu^*(X,Y) = \begin{cases} \frac{card(\pi_1(X \cap Y))}{card(\pi_1(X))} & \text{if } X \neq \emptyset \\ 1 & \text{if } X = \emptyset, \end{cases} \tag{15}$$

for $X, Y \subseteq U \times R_+$.

The lower approximation and upper approximation of Gf in AS are defined by

$$LOW(AS_{IS,\Delta}, Gf) = \bigcup \{I(x) : \nu^*(I(x), Gf) = 1\}, \tag{16}$$

and

$$UPP(AS_{IS,\Delta},Gf) = \bigcup\{I(x):\ \nu^*(I(x),Gf) > 0\}, \tag{17}$$

respectively.

Observe that this definition is different from the standard definition of the lower approximation (Pawlak and Skowron 2007). The defined approximation space is a bit more general than in (Skowron and Stepaniuk 1996), e.g., the values of the uncertainty functions are subsets of $U \times R_+$ instead of U. Moreover, one can easily see that by applying the standard definition of relation approximation to f (Pawlak and Skowron 2007) (this is a special case of relation) the lower approximation of function is almost always equal to the empty set. The new definition is making it possible to express better the fact that a given neighborhood is "well" matching the graph of f (Skowron et al. 2006, Stepaniuk 2008). For expressing this a classical set theoretical inclusion of neighborhood into the graph of f is not satisfactory.

Example 1. We present the first illustrative example of a function approximation. Let $f: U \to R_+$ where $U = \{1,2,3,4,5,6\}$. Let $f(1) = 3$, $f(2) = 2$, $f(3) = 2$, $f(4) = 5$, $f(5) = 5$, $f(6) = 2$.

Let $IS = (U,A)$ be an information system where $A = \{a\}$ and

$$a(x) = \begin{cases} 0 \text{ if } 0 \leq x \leq 2, \\ 1 \text{ if } 2 < x \leq 4, \\ 2 \text{ if } 4 < x \leq 6. \end{cases} \tag{18}$$

Thus the partition $U/IND(A) = \{\{1,2\},\{3,4\},\{5,6\}\}$. The graph of f is defined by $Gf = \{(x, f(x)) : x \in U\} = \{(1,3),(2,2),(3,2),(4,5),(5,5),(6,2)\}$.

We define approximations of Gf given on the sample U of objects.

We obtain $f(U) = \{2,3,5\}$ and let $\Delta = \{\{2,3\},\{5\}\}$ will be a partition of $f(U)$.

We consider an approximation space $AS_{IS,\Delta} = (U, I, \nu^*)$ (relative to given IS and Δ), where

$$I(x) = [x]_{IND(A)} \times \Delta_{Inf_A(x)}, \tag{19}$$

is defined by

$$I(x) = \begin{cases} \{1,2\} \times [1.5,4] \ \text{ if } x \in \{1,2\}, \\ \{3,4\} \times [1.7,4.5] \text{ if } x \in \{3,4\}, \\ \{5,6\} \times [3,4] \quad \text{ if } x \in \{5,6\}. \end{cases} \tag{20}$$

We obtain the lower approximation and upper approximation of Gf in the approximation space $AS_{IS,\Delta}$:

$$\begin{aligned} LOW(AS_{IS,\Delta},Gf) = \bigcup\{I(x):\ \nu^*(I(x),Gf) = 1\} = \\ I(1) \cup I(2) = \{1,2\} \times [1.5,\ 4), \end{aligned} \tag{21}$$

and

$$\begin{aligned} UPP(AS_{IS,\Delta},Gf) = \bigcup\{I(x):\ \nu^*(I(x),Gf) > 0\} = \\ = I(1) \cup I(2) \cup I(3) \cup I(4) = \{1,2\} \times [1.5,\ 4) \cup \{4,5\} \times [1.7,\ 4.5), \end{aligned} \tag{22}$$

respectively.

Example 2. We present the second illustrative example of a function approximation. First, let us recall that an interval is a set of real numbers with the property that any number that lies between two numbers in the set is also included in the set. The closed interval of numbers between v and w ($v,w \in R_+$), including v and w, will be denoted by $[v,w]$.

Let us consider a function $f: R_+ \to R_+$. We have only a partial information about this function given by $G_i = \delta(x_i) \times \varepsilon(f(x_i))$, where $\delta(x_i)$ denotes a closed interval of reals to which belongs x_i, $\varepsilon(f(x_i))$ denotes a closed interval of reals to which belongs $f(x_i)$ and $i = 1,\ldots,n$. A family $\{G_1,\ldots,G_n\}$ is called a partial information about graph $Gf = \{(x,f(x)) : x \in R_+\}$.

Let Nh denotes a family of elements of $P(R_+) \times P(R_+)$ called neighborhoods.

In our example, we consider $Nh = \{X_1,X_2,X_3\}$, where $X_1 = [1,6] \times [0.1,0.4]$, $X_2 = [7,12] \times [1.1,1.4]$ and $X_3 = [13,18] \times [2.1,2.4]$.

Let us recall the inclusion definition between closed intervals:

$$[v_1,w_1] \subseteq [v_2,w_2] \text{ iff } v_2 \le v_1 \;\&\; w_1 \le w_2 \tag{23}$$

We define the new rough inclusion function by

$$\nu(X,\{G_1,\ldots,G_n\}) = \begin{cases} 1 \text{ if } \forall_{i\in\{1,\ldots,n\}}(\pi_1(G_i)\cap\pi_1(X) \neq \emptyset \to G_i \subseteq X) \\ 1/2 \text{ if } \exists_{i\in\{1,\ldots,n\}}(\pi_1(G_i)\cap\pi_1(X) \neq \emptyset \;\&\; G_i\cap X \neq \emptyset). \\ 0 \text{ if } \forall_{i\in\{1,\ldots,n\}}(\pi_1(G_i)\cap\pi_1(X) \neq \emptyset \to G_i\cap X = \emptyset) \end{cases} \tag{24}$$

Let sample values of a function $f: R_+ \to R_+$ are given in Table 5.

Let an approximation space $AS = (R_+,Nh,\nu)$ be given. We define the lower and upper approximation as follows:

$$LOW(AS,\{G_1,\ldots,G_n\}) = \bigcup\{X \in Nh : \; \nu(X,\{G_1,\ldots,G_n\}) = 1\}, \tag{25}$$

Table 5 Sample values of a function f, $\delta(x_i)$, $\varepsilon(f(x_i))$ and G_i

i	x_i	$f(x_i)$	$\delta(x_i)$	$\varepsilon(f(x_i))$	G_i	$\pi_1(G_i)$
1	1.5	0.55	$[1,2]$	$[0.5,0.6]$	$[1,2]\times[0.5,0.6]$	$[1,2]$
2	3.5	0.65	$[3,4]$	$[0.6,0.6]$	$[3,4]\times[0.6,0.6]$	$[3,4]$
3	5.5	0.025	$[5,6]$	$[0,0.5]$	$[5,6]\times[0,0.5]$	$[5,6]$
4	7.5	1.35	$[7,8]$	$[1.3,1.4]$	$[7,8]\times[1.3,1.4]$	$[7,8]$
5	9.5	1.15	$[9,10]$	$[1.1,1.2]$	$[9,10]\times[1.1,1.2]$	$[9,10]$
6	11.5	1.35	$[11,12]$	$[1.3,1.4]$	$[11,12]\times[1.3,1.4]$	$[11,12]$
7	13.5	2.25	$[13,14]$	$[2.2,2.3]$	$[13,14]\times[2.2,2.3]$	$[13,14]$
8	15.5	2.025	$[15,16]$	$[2,2.05]$	$[15,16]\times[2,2.05]$	$[15,16]$
9	17.5	2.475	$[17,18]$	$[2.45,2.5]$	$[17,18]\times[2.45,2.5]$	$[17,18]$

$$UPP(AS,\{G_1,\ldots,G_n\}) = \bigcup\{X \in Nh: \nu(X,\{G_1,\ldots,G_n\}) > 0\}. \quad (26)$$

In our example, we obtain

$$LOW(AS,\{G_1,\ldots,G_9\}) = X_2 = [7,12] \times [1.1,1.4], \quad (27)$$

$$UPP(AS,\{G_1,\ldots,G_9\}) = X_2 \cup X_3 = [7,12] \times [1.1,1.4] \cup [13,18] \times [2.1,2.4]. \quad (28)$$

The above defined approximations are approximations over the set of objects from sample $U \subseteq U^*$. Now, we present an approach for inducing of approximations of the graph Gf^* of function f^* on U^*, i.e., on extension of U. We use an illustrative example to present the approach.

It is worthwhile mentioning that by using boolean reasoning (Pawlak and Skowron 2007) one can generate patterns described by conjunctions of descriptors over IS such that the deviation of f on such patterns in U is less than a given threshold δ. This means that for any such a formula α the set $f(\|\alpha\|_U)$ has diameter less than a given threshold δ, i.e., the image of $\|\alpha\|_U$, i.e., the set $f(\|\alpha\|_U)$, is included into $[y-\delta/2, y+\delta/2)$ for some $y \in U$. Moreover, one can generate such minimal patterns, i.e., formulas α having the above property but no formula obtained by drooping some descriptors from α has that property (Pawlak and Skowron 2007, Bazan et al. 2002). By $PATTERN(A,f,\delta)$ we denote a set of induced patterns with the above properties. One can also assume[3] that $PATTERN(A,f,\delta)$ is extended by adding some shortenings of minimal patterns. For any pattern from $PATTERN(A,f,\delta)$ it is assigned an interval of reals Δ_α such that the deviation of f on $\|\alpha\|_U$ is in Δ_α, i.e., $f(\|\alpha\|_U) \subseteq \Delta_\alpha$.

Note that, for any boolean combination α of descriptors over A, it is also well defined its semantics $\|\alpha\|_{U^*}$ over U^*. However, there is only available information about a part of $\|\alpha\|_{U^*}$ equal to $\|\alpha\|_U = \|\alpha\|_{U^*} \cap U$. Assuming that the patterns from $PATTERN(A,f,\delta)$ are strong (i.e., their support is sufficiently large) one may induce that the following inclusion holds:

$$f^*(\|\alpha\|_{U^*}) \subseteq [y-\delta/2, y+\delta/2). \quad (29)$$

We can now define a generalized approximation space making it possible to extend the approximation of $Gf = \{(x,f(x) : x \in U\}$ over the defined previously approximation space AS to approximation of $Gf^* = \{(x,f(x) : x \in U^*\}$, where $U \subseteq U^*$.

Let us consider a generalized approximation space

$$AS^* = (U, U^*, I^*, \nu^*_{tr}, L^*), \quad (30)$$

where

- tr is a given threshold from the interval [0, 0.5),
- L^* is a language of boolean combinations of descriptors over the information system IS (Pawlak and Skowron 2007) used for construction of patterns from the set $PATTERN(A,f,\delta)$,

[3] Analogously to shortening of decision rules (Pawlak and Skowron 2007).

- $I^*(x) = \{\|\alpha\|_{U^*} \times \Delta_\alpha : \alpha \in PATTERN(A,f,\delta) \;\&\; x \in \|\alpha\|)_{U^*}\}$ for $x \in U^*$, where U^* is an extension of the sample U, i.e., $U \subseteq U^*$,
- for any finite family $\mathscr{X} \subseteq P(U^*) \times \mathscr{I}$, where $P(U^*)$ is the powerset of U^*, $\mathscr{I}$ is a family of intervals of reals of diameter less than δ and for any Y from $U^* \times R_+$ representing the graph of a function from U^* into R_+

$$\nu^*_{tr}(\mathscr{X},Y) = \begin{cases} 1 \text{ if } & Max < tr \\ 1/2 \text{ if } tr \leq Max < 1 - tr \\ 0 \text{ if } & Max \geq 1 - tr, \end{cases} \tag{31}$$

where

1. $Max = max\{\frac{|y^* - mid(\pi_2(Z)|}{max\{y^*, mid(\pi_2(Z)\}} : Z \in \mathscr{X} \;\&\; \nu^*_U(Z,Y) > 0\}$,
2. $\nu^*_U(Z,Y) = \nu^*((\pi_1(Z) \cap U) \times \pi_2(Z), Y \cap (U \times R_+))$, where ν^* is defined by the equation (15),
3. $mid(\Delta) = \frac{a+b}{2}$, where $\Delta = [a,b)$,
4. $$y^* = \frac{1}{c} \sum_{Z \in \mathscr{X}:\; \nu^*_U(Z,Y) > 0} mid(\pi_2(Z)) \cdot card(\pi_1(Z \cap Y) \cap U), \tag{32}$$

 where

 $$c = card\left(\bigcup_{Z \in \mathscr{X}:\; \nu^*_U(Z,Y) > 0} \pi_1(Z) \cap U \right). \tag{33}$$

The lower approximation of Gf^* is defined by

$$\begin{gathered} LOW^*(AS^*, Gf^*) = \\ \{(x,y) : \nu^*_{tr}(I^*(x), Gf^*) = 1 \;\&\; x \in U^* \;\&\; y \in [y^* - \delta/2, y^* + \delta/2)\}, \end{gathered} \tag{34}$$

where y^* is obtained from equation (32) in which $\mathscr{X}$ is substituted by $I(x)$ and Y by Gf^*, respectively.

The upper approximation of Gf^* is defined by

$$\begin{gathered} UPP^*(AS^*, Gf^*) = \\ \{(x,y) : \nu^*_{tr}(I^*(x), Gf^*) > 0 \;\&\; x \in U^* \;\&\; y \in [y^* - \delta/2, y^* + \delta/2)\}, \end{gathered} \tag{35}$$

where y^* is obtained from equation (32) in which $\mathscr{X}$ is substituted by $I(x)$ and Y by Gf^*, respectively.

Let us observe that for $x \in U^*$ the condition $(x,y) \notin UPP^*(AS^*, Gf^*)$ means that $\nu^*_{tr}(I^*(x), Gf^*) = 0 \;\&\; y \in [y^* - \delta/2, y^* + \delta/2)$ or $y \notin [y^* - \delta/2, y^* + \delta/2)$. The first condition describes the set of all pairs (x,y), where the deviation of y from y^* is small (relative to δ) but the prediction of y^* on the set of patterns $I^*(x)$ is very risky.

The values of f^* can be induced by

$$\widehat{f^*}(x) = \begin{cases} [y^* - \delta/2, y^* + \delta/2) & \text{if} \quad \nu^*_{tr}(I^*(x), Gf^*) > 0 \\ \mathit{undefined} & \text{otherwise,} \end{cases} \tag{36}$$

where $x \in U^* \setminus U$ and y^* is obtained from equation (32) in which $\mathscr{X}$ is substituted by $I(x)$ and Y by Gf^*, respectively.

Let us now explain the formal definitions presented above. The value of uncertainty function $I^*(x)$ for a given object x consists all patterns of the form $\|\alpha\|_{U^*} \times \Delta_\alpha$ such that $\|\alpha\|_{U^*}$ is matched by the object x. The condition $x \in \|\alpha\|_{U^*}$ can be verified by checking if the A-signature of x, i.e., $Inf_A(x)$ is matching α (to a satisfactory degree). The deviation on $\|\alpha\|_{U^*}$ is bounded by the interval Δ_α of reals. The degree to which Z is included to Y is estimated by $\nu^*_U(Z,Y)$, i.e., by degree to which the restricted to U projection of the pattern Z is included into Y projected on U. The estimated value for $f^*(x)$ belongs to the interval $[y^* - \delta/2, y^* + \delta/2)$ obtained by fusion of centers of intervals assigned to patterns from $\mathscr{X}$. In this fusion, the weights of these centers are reflecting the strength on U of patterns matching Y to a positive degree. The result of fusion is normalized by c. The degree to which a family of patterns $\mathscr{X}$ is included into Y is measured by the deviation of the value y^* from centers of intervals of patterns Z matching Y to a positive degree (i.e., $\nu^*_U(Z,Y) > 0$). In Figure 3 we illustrate the idea of the presented definition of y^*, where

- $Z_i = \|\alpha_i\|_{U^*} \times \Delta_{\alpha_i}$ for $i = 1,2,3$,
- $I^*(x) = \{Z_1, Z_2, Z_3\}$,
- the horizontal bold lines illustrate projections of sets Z_i ($i = 1,2,3$) on U^*,
- the vertical bold lines illustrate projections of sets Z_i ($i = 1,2,3$) on R_+,
- $y^* = \frac{1}{c}\sum_{t=1}^{3} mid(\Delta_{\alpha_t})card(\|\alpha_t\|_U)$ and c is defined by equation (33),
- $\nu^*_U(Z_i, Gf^*) > 0$ for $i = 1,2,3$ because $(x_1, f(x_1)) \in Z_1$ and $(x_2, f(x_2)) \in Z_2 \cap Z_3$ for $x_1, x_2 \in U$,
- $\nu^*_{tr}(I^*(x), Gf^*) = 1$ means that deviations $|y^* - mid(\Delta_{\alpha_i})|$ are sufficiently small (the exact formula is given by (34)).

The quality of approximations of Gf^* can be estimated using some selected measure defined by a combination of

- closeness between projections on U of approximations of Gf^* and approximations of Gf on sample U (see formulas (21-22)),
- the description lengths of approximations.

The considered approximation space is parameterized by the set of patterns $PATTERN(A, f, \delta)$. The optimization problem for function approximation is defined as the searching problem for a set of patterns optimizing the mentioned above measure based on a version of the minimal length principle (Rissanen 1985). Next, the closeness between the result of optimization and a testing sample of Gf^* can be used for estimation of the approximation quality. Note that one can also tune some parameters of the selected measure for improving the approximation quality. A more detailed discussion on optimization of function approximation will be presented in our next paper.

The presented illustrative method for function approximation based on the rough set approach can be treated as one of many possible ways for inducing function

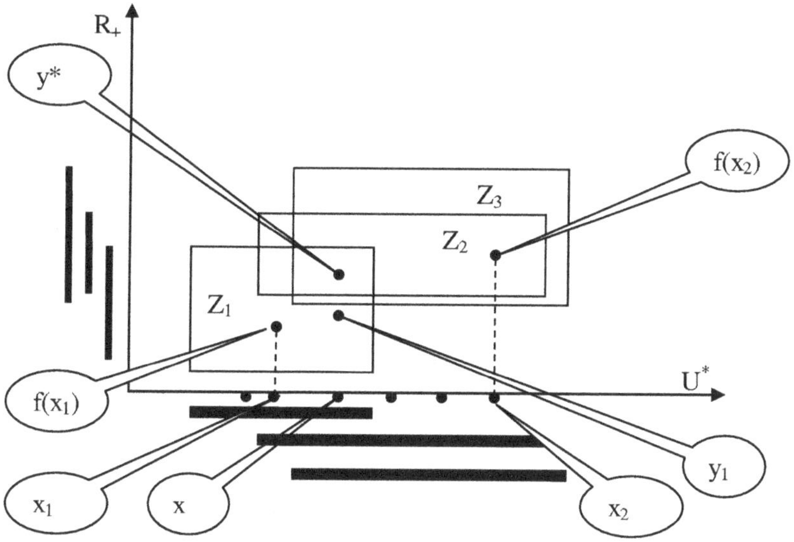

Fig. 3 Inducing the value y^*

approximations from data. For example, some of these methods may use distance functions between objects from U^* or more advanced rules for fusion of votes of matched by objects patterns voting for different approximations of function values.

Example 3. We present a method based on Boolean reasoning for extraction of patterns from data on which the deviation of values of the approximated function is bounded by a given threshold. These patterns are used as left hand sides of decision rules of the following form

if *pattern* **then** *the decision deviation is bounded by a given threshold.*

We consider a decision table $DT = (U, A \cup \{d\})$ such that $U = \{x_1, \ldots, x_5\}$, $A = \{a, b, c\}$ and $V_a = V_b = V_c = \{0, 1\}$ see Table 6 (the last column of the table (labeled by support) means the number of objects described exactly in the same way).

We define the square matrix $[m^{\varepsilon}_{x_i,x_j}]_{x_i,x_j \in U}$ (which is an analogy to discernibility matrix in the standard rough set model) by

$$m^{\varepsilon}_{x_i,x_j} = \{e \in A : e(x_i) \neq e(x_j) \& \mid d(x_i) - d(x_j) \mid > \varepsilon\}. \tag{37}$$

Table 6 Decision Table with Real Valued Decision

U	a	b	c	d	*support*
x_1	0	0	0	160	10
x_2	1	0	0	165	20
x_3	1	0	1	170	30
x_4	0	1	0	175	20
x_5	0	1	1	180	20

Let us observe that for any $x_i, x_j \in U$ $m^{\varepsilon}_{x_i,x_j} \subseteq P(A)$.

Table 7 Square Matrix $[m^5_{x_i,x_j}]_{x_i,x_j \in U}$

	x_1	x_2	x_3	x_4	x_5
x_1	$\emptyset$	$\emptyset$	$\{a,c\}$	$\{b\}$	$\{b,c\}$
x_2	$\emptyset$	$\emptyset$	$\emptyset$	$\{a,b\}$	$\{a,b,c\}$
x_3	$\{a,c\}$	$\emptyset$	$\emptyset$	$\emptyset$	$\{a,b\}$
x_4	$\{b\}$	$\{a,b\}$	$\emptyset$	$\emptyset$	$\emptyset$
x_5	$\{b,c\}$	$\{a,b,c\}$	$\{a,b\}$	$\emptyset$	$\emptyset$

We assume that $\varepsilon = 5$ and the square matrix $[m^5_{x_i,x_j}]_{x_i,x_j \in U}$ is presented in Table 7.

From Table 7 we obtain the boolean function $g(a,b,c)$ with three boolean variables a,b,c corresponding to attributes:

$$g(a,b,c) = (a \vee c) \wedge b \wedge (b \vee c) \wedge (a \vee b) \wedge (a \vee b \vee c) \wedge (a \vee b). \tag{38}$$

This function is an analogy to discernibility function in the standard rough set model. There are two prime implicants of this boolean function g: $a \wedge b$ and $b \wedge c$. Hence we obtain two reducts: $\{a,b\}$ and $\{b,c\}$. Using reducts and values of an object $x \in U$ we construct decision rules of the form

$$\textbf{if} \ldots \textbf{then } d \in (d(x) - \varepsilon, d(x) + \varepsilon). \tag{39}$$

The set of decision rules based on reducts $\{a,b\}$ and $\{b,c\}$ is presented in Table 8.

Let us consider new object $x_{new} \notin U$ described by $a(x_{new}) = 0, b(x_{new}) = 0$ and $c(x_{new}) = 1$. In classification of x_{new} there are applied two decision rules r^1_1 and r^2_3. Using the quality of the rules $Quality(r^1_1) = 10$, $Quality(r^2_3) = 30$ and that $mid(155,165) = 160$, $mid(165,175) = 170$ we obtain that

$$d(x_{new}) = \frac{Quality(r^1_1) mid(155,165) + Quality(r^2_3) mid(165,175)}{Quality(r^1_1) + Quality(r^2_3)} = 167.5. \tag{40}$$

Table 8 Decision rules generated from reducts $\{a,b\}$ and $\{b,c\}$

Object	Decision rule	Quality of the rule
x_1	r^1_1 : **if** $a=0$ **and** $b=0$ **then** $d \in (155,165)$	10
x_1	r^2_1 : **if** $b=0$ **and** $c=0$ **then** $d \in (155,165)$	10
x_2	r^1_2 : **if** $a=1$ **and** $b=0$ **then** $d \in (160,170)$	20
x_2	r^2_2 : **if** $b=0$ **and** $c=0$ **then** $d \in (160,170)$	20
x_3	r^1_3 : **if** $a=1$ **and** $b=0$ **then** $d \in (165,175)$	30
x_3	r^2_3 : **if** $b=0$ **and** $c=1$ **then** $d \in (165,175)$	30
x_4	r^1_4 : **if** $a=0$ **and** $b=1$ **then** $d \in (170,180)$	20
x_4	r^2_4 : **if** $b=1$ **and** $c=0$ **then** $d \in (170,180)$	20
x_5	r^1_5 : **if** $a=0$ **and** $b=1$ **then** $d \in (175,185)$	20
x_5	r^2_5 : **if** $b=1$ **and** $c=1$ **then** $d \in (175,185)$	20

4 Conclusions

We discussed a generalization of approximation spaces based on granular formulas and neighborhoods. We emphasized the fundamental role of approximation spaces in inducing classifiers such as rule based classifiers or function approximations. The approach can be extended for other kinds of classifiers, e.g., knn classifiers, neural networks, or ensembles of classifiers. Efficient strategies searching for relevant approximation spaces for approximation of higher order granules are crucial for application (e.g., in searching for approximation of complex concepts). In our current projects, we are developing such strategies based on some versions of the minimal length principle (Rissanen 1985). The presented approach provides the uniform foundations for implementing diverse strategies searching for (semi)optimal classifiers of different kinds. All of these strategies are based on searching for relevant approximation spaces. In particular, searching for relevant neighborhoods can be supported by feature construction methods based on Boolean reasoning. Searching for relevant approximation spaces can be realized in a network of cooperating strategies searching for classifiers for a given data set. The uniform foundations of our approach can facilitate the cooperation among strategies in such network. We also plan to compare the proposed method for function approximation with the traditional ones such as regression based methods (Hastie et al. 2008) or methods based on the functional data analysis approach (Ramsay 2002).

Acknowledgements

The research has been partially supported by the grants N N516 069235, N N516 368334, and N N516 077837 from Ministry of Science and Higher Education of the Republic of Poland.

References

Bazan, J.G.: Hierarchical classifiers for complex spatio-temporal concepts. In: Peters, J.F., Skowron, A., Rybiński, H. (eds.) Transactions on Rough Sets IX. LNCS, vol. 5390, pp. 474–750. Springer, Heidelberg (2008)

Bazan, J.G., Osmólski, A., Skowron, A., Ślęzak, D., Szczuka, M.S., Wróblewski, J.: Rough Set Approach to the Survival Analysis. In: Alpigini, J.J., Peters, J.F., Skowron, A., Zhong, N. (eds.) RSCTC 2002. LNCS (LNAI), vol. 2475, pp. 522–529. Springer, Heidelberg (2002)

Bazan, J., Skowron, A., Swiniarski, R.: Rough sets and vague concept approximation: From sample approximation to adaptive learning. In: Peters, J.F., Skowron, A. (eds.) Transactions on Rough Sets V. LNCS, vol. 4100, pp. 39–62. Springer, Heidelberg (2006)

Bianucci, D., Cattaneo, G.: Information entropy and granulation co–entropy of partitions and coverings: A summary. In: Peters, J.F., Skowron, A., Wolski, M., Chakraborty, M.K., Wu, W.-Z. (eds.) Transactions on Rough Sets X. LNCS, vol. 5656, pp. 15–66. Springer, Heidelberg (2009)

Greco, S., Matarazzo, B., Słowiński, R.: Dominance-based rough set approach and bipolar abstract rough approximation spaces. In: Chan, C.-C., Grzymala-Busse, J.W., Ziarko, W.P. (eds.) RSCTC 2008. LNCS (LNAI), vol. 5306, pp. 31–40. Springer, Heidelberg (2008)

Grünwald, P.D.: The minimum description length principle. The MIT Press, Cambridge (2007)
Hastie, T., Tibshirani, R., Friedman, J.H.: The Elements of statistical learning: Data mining, inference, and prediction, 2nd edn. Springer, Heidelberg (2008)
Jankowski, A., Skowron, A.: A wistech paradigm for intelligent systems. In: Peters, J.F., Skowron, A., Düntsch, I., Grzymała-Busse, J.W., Orłowska, E., Polkowski, L. (eds.) Transactions on Rough Sets VI. LNCS, vol. 4374, pp. 94–132. Springer, Heidelberg (2007)
Jankowski, A., Skowron, A.: Logic for artificial intelligence: The Rasiowa-Pawlak school perspective. In: Ehrenfeucht, A., Marek, V., Srebrny, M. (eds.) Andrzej Mostowski and Foundational Studies, pp. 106–143. IOS Press, Amsterdam (2008)
Klir, G.J.: Uncertainty and information: Foundations of generalized information theory. John Wiley & Sons, Hoboken (2007)
Łukasiewicz, J.: Die logischen Grundlagen der Wahrscheinlichkeitsrechnung. In: Borkowski, L. (ed.) Jan Łukasiewicz - Selected works, pp. 16–63. North Holland Publishing Company/Polish Scientific Publishers, Amsterdam/Warsaw (1913)
Malyszko, D., Stepaniuk, J.: Adaptive multilevel rough entropy evolutionary thresholding. Inf. Sci. 180(7), 1138–1158 (2010)
Ng, K.S., Lloyd, J.W., Uther, W.T.B.: Probabilistic modelling, inference and learning using logical theories. Ann. of Math. and Artif. Intell. 54(1-3), 159–205 (2008)
Nguyen, H.S.: Approximate Boolean Reasoning: Foundations and Applications in Data Mining. In: Peters, J.F., Skowron, A. (eds.) Transactions on Rough Sets V. LNCS, vol. 4100, pp. 344–523. Springer, Heidelberg (2006)
Pal, S.K., Shankar, B.U., Mitra, P.: Granular computing, rough entropy and object extraction. Pattern Recognition Letters 26(16), 2509–2517 (2005)
Pawlak, Z.: Rough sets: Theoretical aspects of reasoning about data. In: System Theory, Knowledge Engineering and Problem Solving, vol. 9. Kluwer Academic Publishers, Dordrecht (1991)
Pawlak, Z., Skowron, A.: Rudiments of rough sets; Rough sets: Some extensions; Rough sets and boolean reasoning. Inf. Sci. 177(1), 3–27, 28–40, 41–73 (2007)
Pedrycz, W., Skowron, A., Kreinovich, V. (eds.): Handbook of granular computing. John Wiley & Sons, New York (2008)
Peters, J., Skowron, A., Stepaniuk, J.: Nearness of objects: extension of the approximation space model. Fundam. Inform. 79(3,4), 497–512 (2007)
Polkowski, L., Skowron, A.: Rough mereology: A new paradigm for approximate reasoning. Int. J. Approx. Reason. 51, 333–365 (1996)
Ramsay, J.O., Silverman, B.W.: Applied functional data analysis. Springer, Berlin (2002)
Rissanen, J.: Minimum-description-length principle. In: Kotz, S., Johnson, N. (eds.) Encyclopedia of statistical sciences, pp. 523–527. John Wiley & Sons, New York (1985)
Skowron, A., Stepaniuk, J.: Generalized approximation spaces. In: Lin, T.Y., Wildberger, A.M. (eds.) The Third Int. Workshop on Rough Sets and Soft Computing Proceedings (RSSC 1994), pp. 156–163. San Jose State University, San Jose (1994)
Skowron, A., Stepaniuk, J.: Tolerance approximation spaces. Fundam. Inform. 27, 245–253 (1996)
Skowron, A.: Toward intelligent systems: Calculi of information granules. In: Hirano, S., Inuiguchi, M., Tsumoto, S. (eds.) Proc. Int. Workshop On Rough Set Theory And Granular Computing (RSTGC 2001),Matsue, Japan, May 20-22 (2001); Bull. Int. Rough Set Society 5(1-2), 9–30; see also: Terano T., Nishida T., Namatame A., Tsumoto S., Ohsawa Y., Washio T. (eds) New frontiers in Artificial Intelligence, Joint JSAI Workshop Post Proceedings.LNCS(LNAI), vol. 2253, pp. 251–260. Springer, Heidelberg

Skowron, A., Stepaniuk, J., Peters, J., Swiniarski, R.: Calculi of approximation spaces. Fundam. Inform. 72(1-3), 363–378 (2006)
Skowron, A., Szczuka, M.: Toward interactive computations: A rough-granular approach. In: Koronacki, J., Raś, Z.W., Wierzchoń, S.T., Kacprzyk, J. (eds.) Advances in Machine Learning II. SCI, vol. 263, pp. 23–42. Springer, Heidelberg (2010)
Słowiński, R., Vanderpooten, D.: A generalized definition of rough approximations based on similarity. IEEE Trans. Knowl. and Data Eng. 12, 331–336 (2000)
Stepaniuk, J.: Rough–granular computing in knowledge discovery and data mining. Springer, Heidelberg (2008)
Zadeh, L.A.: Fuzzy sets and information granularity. In: Gupta, M., Ragade, R., Yager, R. (eds.) Advances in Fuzzy Set Theory and Applications, pp. 3–18. North-Holland, Amsterdam (1979)
Zadeh, L.A.: Generalized theory of uncertainty (GTU)-principal concepts and ideas. Comput. Stat. and Data Anal. 51, 15–46 (2006)
Zhu, W.: Relationship between generalized rough sets based on binary relation and covering. Inf. Sci. 179(3), 210–225 (2009)
Ziarko, W.: Variable precision rough set model. J. Comput. and Sys. Sci. 46, 39–59 (1993)

Feature Discovery through Hierarchies of Rough Fuzzy Sets

Alfredo Petrosino and Alessio Ferone

Abstract. Rough set theory and fuzzy logic are mathematical frameworks for granular computing forming a theoretical basis for the treatment of uncertainty in many real–world problems, including image and video analysis. The focus of rough set theory is on the ambiguity caused by limited discernibility of objects in the domain of discourse; granules are formed as objects and are drawn together by the limited discernibility among them. On the other hand, membership functions of fuzzy sets enables efficient handling of overlapping classes. The hybrid notion of rough fuzzy sets comes from the combination of these two models of uncertainty and helps to exploit, at the same time, properties like coarseness, by invoking rough sets, and vagueness, by considering fuzzy sets. We describe a model of the hybridization of rough and fuzzy sets, that allows for further refinements of rough fuzzy sets. This model offers viable and effective solutions to some problems in image analysis, e.g. image compression.

Keywords: Rough Fuzzy Sets, Modelling Hierarchies, Feature Discovery, Image Analysis.

1 Introduction

Granular computing is based on the concept of information granule, that is a collection of similar objects which can be considered as indistinguishable. Partition of an universe into granules offers a coarse view of the universe where concepts, represented as subsets, can be approximated by means of granules. In this framework, rough set theory can be regarded to as a family

Alfredo Petrosino · Alessio Ferone
Department of Applied Science
University of Naples "Parthenope"
Centro Direzionale Isola C4
80143 Naples, Italy
e-mail: {alfredo.petrosino,alessio.ferone}@uniparthenope.it

W. Pedrycz and S.-M. Chen (Eds.): Granular Computing and Intell. Sys., ISRL 13, pp. 57–73.
springerlink.com

of methodologies and techniques that make use of granules [9, 10]. The focus of rough set theory is on the ambiguity caused by limited discernibility of objects in the domain of discourse. Granules are formed as objects and are drawn together by the limited discernibility among them. Granulation is of particular interest when a problem involves incomplete, uncertain or vague information. In such cases, precise solutions can be difficult to obtain and hence the use of techniques based on granules can lead to a simplification of the problem at hand.

At the same time, multivalued logic can be applied to handle uncertainty and vagueness present in information systems, the most visible of which is the theory of fuzzy sets [13]. In this framework, uncertainty is modelled by means of functions that define the degree of belonginess of an object to a given concept. Hence membership functions of fuzzy sets enable efficient handling of overlapping classes.

The hybrid notion of rough fuzzy sets comes from the combination of these two models of uncertainty to exploit, at the same time, properties like coarseness, by handling rough sets [9], and vagueness, by handling fuzzy sets [13]. In this combined framework, rough sets embody the idea of indiscernibility between objects in a set, while fuzzy sets model the ill-definition of the boundary of a subclass of this set. Combining both notions leads to consider, as instance, approximation of sets by means of similarity relations or fuzzy partitions. The rough fuzzy synergy is hence adopted to better represent the uncertainty in granular computation.

Nevertheless, some considerations are in order. Classical rough set theory is defined over a given partition, although several equivalence relations, and hence partitions, can be defined over the universe of discourse. Different partitions correspond to a coarser or finer view of the universe, because of different information granules, thus leading to coarser or finer definition of the concept to be provided. Then a substantial interest arises about the possibility of exploiting different partitions and, possibly, rough sets of higher order. Some approaches have been presented to exploit hierarchical granulation [5] where various approximations are obtained with respect to different levels of granulation. Considered as a nested sequence of granulations by a nested sequence of equivalence relations, this procedure leads to a nested sequence of rough set approximations and to a more general approximation structure. Hierarchical representation of the knowledge is also used in [7] to build a sequence of finer reducts so to obtain multiple granularities at multiple layers. The hierarchical reduction can handle problem with coarser granularity at lower level so to avoid incompleteness of data present in finer granularity at deeper layer. A different approach is presented in [8] where authors report a Multi-Granulation model of Rough-Set (MGRS) as an extension of Pawlak's rough set model. Moreover, this new model is used to define the concept of approximation reduct as the smallest attribute subset that preserves the lower approximation and upper approximation of all decision classes in MGRS.

The hybridization of rough and fuzzy sets reported here has been observed to possess a viable and effective solution to some of the most difficult problems in image analysis. The model exhibits a certain advantage of having a new operator to compose rough fuzzy sets, called $\mathcal{RF}$-product, able to produce a sequence of composition of rough fuzzy sets in a hierarchical manner. Theoretical foundations and properties, together with an example of application for image compression are described in the following sections.

2 Rough Fuzzy Sets: A Background

Let us start from the definition of a rough fuzzy set given by Dubois and Prade [6]. Let U be the universe of discourse, X a fuzzy subset of U, such that $\mu_X(u)$ represents the fuzzy membership function of X over U, and R an equivalence relation that induces the partition $U/R = \{Y_1, \ldots, Y_p\}$ (from now on denoted as $\mathcal{Y}$) over U in p disjoint sets, i.e. $Y_i \bigcap Y_j = \emptyset \; \forall i, j = 1, \ldots, p$ and $\bigcup_{i=1}^{p} Y_i = U$. The lower and upper approximation of X by R, i.e. $\underline{R}(X)$ and $\overline{R}(X)$ respectively, are fuzzy sets defined as

$$\mu_{\underline{R}(X)}(Y_i) = \inf\{\mu_X(u) | Y_i = [u]_R\} \tag{1}$$

$$\mu_{\overline{R}(X)}(Y_i) = \sup\{\mu_X(u) | Y_i = [u]_R\} \tag{2}$$

i.e. $[u]_R$ is a set such that (1) and (2) represent the degrees of membership of Y_i in $\underline{R}(X)$ and $\overline{R}(X)$, respectively. The couple of sets $< \underline{R}(X), \overline{R}(X) >$ is called *rough-fuzzy set* denoting a fuzzy concept (X) defined in a crisp approximation space (U/R) by means of two fuzzy sets ($\underline{R}(X)$ and $\overline{R}(X)$). Specifically, identifying $\pi_i(u)$ as the function that returns 1 if $u \in Y_i$ and 0 if $u \notin Y_i$, and considering $Y_i = [u]_R$ and $\pi_i(u) = 1$, the following relationships hold:

$$\mu_{\underline{R}(X)}(Y_i) = \inf_u max\{1 - \pi_i(u), \mu_X(u)\} \tag{3}$$

$$\mu_{\overline{R}(X)}(Y_i) = \sup_u min\{\pi_i(u), \mu_X(u)\} \tag{4}$$

To emphasize that the lower and upper approximations of the fuzzy subset X are, respectively, the infimum and the supremum of the membership functions of the elements of a class Y_i to the fuzzy set X, we can define a rough-fuzzy set as a triple

$$RF_X = (\mathcal{Y}, \mathcal{I}, \mathcal{S}) \tag{5}$$

where $\mathcal{Y} = \{Y_1, \ldots, Y_p\}$ is a partition of U in p disjoint subsets $Y_1, \ldots, Y_p$, and $\mathcal{I}, \mathcal{S}$ are mappings of kind $U \to [0, 1]$ such that $\forall u \in U$,

$$\mathcal{I}(u) = \sum_{i=1}^{p} \underline{\nu_i} \times \mu_{Y_i}(u) \tag{6}$$

$$\mathcal{S}(u) = \sum_{i=1}^{p} \overline{\nu_i} \times \mu_{Y_i}(u) \tag{7}$$

where

$$\underline{\nu}_i = \inf\{\mu_X(u)|u \in Y_i\} \tag{8}$$

$$\overline{\nu}_i = \sup\{\mu_X(u)|u \in Y_i\} \tag{9}$$

for the given subsets $\mathcal{Y} = \{Y_1, \ldots, Y_p\}$ and for every choice of function $\mu : U \to [0,1]$. $\mathcal{Y}$ and μ uniquely define a rough-fuzzy set as stated below

Definition 1. Given a subset $X \subseteq U$, if μ is the membership function μ_X defined on X and the partition $\mathcal{Y}$ is made with respect to an equivalence relation $\mathcal{R}$, i.e. $\mathcal{Y} = U/\mathcal{R}$, then X is a fuzzy set with two approximations $\overline{R}(X)$ and $\underline{R}(X)$, which are again fuzzy sets with membership functions defined as (8) and (9), i.e. $\underline{\nu}_i = \mu_{\underline{R}(X)}$ and $\overline{\nu}_i = \mu_{\overline{R}(X)}$. The pair of sets $< \overline{R}(X), \underline{R}(X) >$ is then a *rough fuzzy set.*

Let us recall the generalized definition of rough set given in [1]. Expressions for the lower and upper approximations of a given set X are

$$\underline{R}(X) = \{(u, \mathcal{I}(u))|u \in U\} \tag{10}$$

$$\overline{R}(X) = \{(u, \mathcal{S}(u))|u \in U\} \tag{11}$$

$\mathcal{I}$ and $\mathcal{S}$ are defined as:

$$\mathcal{I}(u) = \sum_{i=1}^{p} \mu_{Y_i}(u) \times \inf_{\varphi \in U} \max(1 - \mu_{Y_i}(\varphi), \mu_X(\varphi)) \tag{12}$$

$$\mathcal{S}(u) = \sum_{i=1}^{p} \mu_{Y_i}(u) \times \sup_{\varphi \in U} \min(\mu_{Y_i}(\varphi), \mu_X(\varphi)) \tag{13}$$

where μ_{Y_i} is the membership degree of each element $u \in U$ to a granule $Y_i \in U/R$ and μ_X is the membership function associated with X.

If we rewrite (8) and (9) as

$$\underline{\nu}_i = \mu_{\underline{R}(X)}(Y_i) = \inf_{\varphi \in U} \max(1 - \mu_{Y_i}(\varphi), \mu_X(\varphi)) \tag{14}$$

$$\overline{\nu}_i = \mu_{\overline{R}(X)}(Y_i) = \sup_{\varphi \in U} \min(\mu_{Y_i}(\varphi), \mu_X(\varphi)) \tag{15}$$

and considering a Boolean equivalence relation R, we arrive at the same definition of rough fuzzy set as given in (3) and (4). Indeed, considering (14) and the equivalence relation R, $\mu_Y(\varphi)$ takes values in $\{0,1\}$ hence the expression $1 - \mu_Y(\varphi)$ equals 0 if $\varphi \in Y$ or 1 if $\varphi \notin Y$. Furthermore the max operation returns 1 or $\mu_X(\varphi)$ depending on the fact that $\varphi \in Y$ or $\varphi \notin Y$. The operation inf then returns the infimum of such values, that is the minimum value of $\mu_X(\varphi)$. The same applies to (15).

3 Hierarchical Refinement of Rough-Fuzzy Sets

Rough set theory allows to partition the given data into equivalence classes. Nevertheless, given a set U, it is possible to employ different equivalence relations and hence produce different data partitions. This leads to a choice of the partition that represents the data in the best manner. For example, let us consider N-dimensional patterns, with $N = 4$ as in Table 1.

Table 1 Example of data

	A_1	A_2	A_3	A_4
u_1	a	a	b	c
u_2	b	c	c	c
u_3	a	b	c	a
u_4	c	b	a	b

Let $\mathcal{Y}_i$ be the partition obtained applying the equivalence relation R_{A_i} on the attribute A_i. We may get from Table 1 the following four partitions

$$\begin{aligned} \mathcal{Y}^1 &= \{\{u_1, u_3\}, \{u_2\}, \{u_4\}\} \\ \mathcal{Y}^2 &= \{\{u_1\}, \{u_2\}, \{u_3, u_4\}\} \\ \mathcal{Y}^3 &= \{\{u_1\}, \{u_2, u_3\}, \{u_4\}\} \\ \mathcal{Y}^4 &= \{\{u_1, u_2\}, \{u_3\}, \{u_4\}\} \end{aligned} \tag{16}$$

that, without any apriori knowledge, have potentially the same data representation power. To exploit all the possible partitions by means of simple operations, we propose to refine them in a hierarchical manner, so that partitions at each level of the hierarchy retain all the important informations contained into the partitions of the lower levels. The operation employed to perform the hierarchical refinement is called *Rough–Fuzzy product* ($\mathcal{RF}$-product) and is defined by:

Definition 2. Let $RF^i = (\mathcal{Y}^i, \mathcal{I}^i, \mathcal{S}^i)$ and $RF^j = (\mathcal{Y}^j, \mathcal{I}^j, \mathcal{S}^j)$ be two rough fuzzy sets defined, respectively, over partitions $\mathcal{Y}^i = (Y_1^i, \ldots, Y_p^i)$ and $\mathcal{Y}^j = (Y_1^j, \ldots, Y_p^j)$ with $\mathcal{I}^i$ (resp. $\mathcal{I}^j$) and $\mathcal{S}^i$ (resp. $\mathcal{S}^j$) indicating the measures expressed in Eqs. (6) and (7). The *$\mathcal{RF}$–product* between two rough-fuzzy sets, denoted by $\otimes$, is defined as a new rough fuzzy set

$$RF^{i,j} = RF^i \otimes RF^j = (\mathcal{Y}^{i,j}, \mathcal{I}^{i,j}, \mathcal{S}^{i,j})$$

where $\mathcal{Y}^{i,j} = (Y_1^{i,j}, \ldots, Y_{2p-1}^{i,j})$ is a new partition whose equivalence classes are

$$Y_k^{ij} = \begin{cases} \bigcup_{\substack{s=1\\q=h}}^{\substack{s=h\\q=1}} Y_q^i \cap Y_s^j & h = k, \quad k \leq p \\ \\ \bigcup_{\substack{s=h\\q=p}}^{\substack{s=p\\q=h}} Y_q^i \cap Y_s^j & h = k - p + 1, \quad k > p \end{cases} \tag{17}$$

and $\mathcal{I}^{i,j}$ and $\mathcal{S}^{i,j}$ are

$$\mathcal{I}^{i,j}(u) = \sum_{k=1}^{2p-1} \underline{\nu}_k^{i,j} \times \mu_k^{i,j}(u) \tag{18}$$

$$\mathcal{S}^{i,j}(u) = \sum_{k=1}^{2p-1} \overline{\nu}_k^{i,j} \times \mu_k^{i,j}(u) \tag{19}$$

and

$$\underline{\nu}_k^{ij} = \begin{cases} \sup\limits_{\substack{s=1,\ldots,h\\q=h,\ldots,1}} \{\underline{\nu}_q^i, \underline{\nu}_s^j\} & h = k, \quad k \leq p \\ \\ \sup\limits_{\substack{s=h,\ldots,p\\q=p,\ldots,h}} \{\underline{\nu}_q^i, \underline{\nu}_s^j\} & h = k - p + 1, \quad k > p \end{cases} \tag{20}$$

$$\overline{\nu}_k^{ij} = \begin{cases} \inf\limits_{\substack{s=1,\ldots,h\\q=h,\ldots,1}} \{\overline{\nu}_q^i, \overline{\nu}_s^j\} & h = k, \quad k \leq p \\ \\ \inf\limits_{\substack{s=h,\ldots,p\\q=p,\ldots,h}} \{\overline{\nu}_q^i, \overline{\nu}_s^j\} & h = k - p + 1, \quad k > p \end{cases} \tag{21}$$

Let us pick up the example shown in Table 1, and consider partitions $\mathcal{Y}^1$ and $\mathcal{Y}^2$ obtained from equivalence relations R_{A_1} and R_{A_2} defined on U by attributes A_1 and A_2, respectively. In terms of rough–fuzzy sets they are $RF^1 = (\mathcal{Y}^1, \mathcal{I}^1, \mathcal{S}^1)$ and $RF^2 = (\mathcal{Y}^2, \mathcal{I}^2, \mathcal{S}^2)$. Partitions $\mathcal{Y}^1$ and $\mathcal{Y}^2$ are defined as follows

$$\begin{aligned} \{u_4\} &= Y_1^1 & \{u_3, u_4\} &= Y_1^2 \\ \{u_2\} &= Y_2^1 & \{u_2\} &= Y_2^2 \\ \{u_1, u_3\} &= Y_3^1 & \{u_1\} &= Y_3^2 \end{aligned}$$

The refined partition $\mathcal{Y}^{1,2}$ defined on U by both attributes, corresponds to the partition obtained by $\mathcal{RF}$-producting RF^1 and RF^2.

The new partition $\mathcal{Y}^{1,2}$ is obtained by (in matrix notation)

–	–	$(Y_3^1 \cap Y_1^2)$	$(Y_2^1 \cap Y_1^2)$	$(Y_1^1 \cap Y_1^2)$
–	$(Y_3^1 \cap Y_2^2)$	$(Y_2^1 \cap Y_2^2)$	$(Y_1^1 \cap Y_2^2)$	–
$(Y_3^1 \cap Y_3^2)$	$(Y_2^1 \cap Y_3^2)$	$(Y_1^1 \cap Y_3^2)$	–	–

The final partition is obtained by joining sets by column as explained in Eq. 17

$$\begin{aligned} Y_1^{1,2} &= \{(Y_1^1 \cap Y_1^2)\} \\ Y_2^{1,2} &= \{(Y_2^1 \cap Y_1^2) \cup (Y_1^1 \cap Y_2^2)\} \\ Y_3^{1,2} &= \{(Y_3^1 \cap Y_1^2) \cup (Y_2^1 \cap Y_2^2) \cup (Y_1^1 \cap Y_3^2)\} \\ Y_4^{1,2} &= \{(Y_3^1 \cap Y_2^2) \cup (Y_2^1 \cap Y_3^2)\} \\ Y_5^{1,2} &= \{(Y_3^1 \cap Y_3^2)\} \end{aligned}$$

Hence

$$\mathcal{Y}^{1,2} = \{Y_1^{1,2}, Y_2^{1,2}, Y_3^{1,2}, Y_4^{1,2}, Y_5^{1,2}\}$$

and $\mathcal{I}$ and $\mathcal{S}$ of the new rough–fuzzy set, computed as in (18) and (19), are

$$\begin{aligned} \underline{\nu}_1^{1,2} &= \sup\{\inf\{\underline{\nu}_1^1, \underline{\nu}_1^2\}\} \\ \overline{\nu}_1^{1,2} &= \inf\{\sup\{\overline{\nu}_1^1, \overline{\nu}_1^2\}\} \\ \underline{\nu}_2^{1,2} &= \sup\{\inf\{\underline{\nu}_2^1, \underline{\nu}_1^2\}, \inf\{\underline{\nu}_1^1, \underline{\nu}_2^2\}\} \\ \overline{\nu}_2^{1,2} &= \inf\{\sup\{\overline{\nu}_2^1, \overline{\nu}_1^2\}, \sup\{\overline{\nu}_1^1, \overline{\nu}_2^2\}\} \\ \underline{\nu}_3^{1,2} &= \sup\{\inf\{\underline{\nu}_3^1, \underline{\nu}_1^2\}, \inf\{\underline{\nu}_2^1, \underline{\nu}_2^2\}, \inf\{\underline{\nu}_1^1, \underline{\nu}_3^2\}\} \\ \overline{\nu}_3^{1,2} &= \inf\{\sup\{\overline{\nu}_3^1, \overline{\nu}_1^2\}, \sup\{\overline{\nu}_2^1, \overline{\nu}_2^2\}, \sup\{\overline{\nu}_1^1, \overline{\nu}_3^2\}\} \\ \underline{\nu}_4^{1,2} &= \sup\{\inf\{\underline{\nu}_3^1, \underline{\nu}_2^2\}, \inf\{\underline{\nu}_2^1, \underline{\nu}_3^2\}\} \\ \overline{\nu}_4^{1,2} &= \inf\{\sup\{\overline{\nu}_3^1, \overline{\nu}_2^2\}, \sup\{\overline{\nu}_2^1, \overline{\nu}_3^2\}\} \\ \underline{\nu}_5^{1,2} &= \sup\{\inf\{\underline{\nu}_3^1, \underline{\nu}_3^2\}\} \\ \overline{\nu}_5^{1,2} &= \inf\{\sup\{\overline{\nu}_3^1, \overline{\nu}_3^2\}\} \end{aligned} \tag{22}$$

The rough–fuzzy set obtained by $RF^1 \otimes RF^2$ is thus defined by

$$RF^{1,2} = (\mathcal{Y}^{1,2}, \mathcal{I}^{1,2}, \mathcal{S}^{1,2})$$

where

$$\begin{aligned} \mathcal{I}^{1,2}(u) &= \sum_{i=1}^{5} \underline{\nu}_i^{1,2} \times \mu_{Y_i^{1,2}}(u) \\ \mathcal{S}^{1,2}(u) &= \sum_{i=1}^{5} \overline{\nu}_i^{1,2} \times \mu_{Y_i^{1,2}}(u) \end{aligned} \tag{23}$$

In case of partitions of different sizes, it is sufficient to add empty sets to have partitions of the same size.

4 Characterization of $\mathcal{RF}$–product

Let us recall that a partition $\mathcal{Y}$ of a finite set U is a collection $\{Y_1, Y_2, \ldots, Y_p\}$ of nonempty subsets (equivalence classes) such that

$$Y_i \cap Y_j = \emptyset \qquad \forall i, j = 1, \ldots, p \tag{24}$$

$$\bigcup_i Y_i = U \tag{25}$$

Hence, each partition defines an equivalence relation and, conversely, an equivalence relation defines a partition, such that the classes of the partition correspond to the equivalence classes of the relation.

Partitions are partially ordered by reverse refinement $\mathcal{Y}^i \subseteq \mathcal{Y}^j$. We say that $\mathcal{Y}^i$ is finer than $\mathcal{Y}^j$ if every equivalence class of $\mathcal{Y}^i$ is contained in some equivalence class of $\mathcal{Y}^j$, that is, for each equivalence class Y_h^j of $\mathcal{Y}^j$, there are equivalence classes $Y_1^i, \ldots, Y_p^i$ of $\mathcal{Y}^i$ such that $Y_h^j = Y_1^i, \ldots, Y_p^i$. If $E(\mathcal{Y}^i)$ is the equivalence relation defined by the partition $\mathcal{Y}^i$, then $\mathcal{Y}^i \subseteq \mathcal{Y}^j$ iff $\forall u, u' \in U, (u, u') \in E(\mathcal{Y}^i) \implies (u, u') \in E(\mathcal{Y}^j)$, that is, $E(\mathcal{Y}^i) \subseteq E(\mathcal{Y}^j)$.
The set $\Pi(U)$ of partitions of a set U forms a lattice under the partial order of reverse refinement. The minimum is the partition where an equivalence relation is a singleton, while the maximum is the partition composed by one single equivalence relation. The meet $\mathcal{Y}^i \wedge \mathcal{Y}^j \in \Pi(U)$ is the partition whose equivalence classes are given by $Y_k^i \cap Y_h^j \neq \emptyset$, where Y_k^i and Y_h^j are equivalence classes of $\mathcal{Y}^i$ and $\mathcal{Y}^j$, respectively. In terms of equivalence relations

$$R_{\mathcal{Y}^i \wedge \mathcal{Y}^j} = R_{\mathcal{Y}^i} \cap R_{\mathcal{Y}^j} \tag{26}$$

is the largest equivalence relation contained in both $R_{\mathcal{Y}^i}$ and $R_{\mathcal{Y}^j}$. The join $\mathcal{Y}^i \vee \mathcal{Y}^j$ is a partition composed by the equivalence classes of the transitive closure of the union of the equivalence relations defined by $\mathcal{Y}^i$ and $\mathcal{Y}^j$. In terms of equivalence relations

$$\begin{aligned} R_{\mathcal{Y}^i \vee \mathcal{Y}^j} = & R_{\mathcal{Y}^i} \cup R_{\mathcal{Y}^i} \circ R_{\mathcal{Y}^j} \cup R_{\mathcal{Y}^i} \circ R_{\mathcal{Y}^j} \circ R_{\mathcal{Y}^i} \cup \ldots \\ & \cup R_{\mathcal{Y}^j} \cup R_{\mathcal{Y}^j} \circ R_{\mathcal{Y}^i} \cup R_{\mathcal{Y}^j} \circ R_{\mathcal{Y}^i} \circ R_{\mathcal{Y}^j} \cup \ldots \end{aligned} \quad (27)$$

where $R_x \circ R_y$ denotes the composition of the equivalence relations R_x and R_y and is the smallest equivalence relation containing both $R_{\mathcal{Y}^i}$ and $R_{\mathcal{Y}^j}$. $\mathcal{I}$ and $\mathcal{S}$ are defined in (6) and (7). Firstly, we prove that the $\mathcal{RF}$-product yields an equivalence relation

Theorem 1. *Let $R_{\mathcal{Y}^i}$ and $R_{\mathcal{Y}^j}$ be equivalence relations on a set U. Then $E = R_{\mathcal{Y}^i} \otimes R_{\mathcal{Y}^j}$ is an equivalence relation on U.*
Proof. *E is an equivalence relation iff (1) $\forall E_i, E_j \in E, E_i \cap E_j = \emptyset$ and (2) $\cup E = U$.*

1. *Given that $R_{\mathcal{Y}^i}$ and $R_{\mathcal{Y}^j}$ are equivalence relations, $\mathcal{Y}^i$ and $\mathcal{Y}^j \in \Pi(U)$. $\forall u \in U, \exists R \in \mathcal{Y}^i$ and $\exists T \in \mathcal{Y}^j$ such that $u \in R$ and $u \in T$. Then $u \in R \cap T$. If $u \in R \cap T$ then $u \notin P \cap Q, \forall P \in \mathcal{Y}^i (P \neq R)$ and $\forall Q \in \mathcal{Y}^i (Q \neq T)$. The union of the intersections in the $\mathcal{RF}$-product ensure that u belongs to a single equivalence class $E_i \in E$ and hence $\forall E_i E_j \in E, E_i \cap E_j = \emptyset$.*
2. *Given that $\forall u \in U, \exists E_i \in E$ such that $u \in E_i$. Then $\cup E_i = U$.*

The operation $\mathcal{RF}$-product is commutative:

Theorem 2. *Let $R_{\mathcal{Y}^i}$ and $R_{\mathcal{Y}^j}$ be equivalence relations on a set U. Then $R_{\mathcal{Y}^i} \otimes R_{\mathcal{Y}^j} = R_{\mathcal{Y}^j} \otimes R_{\mathcal{Y}^i}$.*

Proof. *The property can be easly proven by first noting that intersection is a commutative operation. The matrix representing the $\mathcal{RF}$–product is built row-wise in $R_{\mathcal{Y}^i} \otimes R_{\mathcal{Y}^j}$ (that is each row is the refinement of an equivalence class of $R_{\mathcal{Y}^j}$ by all equivalence classes of $R_{\mathcal{Y}^i}$), while it is built column-wise in $R_{\mathcal{Y}^j} \otimes R_{\mathcal{Y}^i}$ (that is each column is the refinement of an equivalence class of $R_{\mathcal{Y}^j}$ by all equivalence classes of $R_{\mathcal{Y}^i}$). In both cases the positions considered at the union step are the same, thus yielding the same result.*

Next we prove two theorems which bound the level of refinement of the partitions induced by the $\mathcal{RF}$-product.

Theorem 3. *Let $R_{\mathcal{Y}^i}$ and $R_{\mathcal{Y}^j}$ be equivalence relations on a set U. It holds that $R_{\mathcal{Y}^i} \cap R_{\mathcal{Y}^j} \subseteq R_{\mathcal{Y}^i} \otimes R_{\mathcal{Y}^j}$.*

Proof. *From Eq. 17 it can be easily seen that each equivalence class of $R_{\mathcal{Y}^i} \otimes R_{\mathcal{Y}^j}$ is the union of some equivalence classes of $R_{\mathcal{Y}^i} \cap R_{\mathcal{Y}^j}$. Then each equivalence class of $R_{\mathcal{Y}^i} \cap R_{\mathcal{Y}^j}$ is contained in an equivalence class of $R_{\mathcal{Y}^i} \otimes R_{\mathcal{Y}^j}$. Hence $R_{\mathcal{Y}^i} \cap R_{\mathcal{Y}^j} \subseteq R_{\mathcal{Y}^i} \otimes R_{\mathcal{Y}^j}$.*

Theorem 4. *Let $R_{\mathcal{Y}^i}$ and $R_{\mathcal{Y}^j}$ be equivalence relations on a set U. It holds that $(R_{\mathcal{Y}^i} \otimes R_{\mathcal{Y}^j}) \otimes (R_{\mathcal{Y}^i} \otimes R_{\mathcal{Y}^j}) = R_{\mathcal{Y}^i} \otimes R_{\mathcal{Y}^j}$.*

Proof. *From Theorem 1 $R_{\mathcal{Y}^i} \otimes R_{\mathcal{Y}^j}$ is an equivalence relation. Then $(R_{\mathcal{Y}^i} \otimes R_{\mathcal{Y}^j}) \cap (R_{\mathcal{Y}^i} \otimes R_{\mathcal{Y}^j}) = R_{\mathcal{Y}^i} \otimes R_{\mathcal{Y}^j}$ and from Eq. 17 it derives that each equivalence relation of $R_{\mathcal{Y}^i} \otimes R_{\mathcal{Y}^j}$ is equal to only one equivalence relation of $(R_{\mathcal{Y}^i} \otimes R_{\mathcal{Y}^j}) \cap (R_{\mathcal{Y}^i} \otimes R_{\mathcal{Y}^j})$.*

Another interesting property of the $\mathcal{RF}$-product is that partition $E = R_{\mathcal{Y}^i} \otimes R_{\mathcal{Y}^j}$ can be seen as the coarsest partition with respect to the sequence of operations

$$\begin{aligned} E &= R_{\mathcal{Y}^i} \otimes R_{\mathcal{Y}^j} \\ E' &= E \otimes R_{\mathcal{Y}^j} = (R_{\mathcal{Y}^i} \otimes R_{\mathcal{Y}^j}) \otimes R_{\mathcal{Y}^j} \subseteq E \\ E'' &= E \otimes R_{\mathcal{Y}^i} = (R_{\mathcal{Y}^i} \otimes R_{\mathcal{Y}^j}) \otimes R_{\mathcal{Y}^i} \subseteq E \end{aligned}$$

In other words, at each iteration, the $\mathcal{RF}$-product produces a finer partition with respect to the initial partition. It is worth noting that, at each iteration

$$E = E' \otimes E'' \tag{28}$$

Viewed from another perspective, the $\mathcal{RF}$-product can be seen as a rule generation mechanism. Suppose that it is possible to assign a label to each equivalence class of a partition. Then $R_{\mathcal{Y}^i} \otimes R_{\mathcal{Y}^j}$ represents a partition whose equivalence classes are consistent with the labels of the operands. Consider the following partitions on a set U

$$\begin{aligned} \mathcal{Y}^1 &= \{Y^1_{low}, Y^1_{medium}, Y^1_{high}\} \\ \mathcal{Y}^2 &= \{Y^2_{low}, Y^2_{medium}, Y^2_{high}\} \end{aligned} \tag{29}$$

where $low = 1$ $medium = 2$ $high = 3$ and

$$\begin{aligned} \{u_4\} &= Y^1_{low} & \{u_3, u_4\} &= Y^2_{low} \\ \{u_2\} &= Y^1_{medium} & \{u_2\} &= Y^2_{medium} \\ \{u_1, u_3\} &= Y^1_{high} & \{u_1\} &= Y^2_{high} \end{aligned}$$

Applying $\mathcal{RF}$-product we get

$$\mathcal{Y}^{1,2} = \{Y^{1,2}_{low}, Y^{1,2}_{medium/low}, Y^{1,2}_{medium}, Y^{1,2}_{medium/high}, Y^{1,2}_{high}\} \tag{30}$$

where

$$
\begin{aligned}
&Y^{1,2}_{low} = \{(Y^1_{low} \cap Y^2_{low})\}\\
&Y^{1,2}_{medium/low} = \{(Y^1_{medium} \cap Y^2_{low}) \cup (Y^1_{low} \cap Y^2_{medium})\}\\
&Y^{1,2}_{medium} = \{(Y^1_{high} \cap Y^2_{low}) \cup (Y^1_{medium} \cap Y^2_{medium}) \cup (Y^1_{low} \cap Y^2_{high})\}\\
&Y^{1,2}_{medium/high} = \{(Y^1_{high} \cap Y^2_{medium}) \cup (Y^1_{medium} \cap Y^2_{high})\}\\
&Y^{1,2}_{high} = \{(Y^1_{high} \cap Y^2_{high})\}
\end{aligned}
\tag{31}
$$

Analyzing the new partition we note how the equivalence classes are consistent with the composition of the original ones, i.e.:

a) $u \in U$ belongs to "low" class in $Y^{1,2}$ if it belongs to "low" class in Y^1 and "low" class in Y^2;
b) $u \in U$ belongs to "medium/low" class in $Y^{1,2}$ if it belongs to "low" class in Y^1 and "medium" class in Y^2 or to "medium" class in Y^1 and "low" class in Y^2;
c) $u \in U$ belongs to "medium" class in $\mathcal{Y}^{1,2}$ if it belongs to "medium" class in Y^1 and Y^2 or to "high" class in Y^1 and to "low" class in Y^2 or to "high" class in Y^2 and to "low" class in Y^1;
d) $u \in U$ belongs to "medium/high" class in $Y^{1,2}$ if it belongs to "high" class in Y^1 and "medium" class in Y^2 or to "medium" class in Y^1 and "high" class in Y^2;
e) $u \in U$ belongs to "high" class in $Y^{1,2}$ if it belongs to "high" class in Y^1 and Y^2.

5 Feature Discovery

The basic ideas about how to construct the feature vectors upon the definitions introduced in the previous section are outlined for the case of image analysis.

Let us consider an image I defined over a set $U = [0, ..., H-1] \times [0, ...W-1]$ of picture elements, i.e. $I : u \in U \rightarrow [0,1]$. Let us also consider a grid, superimposed on the image, whose cells Y_i are of dimension $w \times w$, such that all Y_i constitute a partition over I, i.e. eqs (24) and (25) hold and each Y_i^1, for $i = 1 \ldots p$, has dimension $w \times w$ and $p = H/w + W/w$. The size w of each equivalence class will be referred to as *scale*.

Each cell of the grid can be seen as an equivalence class induced by an equivalence relation $\mathcal{R}$ that assigns each pixel of the image to a single cell. Given a pixel u, whose coordinates are u_x and u_y, and a cell Y_i of the grid, whose coordinates of its upper left point are $x(Y_i)$ and $y(Y_i)$, u belongs to Y_i if $x(Y_i) \leq u_x \leq x(Y_i)+w-1$ and $y(Y_i) \leq u_y \leq y(Y_i)+w-1$. In other words, we are defining a partition U/R of the image induced by the relation $\mathcal{R}$, in which each cell represents an equivalence class $[u]_{\mathcal{R}}$. Also suppose that equivalence classes can be ordered in some way, for instance, from left to right.

Moreover, given a subset X of the image, not necessarily included or equal to any $[u]_{\mathcal{R}}$, we define the membership degree $\mu_X(u)$ of a pixel u to X as the normalized gray level value of the pixel.

If we consider different scales, the partitioning scheme yields many partitions of the same image and hence various approximations $\overline{R}(X)$ and $\underline{R}(X)$ of the subset X. For instance, other partitions can be obtained by a rigid translation of $\mathcal{Y}^1$ in the directions of 0°, 45° and 90° of $w-1$ pixels, so that for each partition a pixel belongs to a shifted version of the same equivalence class Y_j^i.

If we consider four equivalence classes, Y_j^1 Y_j^2 Y_j^3 Y_j^4 as belonging to these four different partitions, then there exists a pixel u with coordinates u_x, u_y such that u belongs to the intersection of Y_j^1 Y_j^2 Y_j^3 Y_j^4. Hence each pixel can be seen as belonging to the equivalence class

$$Y_j^{1,2,3,4} = Y_j^1 \cap Y_j^2 \cap Y_j^3 \cap Y_j^4 \tag{32}$$

of the partition obtained by $\mathcal{RF}$-producting the four rough fuzzy sets to which Y_j^i, with $i = 1, \ldots, 4$, belongs, i.e.

$$RF_X^{1,2,3,4} = RF_X^1 \otimes RF_X^2 \otimes RF_X^3 \otimes RF_X^4 \tag{33}$$

The $\mathcal{RF}$-product behaves as a filtering process according to which the image is filtered by a minimum operator over a window $w \times w$ producing $\mathcal{I}$ and by a maximum operator producing $\mathcal{S}$. Iterative application of this procedure consists in applying the same operator to both results $\mathcal{I}$ and $\mathcal{S}$ obtained at the previous iteration.

As instance, X defines the contour or uniform regions in the image. On the contrary, regions appear rather like fuzzy sets of grey levels and their comparison or combination generates more or less uniform partitions of the image. Rough fuzzy sets, as defined in (5), seem to capture these aspects together, trying to extract different kinds of knowledge in data.

This procedure can be efficiently applied to image coding/decoding, getting rise to the method *rough fuzzy vector quantization* (RFVQ)[12]. The image is firstly partitioned in non–overlapping k blocks X_h of dimension $m \times m$, such that $m \geq w$, that is $X = \{X_1, \ldots, X_k\}$ and $k = H/m + K/m$.

Considering each image block X_h, a pixel in the block can be characterized by two values that are the membership degrees to the lower and upper approximation of the set X_h. Hence, the feature extraction process provides two approximations $\underline{R}(X_h)$ and $\overline{R}(X_h)$ characterized by $\mathcal{I}$ and $\mathcal{S}$ as defined in (6) and (7) where

$$\begin{aligned} \underline{\nu}_i &= \mu_{\underline{R}(X_h)}(Y_i) = \inf\{\mu_{X_h}(u) | Y_i = [u]_R\} \\ \overline{\nu}_i &= \mu_{\overline{R}(X_h)}(Y_i) = \sup\{\mu_{X_h}(u) | Y_i = [u]_R\} \end{aligned} \tag{34}$$

and $[u]_R$ is the granule that defines the resolution at which we are observing the block X_h. For a generic pixel $u = (u_x, u_y)$ we can compute the coordinates of the upper left pixel of the four equivalence classes containing u, as shown in Fig. 1:

$$\begin{aligned}
u_x &= x_1 + w - 1 \Rightarrow x_1 = u_x - w + 1\\
u_y &= y_1 + w - 1 \Rightarrow y_1 = u_y - w + 1\\
u_x &= x_2 \Rightarrow x_2 = u_x\\
u_y &= y_2 + w - 1 \Rightarrow y_2 = u_y - w + 1\\
u_x &= x_3 + w - 1 \Rightarrow x_3 = u_x - w + 1\\
u_y &= y_3 \Rightarrow y_3 = u_y\\
u_x &= x_4 \Rightarrow x_4 = u_x\\
u_y &= y_4 \Rightarrow y_4 = u_y
\end{aligned}$$

where the four equivalence classes for pixel u are

$$\begin{aligned}
Y_j^1 &= (x_1, y_1, \underline{\nu}_j^1, \overline{\nu}_j^1)\\
Y_j^2 &= (x_2, y_2, \underline{\nu}_j^2, \overline{\nu}_j^2)\\
Y_j^3 &= (x_3, y_3, \underline{\nu}_j^3, \overline{\nu}_j^3)\\
Y_j^4 &= (x_4, y_4, \underline{\nu}_j^4, \overline{\nu}_j^4)
\end{aligned}$$

For instance, if we choose a granule of dimension $w = 2$ for a generic $j - th$ granule of the $i - th$ partition, equations in (34) become:

$$\begin{aligned}
\underline{\nu}_j^i &= \inf\{(u_x + a, u_y + b)|a, b = 0, 1\}\\
\overline{\nu}_j^i &= \sup\{(u_x + a, u_y + b)|a, b = 0, 1\}
\end{aligned}$$

The compression method performed on each block X_h is composed of three phases: *codebook design*, *coding* and *decoding*. A vector is constructed by retaining the values $\underline{\nu}_j^i$ and $\overline{\nu}_j^i$ at positions u and $u + (w - 1, w - 1)$ in a generic block X_h, or equivalently $\underline{\nu}_j^1, \overline{\nu}_j^1, \underline{\nu}_j^3, \overline{\nu}_j^3$. The vector has hence dimension m^2 consisting of $m^2/2$ inf values and $m^2/2$ sup values. The vectors so constructed and extracted from a training image set are then fed to a quantizer in order to construct the codebook. The aim of vector quantization is the representation of a set of vectors $u \in X \subseteq R^{m^2}$ by a set of C prototypes (or codevectors) $V = \{v_1, v_2, ..., v_C\} \subseteq R^{m^2}$. Thus, vector quantization can also be seen as a mapping from an m^2-dimensional Euclidean space into the finite set $V \subseteq R^{m^2}$, also referred to as the codebook. Codebook design can be performed by clustering algorithms, but it is worth noting that the proposed method relies on the representation capabilities of the vector to be quantized and not on the quantization algorithm, to determine optimal codevectors, i.e. Fuzzy C-Means, Generalized Fuzzy C-Means or any analogous clustering algorithm can be adopted.

The process of coding a new image proceeds as follows. For each block X_h the features extracted are arranged in a vector, following the same scheme used for designing the codebook, and compared with the codewords in the codebook to find the best match, i.e. the closest codeword to the block.

In particular, for each block, inf and sup values are extracted from a window of size 2×2 shifted by one pixel into the block. All the extracted values are arranged in a one-dimensional array, i.e. for block dimension $m \times m$ and a window dimension 2×2 the array is represented by m^2 elements consisting of $m^2/2$ inf values and $m^2/2$ sup values. Doing so, the identificative number (out of C) of the winning codeword, i.e. the best match to the coded block, is saved in place of the generic block X_h.

Given a coded image, the decoding step firstly consists in the substitution of the identificative codeword number with the codeword itself, as reported in the codebook. The codeword consists of $m^2/2$ inf and $m^2/2$ sup values, instead of the original m^2 values of the block. To reconstruct the original block, we apply the theory as follows. As stated above, each pixel can be seen as belonging to the block of the partition obtained by $\mathcal{RF}$-producting the four equivalence classes (32) and (33). Specifically, the blocks contained into the codeword are not the original ones, but those chosen to represent the block, i.e.

$$Y_j^{1,2,3,4} = Y_j^1 \cap Y_j^2 \cap Y_j^3 \cap Y_j^4 \tag{35}$$

where Y_j^i is a set of the partition of the rough-fuzzy set intersecting the generic block of the image X_h. The result of the $\mathcal{RF}$-product operation, with respect to a single block, is represented by another rough fuzzy set, characterized by lower and upper approximations. These values are used to fill the missing values into the decoded block.

In detail, being q_r the codeword corresponding to a generic block, the decoded block $X_{decoded}$ is constructed by filling the missing values, i.e. the original $\underline{\nu}_j^2, \overline{\nu}_j^2, \underline{\nu}_j^4, \overline{\nu}_j^4$ as combination of them, like average, median, etc., yielding $\tilde{\underline{\nu}}_j^2, \tilde{\overline{\nu}}_j^2, \tilde{\underline{\nu}}_j^4, \tilde{\overline{\nu}}_j^4$.

The reconstructed block $X_{decoded}$, again a rough fuzzy set, is obtained by $\mathcal{RF}$-producting the four equivalnce classes Y_j^1 Y_j^2 Y_j^3 Y_j^4, yielding the following

$$\begin{aligned} Y_j^{1,2,3,4} &= Y_j^1 \cap Y_j^2 \cap Y_j^3 \cap Y_j^4 \\ \tilde{\mathcal{I}}^{1,2,3,4}(u) &= \sum_j \tilde{\underline{\nu}}_j^{1,2,3,4} \times \mu_{Y_j}^{1,2,3,4}(u) \\ \tilde{\mathcal{S}}^{1,2,3,4}(u) &= \sum_j \tilde{\overline{\nu}}_j^{1,2,3,4} \times \mu_{Y_j}^{1,2,3,4}(u) \end{aligned}$$

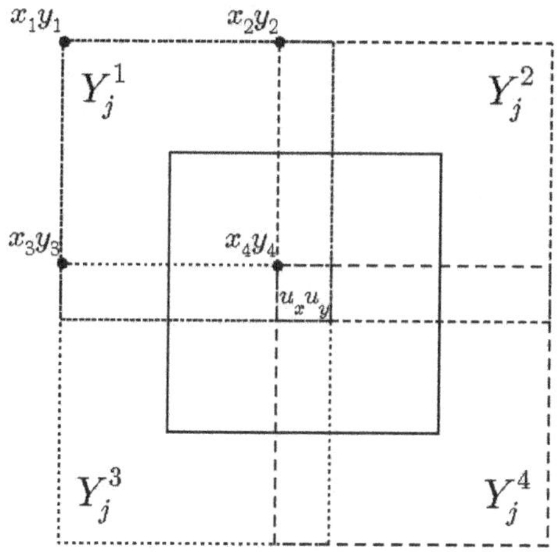

Fig. 1 Equivalence classes coordinates.

where

$$\underline{\tilde{\nu}}_j^{1,2,3,4} = \sup\{\underline{\nu}_j^1, \underline{\tilde{\nu}}^2{}_j, \underline{\nu}_j^3, \underline{\tilde{\nu}}_j^4\} \qquad \tilde{\overline{\nu}}_j^{1,2,3,4} = \inf\{\overline{\nu}_j^1, \tilde{\overline{\nu}}_j^2, \overline{\nu}_j^3, \tilde{\overline{\nu}}_j^4\} \tag{36}$$

Lastly, under the assumption of local smoothness an estimate of the original grey values at the generic position u can be computed composing $\underline{\tilde{\nu}}_j^{1,2,3,4}$ and $\tilde{\overline{\nu}}_j^{1,2,3,4}$, as instance averaging or simply using only one of them.

An example of compression of image *Baboon* is depicted in Figure 2(a) using RFVQ. The codebook has been built by using ISODATA quatizer, a training image set of size 256×256 with 8 bits/pixel: Bird, Bridge, Building, Camera, City, Hat, House, Lena, Mona, Salesman, using the average as combination. Different compression rates are obtained as explained in Table 2.

Table 2 RFVQ parameters at different compression rates.

Compression rate	Number of clusters (C)	Block dimension (m)
0.03	16	4
0.06	256	4
0.14	32	2
0.25	256	2
0.44	16384	2

Analyzing the results shown in Fig.s 2 (b)–(f), we can observe that the proposed method performs well for higher compression rates while it looses efficiency for lower compression rates, reasonably due to the quantization algorithm. Indeed, in order to obtain a compression rate of 0.44 a large number of clusters has to be computed (precisely 16,384), but in this situation the large number of codewords does not ensure that the optimal choice will be performed when selecting the most approximating codeword.

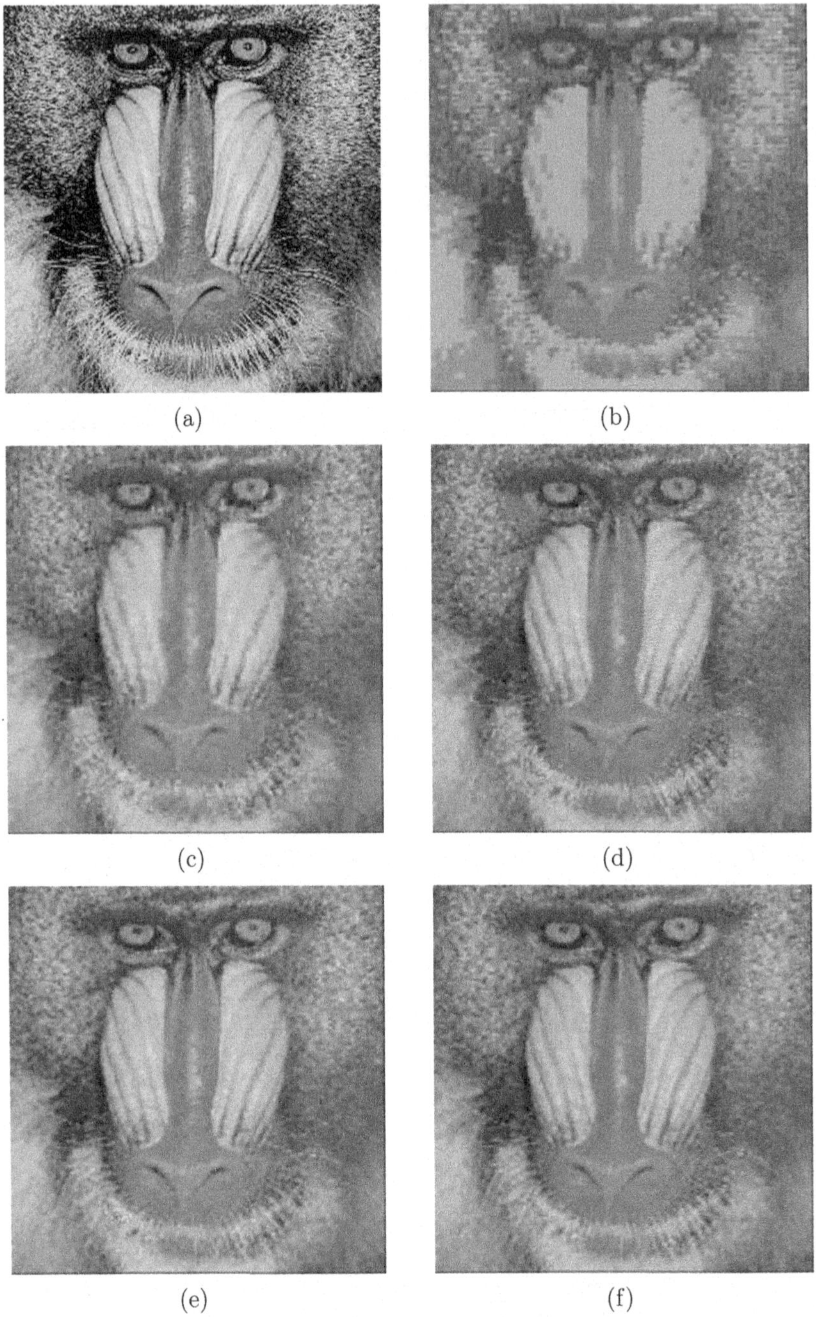

Fig. 2 Image "Baboon" (a) at different compression rates, b) 0.03 (PSNR: 19.10), c) 0.06 (PSNR: 19.89), d) 0.14 (PSNR: 20.61), e) 0.25 (PSNR: 21.08), f) 0.44 (PSNR: 21.36).

6 Concluding Remarks

A model for the hybridization of rough and fuzzy sets has been presented. It is endowed with a new operator, called $\mathcal{RF}$-product, which allows to combine different partitions yielding a refined partition in a hierarchical manner. The model and the $\mathcal{RF}$-operator have been proved to possess peculiar properties effective for feature discovery, useful in applications like image analysis. Here we reported an image compression scheme that exploits the peculiarities of $\mathcal{RF}$-product. Results are quite remarkable, considering that RFVQ does not suffer of the blocking effect, while loosing only a small amount of details. Ongoing work is devoted to exploit the proposed model in other data mining applications and also image processing tasks, like color image segmentation.

References

[1] Sen, D., Pal, S.K.: Generalized Rough Sets, Entropy, and Image Ambiguity Measures. IEEE Trans. on Syst. Man and Cybern. 39, 117–128 (2009)

[2] Caianiello, E.R.: A calculus of hierarchical systems. In: Proc. Internat. Conf. on Pattern Recognition, pp. 1–5 (1973)

[3] Caianiello, E.R., Petrosino, A.: Neural networks, fuzziness and image processing. In: Cantoni, V. (ed.) Machine and Human Perception: Analogies and Divergencies, Plenum Press, NewYork (1994)

[4] Caianiello, E.R., Ventre, A.: A model for C-calculus. International Journal of General Systems 11, 153–161 (1985)

[5] Yao, Y.Y.: Information granulation and rough set approximation. International Journal of Intelligent Systems 16, 87–104 (2001)

[6] Dubois, D., Prade, H.: Rough fuzzy sets and fuzzy rough sets. Int. J. Gen. Sys. 17, 191–209 (1990)

[7] Li, Y., Qiao, B.: Hierarchical Reduction Algorithm of Rough Sets. In: Sixth International Conference on Intelligent Systems Design and Applications, vol. 1, pp. 497–502 (2006)

[8] Qian, Y., Liang, J., Yao, Y., Dang, C.: MGRS: A multi-granulation rough set. Information Sciences 180, 949–970 (2010)

[9] Pawlak, Z.: Rough sets. International Journal of Computer and Information Sciences 11, 341–356 (1982)

[10] Pawlak, Z.: Granularity of knowledge, indiscernibility and rough sets. In: Proceedings of IEEE International Conference on Fuzzy Systems, pp. 106–110 (1998)

[11] Petrosino, A., Salvi, G.: Rough fuzzy set based scale space transforms and their use in image analysis. International Journal of Approximate Reasoning 41, 212–228 (2006)

[12] Petrosino, A., Ferone, A.: Rough fuzzy set-based image compression. Fuzzy Sets and Systems 160, 1485–1506 (2009)

[13] Zadeh, L.: Fuzzy sets. Information and Control, pp. 338–353 (1964)

Comparative Study of Fuzzy Information Processing in Type-2 Fuzzy Systems

Oscar Castillo and Patricia Melin

Abstract. Fuzzy information processing in type-2 fuzzy systems has been implemented in most cases based on the Karnik and Mendel (KM) and Wu-Mendel (WM) approaches. However, both of these approaches are time consuming for most real-world applications, in particular for control problems. For this reason, a more efficient method based on evolutionary algorithms has been proposed by Castillo and Melin (CM). This method is based on directly obtaining the type reduced results by using an evolutionnary algorithm (EA). The basic idea is that with an EA the upper and lower membership functions in the output can be obtained directly based on experimental data available for a particular problem. A comparative study (in control applications) of the three methods, based on accuracy and efficiency is presented, and the CM method is shown to outperform both the KM and WM methods in efficiency while accuracy produced by this method is comparable.

Keywords: Intelligent Control, Type-2 Fuzzy Logic, Interval Fuzzy Logic, Hybrid Intelligent Systems, Evolutionary Algorithm, Hardware Implementation, Fuzzy Controllers, Type Reduction.

1 Introduction

Uncertainty affects decision-making and appears in a number of different forms. The concept of information is fully connected with the concept of uncertainty. The most fundamental aspect of this connection is that the uncertainty involved in any problem-solving situation is a result of some information deficiency, which may be incomplete, imprecise, fragmentary, not fully reliable, vague, contradictory, or deficient in some other way. Uncertainty is an attribute of information (Zadeh 2005). The general framework of fuzzy reasoning allows handling much of this uncertainty and fuzzy systems that employ type-1 fuzzy sets represent uncertainty by numbers in the range [0, 1]. When a phenomenon is uncertain, like a measurement, it is difficult to determine its exact value, and of course type-1 fuzzy sets

Oscar Castillo and Patricia Melin
Tijuana Institute of Technology, Division of Graduate Studies Tijuana, Mexico
e-mail: {ocastillo,pmelin}@tectijuana.mx

W. Pedrycz and S.-M. Chen (Eds.): Granular Computing and Intell. Sys., ISRL 13, pp. 75–93.
springerlink.com

make more sense than using sets (Zadeh 1975). However, it is not reasonable to use an accurate membership function for something uncertain, so in this case what we need is higher order fuzzy sets, those which are able to handle these uncertainties, like the so called type-2 fuzzy sets (Mendel 2004) (Mizumoto and Tanaka 1976). So, the amount of uncertainty can be managed by using type-2 fuzzy logic because it offers better capabilities to handle linguistic uncertainties by modeling vagueness and unreliability of information (Wagenknecht and Hartmann 1988).

Recently, we have seen the use of type-2 fuzzy sets in Fuzzy Logic Systems (FLS) in different areas of application (Castillo and Melin 2008). A novel approach for realizing the vision of ambient intelligence in ubiquitous computing environments (UCEs), is based on intelligent agents that use type-2 fuzzy systems which are able to handle the different sources of uncertainty in UCEs to give a good response (Doctor et al. 2005). There are also papers with emphasis on the implementation of type-2 FLS (Karnik et al. 1999) and in others, it is explained how type-2 fuzzy sets let us model the effects of uncertainties in rule-base FLS (Mendel and John 2002). In industry, type-2 fuzzy logic and neural networks was used in the control of non-linear dynamic plants (Melin and Castillo 2004); also we can find studies in the field of mobile robots (Astudillo et al. 2006) (Hagras 2004). In this paper we deal with the application of interval type-2 fuzzy control to non-linear dynamic systems. It is a well known fact, that in the control of real systems, the instrumentation elements (instrumentation amplifier, sensors, digital to analog, analog to digital converters, etc.) introduce some sort of unpredictable values in the information that has been collected (Castillo and Melin 2001). The controllers designed under idealized conditions tend to behave in an inappropriate manner (Castillo and Melin 2003). For this reason, type-2 fuzzy controllers, which can cope better with uncertainty, may have better performance under non-ideal conditions (Castillo and Melin 2004).

Fuzzy information processing in interval type-2 fuzzy systems has been implemented in most cases based on the Karnik and Mendel (KM) and Wu-Mendel (WM) approaches (Karnik and Mendel 2001). However, both of these approaches are time consuming for most real-world applications, in particular for control problems (Coupland and John 2008) (Starzewski 2009) (Martinez et al. 2009). For this reason, a more efficient method based on evolutionary algorithms (Sepulveda et al. 2007) has been proposed by Castillo and Melin (CM). This method is based on directly obtaining the type reduced results by searching the space of possible results using an evolutionary algorithm (Montiel et al. 2007). The basic idea is that with an EA the upper and lower membership functions in the output can be obtained directly based on experimental data for a particular problem. In this paper, a comparative study (in control applications) of the three methods, based on accuracy and efficiency is presented. The CM method is shown to outperform both the KM and WM methods in efficiency while accuracy is comparable. This fact makes the CM method a good choice for real-world control applications in which efficiency is of fundamental importance.

2 Fuzzy Logic Systems

In this section, a brief overview of type-1 and type-2 fuzzy systems is presented. This overview is considered to be necessary to understand the basic concepts needed to develop the methods and algorithms presented later in the paper.

2.1 Type-1 Fuzzy Logic Systems

Soft computing techniques have become an important research topic, which can be applied in the design of intelligent controllers, which utilize the human experience in a more natural form than the conventional mathematical approach (Zadeh 1971) (Zadeh 1973). A FLS, described completely in terms of type-1 fuzzy sets is called a type-1 fuzzy logic system (type-1 FLS). In this paper, the fuzzy controller has two input variables, which are the error $e(t)$ and the change of error $\Delta e(t)$,

$$e(t) = r(t) - y(t) \tag{1}$$

$$\Delta e(t) = e(t) - e(t-1) \tag{2}$$

The control system can be represented as in Figure 1.

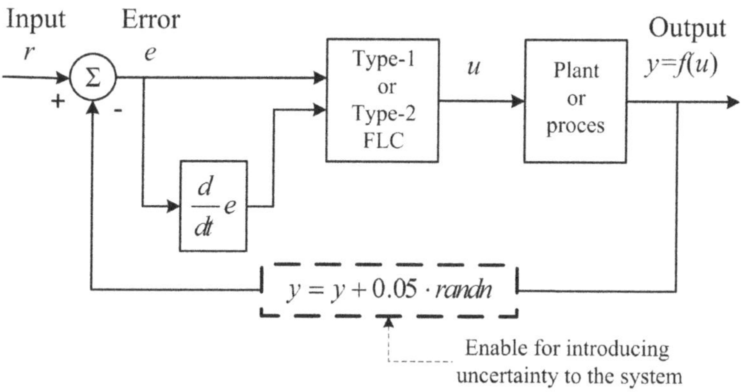

Fig. 1 System used for obtaining the experimental results.

2.2 Type-2 Fuzzy Logic Systems

If for a type-1 membership function, as in Figure 2, we blur its values to the left and to the right, as illustrated in Figure 3, then a type-2 membership function is obtained. In this case, for a specific value x', the membership function (u'), takes on different values, which are not all weighted the same, so we can characterize them by a distribution of membership values.

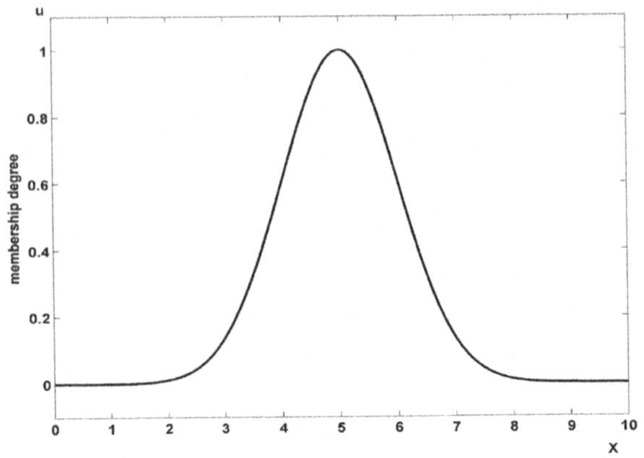

Fig. 2 Type-1 membership function.

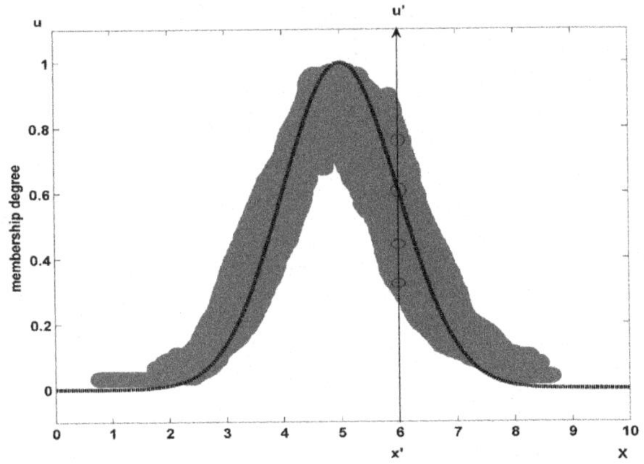

Fig. 3 Blurred type-1 membership function.

A type-2 fuzzy set $\tilde{A}$, is characterized by the membership function (Mendel 2001) (Mendel and Mouzouris 1999):

$$\tilde{A} = \{((x,u), \mu_{\tilde{A}}(x,u)) \mid \forall x \in X, \forall u \in J_x \subseteq [0,1]\} \tag{3}$$

in which $0 \leq \mu_{\tilde{A}}(x,u) \leq 1$. Another expression for $\tilde{A}$ reads as,

$$\tilde{A} = \int_{x \in X} \int_{u \in J_x} \mu_{\tilde{A}}(x,u) / (x,u) \qquad J_x \subseteq [0,1] \tag{4}$$

Where $\int\int$ denotes the union over all admissible input variables x and u. For discrete universes of discourse, the symbol $\int$ is replaced by $\sum$ (Mendel and

John 2002). In fact $J_x \subseteq [0,1]$ represents the primary membership of *x*, and $\mu_{\tilde{A}}(x,u)$ is a type-1 fuzzy set known as the secondary set. Hence, a type-2 membership grade can be any subset in [0,1], the primary membership, and corresponding to each primary membership, there is a secondary membership (which can also be in [0,1]) that defines the possibilities for the primary membership. Uncertainty is represented by a region, which is called the footprint of uncertainty (FOU). When $\mu_{\tilde{A}}(x,u)=1, \forall u \in J_x \subseteq [0,1]$ we have an interval type-2 membership function, as shown in Figure 4. The uniform shading for the FOU represents the entire interval type-2 fuzzy set and it can be described in terms of an upper membership function $\bar{\mu}_{\tilde{A}}(x)$ and a lower membership function $\underline{\mu}_{\tilde{A}}(x)$.

A FLS described using at least one type-2 fuzzy set is called a type-2 FLS. Type-1 FLSs are unable to directly handle rule uncertainties, because they use type-1 fuzzy sets that are certain (Castro et al. 2009). On the other hand, type-2 FLSs, are very useful in circumstances where it is difficult to determine an exact membership function, and there are measurement uncertainties (Mendel 2001) (Li and Zhang 2006).

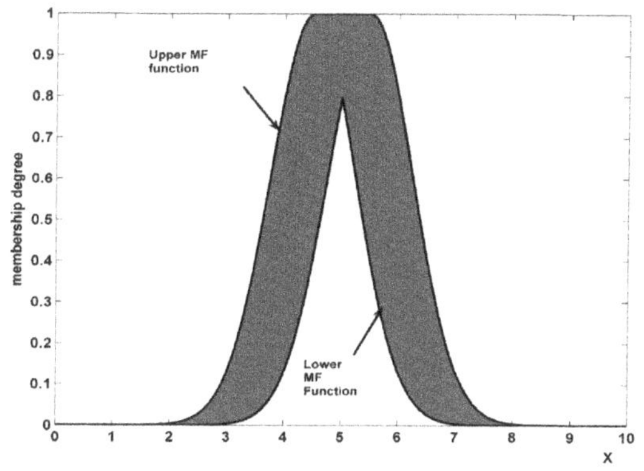

Fig. 4 Interval type-2 membership function.

A type-2 FLS is again characterized by IF-THEN rules, but its antecedent or consequent sets are now of type-2. Similar to a type-1 FLS, a type-2 FLS includes a fuzzifier, a rule base, fuzzy inference engine, and an output processor, as we can see in Figure 5. The output processor includes type-reducer and defuzzifier; it generates a type-1 fuzzy set output (type-reducer) or a crisp number (defuzzifier) (Karnik and Mendel 2001).

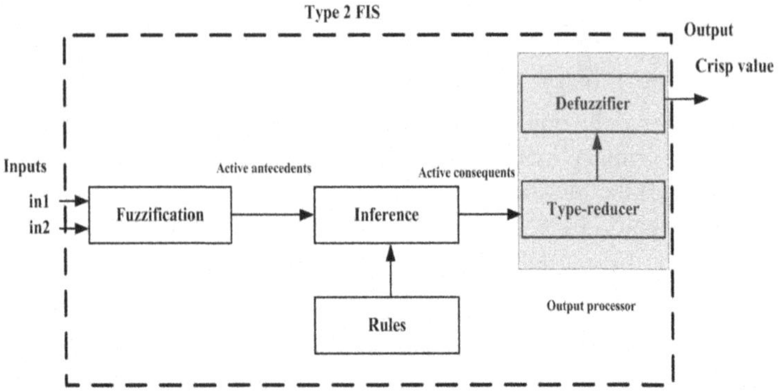

Fig. 5 Type-2 Fuzzy Logic System

2.2.1 Fuzzifier

The fuzzifier maps a point $\mathbf{x}=(x_1,\ldots,x_p)^T \in X_1xX_2x\ldots xX_p \equiv \mathbf{X}$ into a type-2 fuzzy set $\tilde{A}_x$ in **X**, interval type-2 fuzzy sets in this case. We will use type-2 singleton fuzzifier, in a singleton fuzzification, the input fuzzy set has only a single point with nonzero membership (Mendel 2001). $\tilde{A}_x$ is a type-2 fuzzy singleton if $\mu_{\tilde{A}_x}(x) = 1/1$ for **x=x'** and $\mu_{\tilde{A}_x}(x) = 1/0$ for all other **x≠x'**.

2.2.2 Rules

The structure of rules in a type-1 FLS and a type-2 FLS is the same, but in the latter the antecedents and the consequents will be represented by type-2 fuzzy sets. So for a type-2 FLS with p inputs $x_1 \in X_1,\ldots,x_p \in X_p$ and one output $y \in Y$, Multiple Input Single Output (MISO), if we assume there are M rules, the lth rule in the type-2 FLS can be written as follows (Mendel 2001):

$$R^l\text{: IF } x_1 \text{ is } \tilde{F}_1^{\,l} \text{ and } \cdots \text{and } x_p \text{ is } \tilde{F}_p^{\,l}\text{ , THEN } y \text{ is } \tilde{G}^l \qquad l=1,\ldots,M \tag{5}$$

2.2.3 Inference

In the type-2 FLS, the inference engine combines rules and gives a mapping from input type-2 fuzzy sets to output type-2 fuzzy sets. It is necessary to compute the join $\sqcup$, (unions) and the meet $\sqcap$ (intersections), as well as use the extended sup-star compositions (sup star compositions) of type-2 relations. If $\tilde{F}^l{}_1 \times\cdots\times \tilde{F}^l{}_p = \tilde{A}^l$, expression (5) can be re-written as

$$R^l : \tilde{F}^l{}_1 \times\cdots\times \tilde{F}^l{}_p \rightarrow \tilde{G}^l = \tilde{A}^l \rightarrow \tilde{G}^l \qquad l=1,\ldots,M \tag{6}$$

R^l is described by the membership function $\mu_{R^l}(\mathbf{x}, y) = \mu_{R^l}(x_1,\ldots,x_p, y)$,

where

$$\mu_{R^l}(\mathbf{x}, y) = \mu_{\tilde{A}^l \to \tilde{G}^l}(\mathbf{x}, y) \tag{7}$$

can be written as (Mendel 2001):

$$\mu_{R^l}(\mathbf{x}, y) = \mu_{\tilde{A}^l \to \tilde{G}^l}(\mathbf{x}, y) = \mu_{\tilde{F}_1^l}(x_1) \sqcap \cdots \sqcap \mu_{\tilde{F}_p^l}(x_p) \sqcap \mu_{\tilde{G}^l}(y)$$
$$= [\sqcap_{i=1}^{p} \mu_{\tilde{F}^l_i}(x_i)] \sqcap \mu_{\tilde{G}^l}(y) \tag{8}$$

In general, the p-dimensional input to R^l is given by the type-2 fuzzy set $\tilde{A}_x$ whose membership function is

$$\mu_{\tilde{A}_x}(\mathbf{x}) = \mu_{\tilde{x}_1}(x_1) \sqcap \cdots \sqcap \mu_{\tilde{x}p}(x_p) = \sqcap_{i=1}^{p} \mu_{\tilde{x}i}(x_i) \tag{9}$$

where $\tilde{X}_i (i = 1,..., p)$ are the labels of the fuzzy sets describing the inputs. Each rule R^l determines a type-2 fuzzy set $\tilde{B}^l = \tilde{A}_x \circ R^l$ such that:

$$\mu_{\tilde{B}^l}(y) = \mu_{\tilde{A}_x \circ R^l} = \sqcup_{\mathbf{x} \in \mathbf{X}} \left[\mu_{\tilde{A}_x}(\mathbf{x}) \sqcap \mu_{R^l}(\mathbf{x}, y) \right] \quad y \in Y \;\; l=1,\ldots,M \tag{10}$$

This equation is the input/output relation in Figure 5 between the type-2 fuzzy set that activates one rule in the inference engine and the type-2 fuzzy set at the output of that engine. In the FLS we used interval type-2 fuzzy sets and meet under product t-norm, so the result of the input and antecedent operations, which are contained in the firing set $\sqcap_{i=1}^{p} \mu_{\tilde{F}_{i_i}}(x'_i \equiv F^l(\mathbf{x}')$, is an interval type-1 set (Mendel 2001),

$$F^l(\mathbf{x}') = \left[\underline{f}^l(\mathbf{x}'), \bar{f}^l(\mathbf{x}') \right] \equiv \left[\underline{f}^l, \bar{f}^l \right] \tag{11}$$

where

$$\underline{f}^l(\mathbf{x}') = \underline{\mu}_{\tilde{F}_1^l}(x_1') * \cdots * \underline{\mu}_{\tilde{F}_p^l}(x_p') \tag{12}$$

$$\bar{f}^l(\mathbf{x}') = \bar{\mu}_{\tilde{F}_1^l}(x_1') * \cdots * \bar{\mu}_{\tilde{F}_p^l}(x_p') \tag{13}$$

where * is the product operation.

2.2.4 Type Reducer

The type-reducer generates a type-1 fuzzy set output, which is then converted in a crisp output through the defuzzifier. This type-1 fuzzy set is also an interval set, for the case of our FLS we used center of sets (cos) type reduction, $Y_{\cos}$ which is expressed as (Mendel 2001):

$$Y_{\cos}(\mathbf{x}) = [y_l, y_r] = \int_{y^1\in[y_l^1, y_r^1]} \cdots \int_{y^M\in[y_l^M, y_r^M]} \int_{f^1\in[\underline{f}^1, \bar{f}^1]} \cdots \int_{f^M\in[\underline{f}^M, \bar{f}^M]} 1 \Big/ \frac{\sum_{i=1}^{M} f^i y^i}{\sum_{i=1}^{M} f^i} \tag{14}$$

this interval set is determined by its two end points, y_l and y_r, which corresponds to the centroid of the type-2 interval consequent set $\tilde{G}^i$,

$$C_{\tilde{G}^i} = \int_{\theta_1\in J_{y1}} \cdots \int_{\theta_N\in J_{yN}} 1 \Big/ \frac{\sum_{i=1}^{N} y_i\theta_i}{\sum_{i=1}^{N} \theta_i} = [y_l^i, y_r^i] \tag{15}$$

before the computation of $Y_{\cos}$ ($\mathbf{x}$), we must evaluate equation (15), and its two end points, y_l and y_r. If the values of f_i and y_i that are associated with y_l are denoted f_l^i and y_l^i, respectively, and the values of f_i and y_i that are associated with y_r are denoted f_r^i and y_r^i, respectively, from 14, we have

$$y_l = \frac{\sum_{i=1}^{M} f_l^i y_l^i}{\sum_{i=1}^{M} f_l^i} \tag{16}$$

$$y_r = \frac{\sum_{i=1}^{M} f_r^i y_r^i}{\sum_{i=1}^{M} f_r^i} \tag{17}$$

2.2.5 Defuzzifier

From the type-reducer we obtain an interval set $Y_{\cos}$, to defuzzify it we use the average of y_l and y_r, so the defuzzified output of an interval singleton type-2 FLS is (Mendel 2001)

$$y(\mathbf{x}) = \frac{y_l + y_r}{2} \tag{18}$$

3 Average Type-2 FIS (CM Method)

In cases where the performance of an IT2FIS is important, especially in real time applications, an option to avoid the computational delay of type-reduction, is the Wu-Mendel method (Mendel 2001), which is based on the computation of inner and outer bound sets. Another option to improve computing speed in an IT2FIS, is to use the average of two type-1 FIS method (CM method), which was proposed for systems where the type-2 MFs of the inputs and output, have no uncertainty in the mean or center; it is achieved by substituting the IT2FIS with two type-1 FIS, located adequately at the upper and lower footprint of uncertainty (FOU) of the type-2 MFs (Sepulveda et al. 2007).

For the average (CM) method the fuzzification, the inference and the defuzzification stages for each FIS remain identical, the difference is at the output because the crisp value is calculated by taking the arithmetic average of the crisp output of each type-1 FIS, as it is shown in Figure 6, using the height method to calculate the defuzzified crisp output. In the average (CM) method, to achieve the defuzzification, one type-1 FIS is used for the upper bound of uncertainty, and the second FIS for the lower bound of uncertainty. So, as it was explained in Section 2, the defuzzification of a type-1 FIS is used in the average (CM) method and it is illustrated in Figure 6.

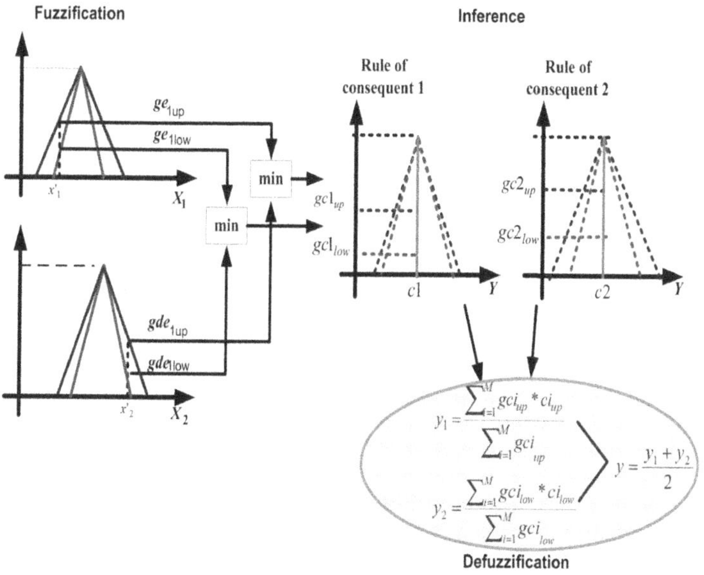

Fig. 6 The fuzzification, the inference and the defuzzification stages in the average (CM) method uses two type-1 FIS.

4 Experimental Results for Intelligent Control

The experimental results are presented here to show a comparison in the system's response in a feedback controller when using a type-1 FLC or a type-2 FLC. A set of five experiments is described in this section. The first two experiments were performed in ideal conditions, i.e., without any kind of disturbance. In the last three experiments, Gaussian noise was added to the feedback loop with the purpose of simulating, in a global way, the effects of uncertainty from several sources. Figure 1 shows the feedback control system that was used for obtaining the simulation results. The complete system was simulated, and the controller was designed to follow the input as closely as possible. The plant is a nonlinear system modeled with equation:

$$y(i) = 0.2 \cdot y(i-3) \cdot 0.07 y(i-2) + 0.9 \cdot y(i-1) + 0.05\, u(i-1) + 0.5 \cdot u(i-2) \quad (19)$$

To illustrate the dynamics of the system, two different inputs are applied, first the input of equation is given by:

$$u(i) = \begin{cases} 0 & 1 \le i < 5 \\ .1 & 5 \le i < 10 \\ .5 & 10 \le i < 15 \\ 1 & 15 \le i < 20 \\ .5 & 20 \le i < 25 \\ 1 & 25 \le i < 30 \\ 0 & 30 \le i < 35 \\ 1.47 & 35 \le i < 40 \end{cases} \quad (20)$$

Now, for a slightly different input given by equation:

$$u(i) = \begin{cases} 0 & 1 \le i < 5 \\ .1 & 5 \le i < 10 \\ .5 & 10 \le i < 15 \\ 1 & 15 \le i < 20 \\ .5 & 20 \le i < 25 \\ 1 & 25 \le i < 30 \\ 0 & 30 \le i < 35 \\ 1.4 & 35 \le i < 40 \end{cases} \quad (21)$$

Going back to the control problem, this system given by equation (19) was used in Figure 1, under the name of plant or process, in this figure we can see that the controller's output is applied directly to the plant's input. Since we are interested in comparing the performance between type-1 and type-2 FLC systems, the controller was tested in two ways:

1. Considering the system as ideal. We have not introduced in the modules of the control system any source of uncertainty (experiments 1 and 2).
2. Simulation of the effects of uncertain modules (subsystems) response introducing some uncertainty (experiments 3, 4 and 5).

For both cases, as it is shown in Figure 1, the system's output is directly connected to the summing junction, but in the second case, the uncertainty was simulated introducing random noise with normal distribution (the dashed square in Figure 1). We added noise to the system's output $y(i)$ using a function "randn", which generates random numbers with a Gaussian distribution. The signal and the added noise in turn, were obtained by using the expression (22), the result $y(i)$ was introduced to the summing junction of the controller system. Note that in expression (22) we are using the value 0.05, for experiments 3 and 4, but in the set of tests for experiment 5, we varied this value to obtain different SNR values.

$$y(i) = y(i) + 0.05 \cdot randn \tag{22}$$

The system was tested using as input, a unit step sequence free of noise, $r(i)$. For evaluating the system's response and comparing between type 1 and type 2 fuzzy controllers, the performance criteria of Integral of Squared Error (ISE), Integral of Absolute Value of Error (IAE), and Integral of Time per Absolute Value of Error (ITAE) were used. In Table 3, we summarize the values obtained in an ideal system for each criterion considering 400 units of time. For calculating ITAE a sampling time of $T_s = 0.1$ sec. was considered. In Experiment 5, we tested the systems, type-1 and type-2 FLCs, introducing different values of noise η. This was done by modifying the signal to noise ratio SNR (Proakis and Manolakis 1996),

$$SNR = \frac{\sum |s|^2}{\sum |\eta|^2} = \frac{P_{signal}}{P_{noise}} \tag{23}$$

Because many signals have a very wide dynamic range, SNRs are usually expressed in the logarithmic decibel scale in SNR(db),

$$SNR(db) = 10\log_{10}\left(\frac{P_{signal}}{P_{noise}}\right) \tag{24}$$

In Table 4, we show, for different values of SNR(db), the behavior of the errors ISE, IAE, ITAE for type-1 and type-2 FLCs. In all the cases the results for type-2 FLC are better than type-1 FLC. In the type-1 FLC, Gaussian membership functions (Gaussian MFs) for the inputs and for the output were used. A Gaussian MF is specified by two parameters $\{c,\sigma\}$:

$$\mu_A(x) = e^{-\frac{1}{2}\left(\frac{x-c}{\sigma}\right)^2} \tag{25}$$

c represents the MFs center and σ determines the spread of the MFs.

For each of the inputs of the type-1 FLC, three Gaussian MFs were defined as: negative, zero, positive. The universe of discourse for these membership functions is in the range [-10 10]. For the output of the type-1 FLC, we have five Gaussian MFs denoted by NG, N, Z, P and PG. Table 1 illustrates the characteristics of the MFs of the inputs and output of the type-1 FLC.

Table 1 Characteristics of the Inputs and Output of the Type-1 FLC.

Variable	Term	Center c	Standard deviation σ
Input e	negative	-10	4.2466
	zero	0	4.2466
	positive	10	4.2466
Input Δe	Negative	-10	4.2466
	Zero	0	4.2466
	positive	10	4.2466
Output *cde*	NG	-10	2.1233
	N	-5	2.1233
	Z	0	2.1233
	P	5	2.1233
	PG	10	2.1233

In experiments 2, 4, and 5, for the type-2 FLC, as in type-1 FLC, we also selected Gaussian MFs for the inputs and for the output, but in this case we have interval type-2 Gaussian MFs with a fixed center, c, and some spread σ, i.e.,

$$\mu_A(x) = e^{-\frac{1}{2}\left(\frac{x-c}{\sigma}\right)^2} \tag{26}$$

In terms of the upper and lower membership functions, we have for $\mu_{\tilde{A}}(x)$,

$$\overline{\mu}_{\tilde{A}}(\mathrm{x}) = \mathrm{N}(\mathrm{c}, \sigma_2; \mathrm{x}) \tag{27}$$

and for the lower membership function $\underline{\mu}_{\tilde{A}}(x)$,

$$\underline{\mu}_{\tilde{A}}(\mathrm{x}) = \mathrm{N}(\mathrm{c}, \sigma_1; \mathrm{x}) \tag{28}$$

where $\mathrm{N}(\mathrm{c}, \sigma_2, \mathrm{x}) \equiv \mathrm{e}^{-\frac{1}{2}\left(\frac{\mathrm{x}-\mathrm{c}}{\sigma_2}\right)^2}$, and $\mathrm{N}(\mathrm{c}, \sigma_1, \mathrm{x}) \equiv \mathrm{e}^{-\frac{1}{2}\left(\frac{\mathrm{x}-\mathrm{c}}{\sigma_1}\right)^2}$, (Mendel 2001). Hence, in the type-2 FLC, for each input we defined three-interval type-2 fuzzy Gaussian MFs: negative, zero, positive in the interval [-10 10], as illustrated in Figures 7 and 8. For computing the output we have five interval type-2 fuzzy Gaussian MFs, which are NG, N, Z, P and PG, in the interval [-10 10], as can be seen in Figure 9. Table 2 shows the characteristics of the inputs and output of the type-2 FLC.

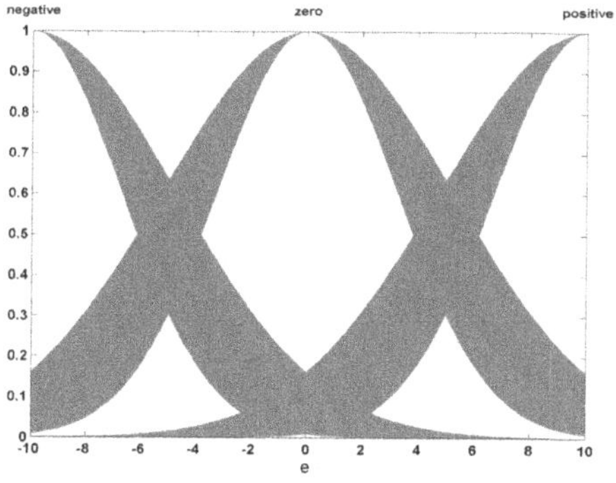

Fig. 7 Input e membership functions for the type-2 FLC.

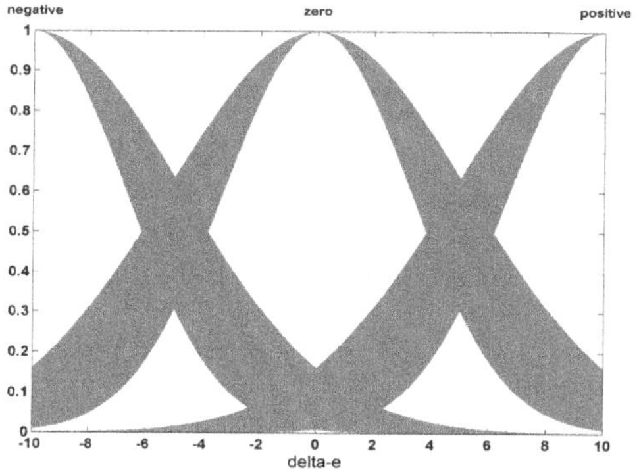

Fig. 8 Input Δe membership functions for the type-2 FLC.

In all experiments, we have a dash-dot line for illustrating the system's response and behavior of type-1 FLC, in the same sense, a continuous line for type-2 FLC. The reference r is shown with a dot line.

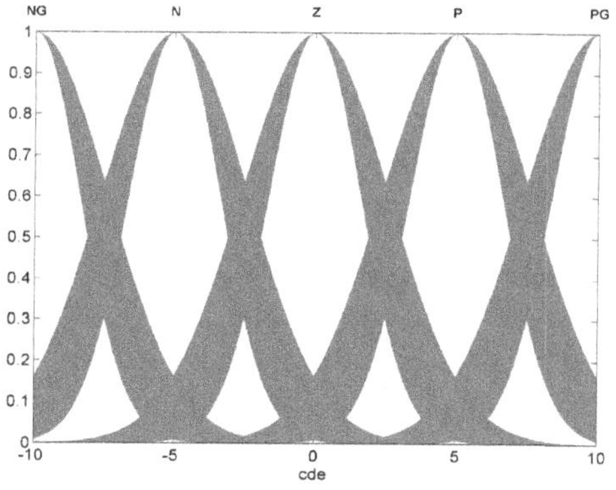

Fig. 9 Output cde membership functions for the type-2 FLC.

Table 2 Input and Output Parameters of the Type-2 FLC.

Variable	Term	Center c	Standard deviation σ_1	Standard deviation σ_2
Input e	negative	-10	5.2466	3.2466
	zero	0	5.2466	3.2466
	positive	10	5.2466	3.2466
Input Δe	Negative	-10	5.2466	3.2466
	Zero	0	5.2466	3.2466
	positive	10	5.2466	3.2466
Output *cde*	NG	-10	2.6233	1.6233
	N	-5	2.6233	1.6233
	Z	0	2.6233	1.6233
	P	5	2.6233	1.6233
	PG	10	2.6233	1.6233

Experiment 1: Simulation of an ideal system with a type-1 FLC.

In this experiment, uncertainty data was not added to the system, and the system response produced a settling time of about 140 units of time; i.e., the system tends to stabilize with time and the output will follow accurately the input. In Table 3, we listed the values of ISE, IAE, and ITAE for this experiment.

Table 3 Performance Criteria for Type-1 and Type-2 Fuzzy Controllers for 20 dB Signal to Noise Ratio (After 200 Samples).

Performance Criteria	**Type-1 FLC**		**Type-2 FLC**	
	Ideal System	Syst. with uncertainty	Ideal System	Syst. with uncertainty
ISE	7.65	19.4	6.8	18.3
IAE	17.68	49.5	16.4	44.8
ITAE	62.46	444.2	56.39	402.9

Experiment 2: Simulation of an ideal system using the type-2 FLC.

Here, the same test conditions of Experiment 1 were used, but in this case, we implemented the controller with type-2 fuzzy logic. The corresponding performance criteria are listed in Table 3. We can observe that when using a type-2 FLC we obtained the lower errors.

Experiment 3: System with uncertainty using a type-1 FLC.

In this case, expression (25) was used to simulate the effects of uncertainty introduced to the system by transducers, amplifiers, and any other element that in real world applications affects expected values. In this experiment the noise level was assumed to be in the range of 20 dB of SNR ratio.

Experiment 4: System with uncertainty using a type-2 FLC.

In this experiment, uncertainty was introduced in the system, in the same way as in Experiment 3. In this case, a type-2 FLC was used and the results obtained with a type-1 FLC (Experiment 3) were improved.

Experiment 5. Varying the Signal to Noise Ratio (SNR) in type-1 and type-2 FLCs.

To test the robustness of the type-1 and type-2 FLCs, we repeated experiments 3 and 4 giving different noise levels, going from 30 db to 8 db of SNR ratio in each experiment. In Table 4, we summarized the values for ISE, IAE, and ITAE considering 200 units of time with a P_{signal} of 22.98 dB in all cases. As it can be seen in Table 4, in presence of different noise levels, the behavior of type-2 FLC is in general better than type-1 FLC.

Table 4 Behavior of theType-1 and Type-2 Fuzzy Logic Controllers after Variation of Signal to Noise Ratio (Values Obtained for 200 Samples).

Noise variation				Type-1 FLC			Type-2 FLC		
SNR (dB)	SNR	Sum-Noise	Sum-Noise (dB)	ISE	IAE	ITAE	ISE	IAE	ITAE
8	6.4	187.42	22.72	321.1	198.1	2234.1	299.4	194.1	2023.1
10	10.05	119.2	20.762	178.1	148.4	1599.4	168.7	142.2	1413.5
12	15.86	75.56	18.783	104.7	114.5	1193.8	102.1	108.8	1057.7
14	25.13	47.702	16.785	64.1	90.5	915.5	63.7	84.8	814.6
16	39.88	30.062	14.78	40.9	72.8	710.9	40.6	67.3	637.8
18	63.21	18.967	12.78	27.4	59.6	559.1	26.6	54.2	504.4
20	100.04	11.984	10.78	19.4	49.5	444.2	18.3	44.8	402.9
22	158.54	7.56	8.78	14.7	42	356.9	13.2	37.8	324.6
24	251.3	4.77	6.78	11.9	36.2	289	10.3	32.5	264.2
26	398.2	3.01	4.78	10.1	31.9	236.7	8.5	28.6	217.3
28	631.5	1.89	2.78	9.1	28.5	196.3	7.5	25.5	180.7
30	1008	1.19	0.78	8.5	25.9	164.9	7	23.3	152.6

From Table 4, considering two examples, the extreme cases; we have for an SNR ratio of 8 dB, in type-1 FLC the following performance values ISE=321.1, IAE=198.1, ITAE=2234.1; and for the same case, in type-2 FLC, we have ISE=299.4, IAE=194.1, ITAE=2023.1. For 30 db of SNR ratio, we have for the type-1 FLC, ISE=8.5, IAE=25.9, ITAE=164.9, and for the type-2 FLC, ISE=7, IAE=23.3, ITAE=152.6. These values indicate a better performance of the type-2 FLC than type-1 FLC, because they are a representation of the errors, and as the error increases the performance of the system goes down.

Finally in Table 5 we show the values obtained in the optimization process of the optimal parameters for the MFs after 30 tests of: the variance, the Standard deviation, best ISE value, average ISE obtained with the optimized interval type-2 FLC, and with the average of two optimized type-1 FLCs (CM method).

Table 5 Comparison of the Variance, Standard Deviation, Best ISE value, ISE average, obtained with the Optimized Interval Type-2 FLC and the optimized Average of two Type-1 FLCs

Parameters	Type-2 FLC (WM Method)	Average of two Type-1 FLCs (CM)
Search Interval	2.74 to 5.75	2.74 to 5.75
Best ISE value	4.3014	4.1950
ISE Average	4.4005	4.3460
Standard deviation	0.1653	0.1424
Variance	0.0273	0.0203

In order to know which system behaves in a better way in the experiments, where uncertainty was simulated through different noise levels, first we compare the values of the ISE, IAE and ITAE errors obtained with the optimized parameters of the MFs of the interval type-2 FLC and the average of the two type-1 FLCs. The second comparison is made with the values of standard deviation and the variance obtained in each optimization process to get the optimal parameters of the MFs for the minimal ISE, IAE and ITAE errors. Figure 10 shows a comparison between the ISE values of the type-2 FLC based on the WM method (ISE T2) and the ISE of the type-2 FLC based on CM method (ISE PROM), which uses the average of two type-1 fuzzy systems. In this case, the ISE values are consistently lower for the CM method.

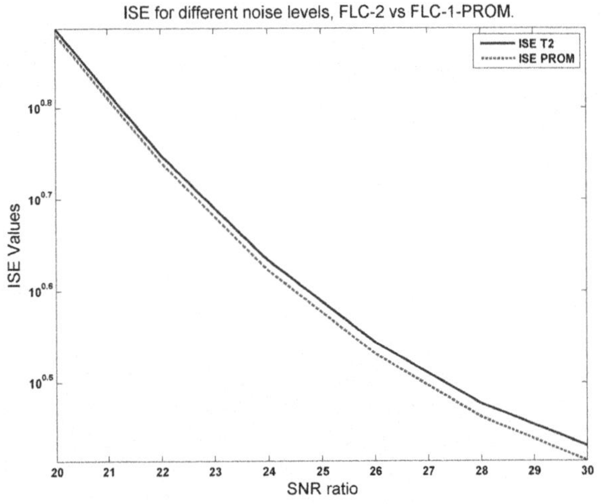

Fig. 10 Comparison of the ISE errors of optimized interval type-2 FLC and the optimized average of two type-1 FLCs, for different noise levels.

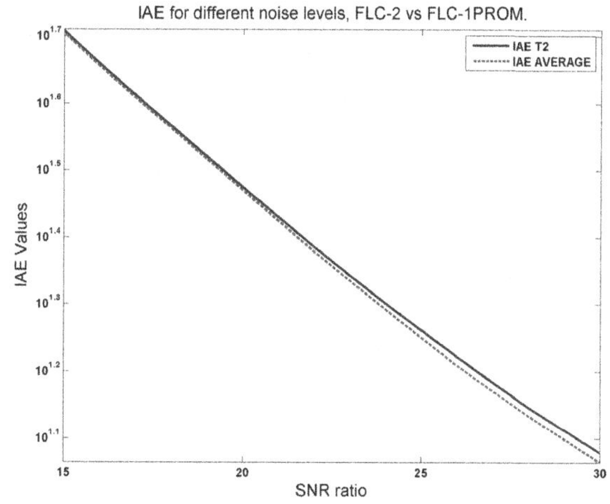

Fig. 11 Comparison of the IA errors of optimized interval type-2 FLC and the optimized average of two type-1 FLC, for different noise levels.

We can see in Tables 4 and 5 that with the average of two type-1 FLC optimized under certain FOU, it was obtained a minimum advantage in the values of ISE, IAE and ITAE errors than with the interval type-2 FLC optimized under the same conditions than the average of two type-1 FLC. In Figures 10, 11 and 12 it is shown that this advantage is notorious for low noise level.

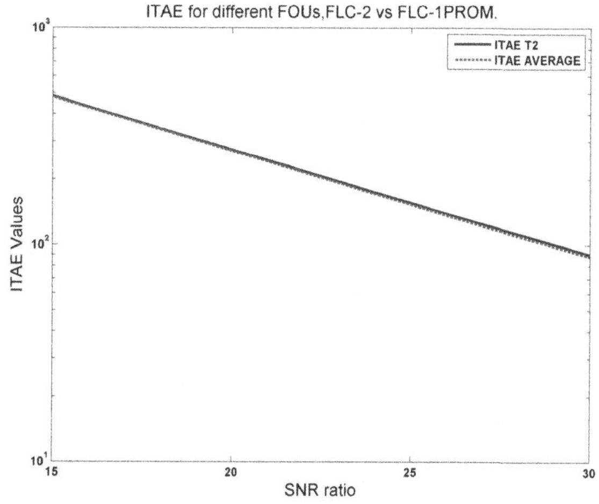

Fig. 12 Comparison of the ITAE errors of optimized interval type-2 FLC and the optimized average of two type-1 FLC, for different noise levels. Practically they behave in the same manner.

In this paper, an improved type-2 inference engine with the CM method was proposed and implemented into an FPGA. The type-2 engine process all the rules in parallel providing high speed computations, the processing time of the whole inference engine is just one clock cycle, approximately 0.02 microseconds for the Spartan 3 FPGA (Montiel et al. 2008). The processing time of a type-2 system implemented with the type-1 inference engine will not grow up since both inference engines (of the two type-1 fuzzy systems) are connected in parallel, hence the processing time remains almost the same for this stage. On the other hand, using KM or WM the times required for type-2 processing would be at least 1000 times more than with the CM method. This makes the proposed CM method of fundamental importance for real-world type-2 fuzzy logic applications, in particular for intelligent control.

5 Conclusions

We have presented the study of the controllers' design for nonlinear control systems using type-1 and type-2 fuzzy logic. We presented five experiments where we simulated the systems' responses with and without uncertainty presence. In the experiments, a quantification of errors was achieved and documented in detail for different criteria such as ISE, IAE, and ITAE. It was also shown that the lower overshoot and the best settling times were obtained using a type-2 FLC. Based on the experimental results, we can say that the best results are obtained using type-2 fuzzy systems. A comparative study of the three methods, based on accuracy and efficiency is presented, and the CM is shown to outperform both the KM and WM methods in efficiency while accuracy is comparable. In our opinion, this is because the lower and upper membership functions' estimations, of the outputs, are more easily found by directly obtaining them using an optimization method, like an evolutionary algorithm. This fact makes the CM method a good choice for real-world control applications in which efficiency is of fundamental importance.

References

Astudillo, L., Castillo, O., Aguilar, L.T.: Intelligent control of an autonomous mobile robot using type-2 fuzzy logic. In: Proceedings of the International Conference on Artificial Intelligence, Las Vegas Nevada (2006)

Castillo, O., Melin, P.: Soft computing for control of non-linear dynamical systems. Springer, Heidelberg (2001)

Castillo, O., Melin, P.: Soft computing and fractal theory for intelligent manufacturing. Springer, Heidelberg (2003)

Castillo, O., Melin, P.: A new approach for plant monitoring using type-2 fuzzy logic and fractal theory. Int. J. Gen. Syst. 33, 305–319 (2004)

Castillo, O., Melin, P.: Type-2 fuzzy logic: theory and applications. Springer, Heidelberg (2008)

Castro, J.R., Castillo, O., Melin, P., Rodriguez-Diaz, A.: A hybrid learning algorithm for a class of interval type-2 fuzzy neural networks. Inf. Sci. 179, 2175–2193 (2009)

Coupland, S., John, R.I.: New geometric inference techniques for type-2 fuzzy sets. Int. J. Approx. Rea. 49, 198–211 (2008)
Doctor, F., Hagras, H., Callaghan, V.: A type-2 fuzzy embedded agent to realize ambient intelligence in ubiquitous computing environments. Inf. Sci. 171, 309–334 (2005)
Hagras, H.: Hierarchical type-2 fuzzy logic control architecture for autonomous mobile robots. IEEE Trans. on Fuzzy Sys. 12, 524–539 (2004)
Karnik, N.N., Mendel, J.M., Liang, Q.: Type-2 fuzzy logic systems. IEEE Trans. Fuzzy Syst. 7, 643–658 (1999)
Karnik, N.N., Mendel, J.M.: Operations on type-2 fuzzy sets. Fuzzy Sets Syst. 122, 327–348 (2001)
Karnik, N.N., Mendel, J.M.: Centroid of a type-2 fuzzy set. Inf. Sci. 132, 195–220 (2001)
Li, S., Zhang, X.: Fuzzy logic controller with interval-valued inference for distributed parameter system. Int. J. Innovat. Comput. Inf. Control 2, 1197–1206 (2006)
Martinez, R., Castillo, O., Aguilar, L.T.: Optimization of interval type-2 fuzzy logic controllers for a perturbed autonomous wheeled mobile robot using Genetic Algorithms. Inf.Sci. 179, 2158–2174 (2009)
Melin, P., Castillo, O.: A new method for adaptive control of non-linear plants using type-2 fuzzy logic and neural networks. Int. J. Gen. Syst. 33, 289–304 (2004)
Mendel, J.M.: Uncertain rule-based fuzzy logic systems: introduction and new directions. Prentice Hall, New Jersey (2001)
Mendel, J.M.: Computing derivatives in interval type-2 fuzzy logic systems. IEEE Trans. Fuzzy Syst. 12, 84–98 (2004)
Mendel, J.M., John, R.I.: Type-2 Fuzzy Sets Made Simple. IEEE Trans. Fuzzy Syst. 10, 117–127 (2002)
Mendel, J.M., Mouzouris, G.C.: Type-2 fuzzy logic systems. IEEE Trans. Fuzzy Syst. 7, 643–658 (1999)
Mizumoto, M., Tanaka, K.: Some properties of fuzzy sets of type-2. Inform Control 31, 312–340 (1976)
Montiel, O., Castillo, O., Melin, P., Rodriguez-Diaz, A., Sepulveda, R.: Human evolutionary model: a new approach to optimization. Inf. Sci. 177, 2075–2098 (2007)
Montiel, O., Maldonado, Y., Sepulveda, R., Castillo, O.: Simple tuned fuzzy controller embedded into an FPGA. In: Proceedings of the 2008 NAFIPS Conference, New York, USA (2008)
Proakis, J.G., Manolakis, D.G.: Digital signal processing principles, algorithms and applications. Prentice Hall, New Jersey (1996)
Sepulveda, R., Castillo, O., Melin, P., Rodriguez-Diaz, A., Montiel, O.: Experimental study of intelligent controllers under uncertainty using type-1 and type-2 fuzzy logic. Inf. Sci. 177, 2023–2048 (2007)
Starczewski, J.T.: Efficient triangular type-2 fuzzy logic systems. Int. J. of Approx Rea. 50, 799–811 (2009)
Wagenknecht, M., Hartmann, K.: Application of fuzzy sets of type 2 to the solution of fuzzy equations systems. Fuzzy Sets Syst. 25, 183–190 (1988)
Zadeh, L.A.: Similarity relations and fuzzy ordering. Inf. Sci. 3, 177–206 (1971)
Zadeh, L.A.: Outline of a new approach to the analysis of complex systems and decision processes. IEEE Trans. Syst. Man and Cyber. 3, 28–44 (1973)
Zadeh, L.A.: The concept of a linguistic variable and its application to approximate reasoning. Part 1. Inf. Sci. 8, 199–249 (1975)
Zadeh, L.A.: Toward a generalized theory of uncertainty (GTU)-an outline. Inf. Sci. 172, 1–40 (2005)

Type-2 Fuzzy Similarity in Partial Truth and Intuitionistic Reasoning

Chung-Ming Own

Abstract. Representing and manipulating the vague concepts of partially true knowledge pose a major challenge to the development of machine intelligence. In particular, the issue of how to extract approximate facts from vague and partially true statements has received considerable attention in the field of fuzzy information processing. However, vagueness is often due to a lack of available information, making it impossible to satisfactorily evaluate membership. Atanassov (1996) demonstrated the feasibility of mapping intuitionistic fuzzy sets to historical fuzzy sets. Intuitionistic fuzzy sets are isomorphic to interval valued fuzzy sets, while interval valued fuzzy sets have been regarded as unique value among type-2 fuzzy sets. This study presents a theoretical method to represent and manipulate partially true knowledge. The proposed method is based on the measurement of similarity among type-2 fuzzy sets, which are used directly to handle rule uncertainty that type-1 fuzzy sets are unable to deal with. Moreover, the switching relationship between type-2 fuzzy sets and intuitionist fuzzy sets is defined axiomatically. Results of this study demonstrate the effectiveness of the proposed theoretical method in pattern recognition and reasoning with regard to medical diagnosis.

Keywords: Type-2 fuzzy sets, Intuitionistic fuzzy sets, Fuzzy similarity, Partial truth.

1 Introduction

In order to distinguish between similar entities or groups of entities in daily life, one must determine the degree of similarity between them. Fuzzy set theory was developed by Zadeh, and ushered in an era of research into the measurement of similarity between fuzzy sets. These fuzzy set-based developments are applicable in data preprocessing, data mining for identifying dependency relationships between concepts, inference reasoning (Li et al. 2002, Tianjiang et al. 2002, Li et al. 2005, Song 2008, Fu X and Shen 2010, Petry and Yager 2010), and pattern

Chung-Ming Own
Department of Computer and Communication Engineering,
St. John's Univeristy

W. Pedrycz and S.-M. Chen (Eds.): Granular Computing and Intell. Sys., ISRL 13, pp. 95–115.
springerlink.com

recognition (Dengfeg and Chuntian 2002, Mitchell 2003, Zhizhen and Pengfei 2003, Hung and Yang 2004, Tang 2008, Sledge et al. 2010). Among other recently developed ways of measuring similarity, intuitionistic fuzzy sets are regarded as a distinct category. For examples of intuitionistic fuzzy sets, please refer to Chen (1995, 1997), Hong and Kim (1999) and Fan and Zhangyan (2001), Xu (2007), Mushrif and Ray (2009), and Chaira (2010).

In this chapter, we analyze existing measures of similarity among type-2 fuzzy sets based on counter-intuitive examples of partially true knowledge. We axiomatically define how type-2 fuzzy sets and intuitionistic fuzzy sets are related. Finally, we illustrate the usefulness of our proposed conversion in the application to reasoning in medical diagnosis.

1.1 Partially Truth

Reasoning systems are altered to handle incomplete or partially true knowledge by splitting each partially true statement into two components: a proposition component; and an associated truth-value component (Dubois and Prade 1980). The truth-value component provides an effective means of modifying the significance of the original proposition. A proposition such as "*x* is *F*" is expressed as "*x* is *F* is τ", where τ denotes a linguistic value of partial truth qualification, defined as the degree of compatibility of the situation with the proposition "*x* is *F*". Therefore, "*x* is *F*" denotes a proposition component, and a linguistic truth value, τ, denotes an associated truth-value component. Equivalent statements in natural language include:

" 'David is healthy' is quite true."

" 'The speed is moderate' is absolutely true."

The unit interval [0, 1] is taken as a set of partially true values. Any vague definition related to truth can be represented by a fuzzy set on [0, 1].

A simple vague proposition about the truth-value, such as "This truth value represents 'mostly true'", can be translated into a rule in the form of a general fuzzy set:

"Truth value is mostly true"

$$= \sum_{x \in X} \mu_{mostly\ true}(x)/x$$

$$= \frac{0.4}{0.65} + \frac{0.5}{0.7} + \frac{0.75}{0.75} + \frac{1}{0.8} + \frac{0.75}{0.9} + \frac{0.26}{0.95}.$$

The value of $\mu_{mostly\ true}(x)$ does not change the meaning of the proposition, but represents a subjective opinion concerning the meaning of the proposition. That is, when $\mu_{mostly\ true}(x) = 0$, the truth value certainly differs from *x*, and when $\mu_{mostly\ true}(x) = 1$, the truth value equals *x*. Notably, $\mu_{mostly\ true}(x)$ reveals the uncertainty of the original knowledge.

Assuming that this partial truth qualification is local rather than absolute, Bellman and Zadeh obtained a true statement based on a partially true statement, and derived the corresponding fuzzy set as a representation of such a statement (Zadeh 1979, Bellman and Zadeh 1977). Baldwin proposed implied statements, consisting of a fuzzy truth value restricted to a Lukasiewicz logical implication related to a fuzzy truth space (Baldwin 1979). Based on set-theoretical considerations, that study also obtained constraints to fuzzy truth values on truth value restrictions from conditional fuzzy linguistic statements, by applying an inverse procedure to modify the truth functions.

Accordingly, Raha and Ray proposed a theoretical method for reasoning with a partial truth value associated with a vague sentence (Raha and Ray 1999, 1997, 2000). The partial truth values were defined by fuzzy sets on the universe of discourse [0,1], which is a unit interval. This vague proposition is presented as a possibility distribution, in which each possibility distribution is assigned to and manipulated by a fuzzy set/relation.

In contrast with the above approaches, the theoretical method developed in this study attempts to eliminate the deficiencies involved in the representation of partially true knowledge. Despite associating and manipulating the partial truth value according to the proposition, previous methods have denoted and estimated the corresponding fuzzy set and qualification of partial truth, separately. In other words, set-theoretical considerations cannot be used to derive partially true knowledge, as long as partially true statements are not associated with the existing proposition. The proposed theoretical method used to represent and manipulate such partially true knowledge is therefore based on the type-2 fuzzy set theory.

1.2 Type-2 Fuzzy Sets

Type-2 fuzzy sets were initially defined by Zadeh (1979), and characterized by a fuzzy membership. The membership value for each element of this set is a fuzzy set in [0,1], whereas the membership grade of a type-1 fuzzy set is a numeric value in [0,1]. To clarify the above statement, the fuzzy set '*tall*' is represented as:

$$tall = \frac{0.95}{Michael} + \frac{0.4}{Danny} + \frac{0.6}{Robert}.$$

Conversely, the interpretation of type-2 fuzzy sets is

$$tall = \frac{High}{Michael} + \frac{Low}{Danny} + \frac{Medium}{Robert},$$

where membership functions of High, Low, and Medium, themselves are fuzzy sets. The former set is measured by one condition for one element, while the latter set is evaluated by several conditions for one element. Type-2 fuzzy sets are useful when the exact membership function for a type-1 fuzzy set cannot be easily determined. For this reason, Ttype-2 fuzzy sets are advantageous for the incorporation of uncertainty.

According to Mendel (2000, 2006), type-2 fuzzy sets are defined as follows. For simplicity, the universe of discourse is assumed to be a finite set, although the definition is also applicable for infinite universes of discourse.

Definition 1 (Mendel 2000, 2006). Type-2 fuzzy set, $\tilde{A}$, is characterized by a type-2 membership function, $\mu_{\tilde{A},}(x)$, where X is the universal set, $x \in X$ and $u \in J_x \subseteq [0,1]$, J_x is the possible membership collection for every x. Meanwhile, the amplitude of the secondary membership function is called a secondary grade, and $f_x(u)$ is a secondary grade. That is,

$$\tilde{A} = \left\{\left(x, \mu_{\tilde{A},}(x)\right) \middle| \forall x \in X\right\},$$

or, as

$$\tilde{A} = \sum_{x \in X} \mu_{\tilde{A}}(x)/x = \sum_{x \in X} [\sum_{x \in J_x} f_x(u)/u]/x.$$

1.3 Intuitionistic Fuzzy Sets

Assume that X denotes the universe of discourse, $X = \{x_1, x_2, \cdots, x_n\}$. Ordinary fuzzy set theory lacks an effective means of incorporating that hesitation into the degree of membership. Atanassov (1986) developed intuitionistic fuzzy sets, along with the ability to model hesitation and uncertainty by using an additional degree. Each intuitionistic fuzzy set $\breve{A}$ allots a membership degree $\mu_{\breve{A}}(x)$ and a non-membership degree $v_{\breve{A}}(x)$ to each element x of the universe X, note that $\mu_{\breve{A}}(x) \in [0,1]$, $v_{\breve{A}}(x) \in [0,1]$ and $\mu_{\breve{A}}(x) + v_{\breve{A}}(x) \leq 1$. The value $\pi(x) = 1 - (\mu_{\breve{A}}(x) + v_{\breve{A}}(x))$ is called the hesitation part, which is the hesitancy degree of whether x belongs to $\breve{A}$. The set of all the intuitionistic fuzzy sets in X is representing as $IFS(X)$.

Definition 2 (Atanassov 1986). When the universe of discourse X is discrete, an intuitionistic fuzzy set $\breve{A}$ is denoted as follows:

$$\breve{A} = \sum_{i=1}^{n} [x, \mu_{\breve{A}}(x), v_{\breve{A}}(x)], \ \forall x_i \in X.$$

For the sake of simplicity, the universe of discourse is assumed to be a finite set, although the definition can be applied for infinite sets.

The following properties are expressed for all $\breve{A}$ and $\breve{B}$ belonging to IFSs(X) in (Pappis and Karacpailidis 1993),

(a). $\breve{A} \leq \breve{B}$ *if and only if* $\mu_{\breve{A}}(x) \leq \mu_{\breve{B}}(x)$ *and* $v_{\breve{A}}(x) \geq v_{\breve{B}}(x)$ *for all* $x \in X$.

(b). $\breve{A} = \breve{B}$ *if and only if* $\breve{A} \leq \breve{B}$ *and* $\breve{A} \geq \breve{B}$.

(c). $\breve{A}_c = \sum_{i=1}^{n} [x, v_{\breve{A}}(x), \mu_{\breve{A}}(x)], \ \forall x_i \in X$.

1.4 Type-2 Fuzzy Similarity

In our study, the universe of discourse denotes a finite set, the type-2 fuzzy set $\tilde{A}$ is expressed as:

$$\tilde{A} = \sum_{x \in X} [\sum_{x \in J_x} f_x(u)/u]/x = \sum_{i=1}^{N} [\sum_{x \in J_{x_i}} f_x(u)/u]/x_i$$
$$= [\sum_{k=1}^{M_1} f_{x_1}(u_{1k})/u_{1k}]/x_1 + \cdots + [\sum_{k=1}^{M_N} f_{x_N}(u_{Nk})/u_{Nk}]/x_N.$$

Assume that x has been incorporated into N values, with each value u discretized into M_i values. Many choices are possible for these secondary membership functions. The secondary membership function can be treated as a type-1 membership function along each x. Hence, the similarity between type-2 fuzzy sets is

$$\tilde{S}(\tilde{A}, \tilde{B}) = 1/N \sum_{i=1}^{N} S(\mu_{\tilde{A}}(x_i), \mu_{\tilde{B}}(x_i)),$$

where $\tilde{S}(\cdot)$ can be any traditional similarity index for the general fuzzy sets. Note that, $\mu_{\tilde{A}}(x_i)$ and $\mu_{\tilde{B}}(x_i)$ are two secondary membership functions. For instance, the proposed similarity methods of Rahaet al. (2002) and Pappiset al. (1993) are:

$$S(A, B) = \sum_{x \in X} \{\mu_A(x) \cdot \mu_B(x)\} / \sum_{x \in X} max\{\mu_A(x), \mu_B(x)\}^2, \quad (1)$$

or

$$S(A, B) = |A||B|\cos(\theta)/(\max(|A|^2, |B|^2)),$$

where A and B are two type-1 fuzzy sets; $|A|$ is the length of the vector A, and $\cos(\theta)$ is the cosine of the angle between the two vectors. An important consideration is to select the similarity index of type-2 fuzzy sets such that the index exhibits the properties of similarity. Expression(1) is adopted in this study, and the similarity index of type-2 fuzzy sets is formulated as follows:

$$\tilde{S}(\tilde{A}, \tilde{B}) = 1/N \sum_{i=1}^{N} S(\mu_{\tilde{A}}(x_i), \mu_{\tilde{B}}(x_i))$$
$$= 1/N \sum_{i=1}^{N} \frac{\sum_u \{f_{x_i}(u), g_{x_i}(u)\}}{\sum_u [\max\{f_{x_i}(u), g_{x_i}(u)\}^2]}, \quad (2)$$

where $x_i \in X$ and $u \in J_x$. In addition, when we defined $\sum_u[\max\{f_{x_i}(u), g_{x_i}(u)\}^2=0$, then $S(\mu_{\tilde{A}}(x_i), \mu_{\tilde{B}}(x_i)) = 1$, that is, $\tilde{S}(\tilde{A}, \tilde{B}) = 1$.

Notably, because the primary membership values may not be the same for a specific value of x, that is J_x on $\tilde{A}$ and J_x on $\tilde{B}$ cannot be exactly computed in some cases. Because the notion of a zero membership value generalizes in fuzzy set theory to the situation in which a non-zero membership is not clearly stated, this concept has also been executed in this study. The value '0' in the secondary grade denotes this item as useless in the deterministic process. The value '0' is applied as the appended secondary grade to those missing items in the minus between J_x on $\tilde{A}$ and J_x on $\tilde{B}$ for the computation of similarity.

2 Reasoning with Type-2 Similarity

A type-1 fuzzy inference engine normally combines rules and provides a mapping from input type-1 fuzzy sets to output type-1 fuzzy sets. Additionally, the inference process is very similar in the case of type-2 inference. The inference engine combines rules and provides a mapping from input type-2 fuzzy sets to output type-2 fuzzy sets. Another difference is in the defuzzification. In the type-2 cases, the output sets are type-2; the extended defuzzification operation in the type-2 case gives the type-1 fuzzy sets as the output. This operation is called a "type reducer", and the type-1 fuzzy set is obtained as a "type reduced set", which may then be defuzzified to obtain a single crisp number.

This study presents a new reasoning method, involving the measure of similarity between type-2 fuzzy sets as an inference methodology. Consider two type-2 fuzzy sets $\tilde{A}$ and $\tilde{A}'$ defined in the same universe of discourse X. Another two type-2 fuzzy sets $\tilde{B}$ and $\tilde{B}'$ are defined over the same universe of discourse Y. Two corresponding linguistic variables x and y are also defined, and the typical propositions is presented as:

$$\text{Rule: IF } x \text{ is } \tilde{A} \text{ then } y \text{ is } \tilde{B}$$

$$\text{Fact: } x \text{ is } \tilde{A}'$$

$$\Rightarrow \text{Conclusion: } y \text{ is } \tilde{B}'$$

Let $\tilde{S}(\tilde{A}, \tilde{A}')$ denote the measure of similarity between two type-2 fuzzy sets $\tilde{A}$ and $\tilde{A}'$. Existing methods use the measure of similarity to directly compute the inference without considering the induced relationship. In the proposed method, the authors translate the conditional statement into a fuzzy relationship. The similarity between the fact and the antecedent of the rule is used to modify the derived relationship. That is, every change in the conditional premise and in the fact is incorporated into the induced fuzzy relationship. Accordingly, a conclusion can be derived using the sub-projection operation. Thus, the conclusion is influenced by the modification of the fact and the antecedent of the rule fired.

Two modification procedures are proposed to modify the derived relationship. They are listed as follows:

$$\text{expansion form: } \mu_{\tilde{Q}'}(x,y) = m_1(\mu_{\tilde{Q}}(x,y), \tilde{S}(\tilde{A},\tilde{A}'))$$
$$= \sum_{u \in J_{x,y}} \left(\frac{f_{x,y}(u)}{\tilde{S}(\tilde{A},\tilde{A}')}\right) / \left(\frac{u}{\tilde{S}(\tilde{A},\tilde{A}')}\right),$$

and

$$\text{reduction form: } \mu_{\tilde{Q}'}(x,y) = m_2(\mu_{\tilde{Q}}(x,y), \tilde{S}(\tilde{A},\tilde{A}'))$$
$$= \sum_{u \in J_{x,y}} (f_{x,y}(u) \cdot \tilde{S}(\tilde{A},\tilde{A}')) / (u \cdot \tilde{S}(\tilde{A},\tilde{A}')),$$

where $\tilde{Q}$ is a fuzzy relation in the Cartesian product space $X \times Y$, $m_1(\cdot)$ and $m_2(\cdot)$are two modification functions for the expansion and reduction forms; and $f_{x,y}(u)$ is the secondary grade of the fuzzy relationship. In this study, the proposed method focused on the significant difference between $\tilde{A}$ and $\tilde{A}'$ to make the conclusion $\tilde{B}'$ less specific, and then choosing the expansion form. Hence, with a decrease in similarity, occurring at a significant difference between $\tilde{A}$ and $\tilde{A}'$, the inferred conclusion would be close to Y. Conversely, when $\tilde{A} = \tilde{A}'$ the inferred conclusion is obtained as $\tilde{B} = \tilde{B}'$. Notably, when $\tilde{S}(\tilde{A},\tilde{A}') = 0$, nothing can be concluded when $\tilde{A}$ and $\tilde{A}$ are dissimilar, and $\tilde{B} = \tilde{B}^c$ is obtained.

Subsequently, assume that k linguistic variables $x_1, \ldots, x_k$ defined on the universes of discourses $X_1, \ldots, X_k$. These typical propositions are listed:

Rule: IF x is $\tilde{A}_1$ and ... and x is $\tilde{A}_k$ then y is $\tilde{B}$

Fact: IF x is $\tilde{A}_1'$ and ... and x is $\tilde{A}_k'$

$\Rightarrow$ Conclusion: y is $\tilde{B}'$

3 Truth-Qualified Proposition

Accordingly, the partially truth-qualified statements of the form illustrated as follows:

" ' David is healthy' is quite true," or
" 'The temperature is moderate' is mostly true."

Simple statements are of the general propositional form,

"x is F; t is Q"

where x and t are two linguistic variables, and t denotes the truth value. F represents the vague descriptions of the object x, and Q denotes the truth of proposition "x is F". Restated, F and Q are type-1 fuzzy sets. Consequently, the previous general propositional form can be translated into a type-2 fuzzy statement,

"x is F"

where

$$\tilde{F} = \sum \mu_{\tilde{F}}(x)/x = \sum [Q_x(u)/u]/x, \quad u \epsilon J_x \subseteq U = [0,1].$$

Notably, a secondary grade, $Q_x(u)$, is applied to state the truth value.

The partially truth-qualified statement is represented as the type-2 fuzzy set. Consequently, the truth value of a composite proposition is computed as follows:

$$(x \text{ is } F; t \text{ is } Q) \wedge (x \text{ is } G; t \text{ is } R)$$
$$= (x \text{ is } F) \wedge (x \text{ is } G) \Rightarrow \mu_{\tilde{F}}(x) \cap \mu_{\tilde{G}}(x),$$

$$(x \text{ is } F; t \text{ is } Q) \vee (x \text{ is } G; t \text{ is } R)$$
$$= (x \text{ is } F) \vee (x \text{ is } G) \Rightarrow \mu_{\tilde{F}}(x) \cup \mu_{\tilde{G}}(x),$$

and

$$\neg(x \text{ is } F; t \text{ is } Q) = (x \text{ is } F^c) \Rightarrow \mu_{\tilde{F}^c}(x).$$

Accordingly, the deductive processes are introduced on the similarity measure among type-2 fuzzy sets. In the following, k linguistic variables $x_1, \ldots, x_k$ are defined on the universe of discourses $X_1, \ldots, X_k$. t denotes as the truth of the proposition. These typical propositions listed as:

$$rule\text{: if } x_1 \text{ is } A_1 \text{ and } \ldots \text{ and } x_k \text{ is } A_k \text{ then } y \text{ is } B; t \text{ is } C_{true}$$
$$fact\text{: if } x_1 \text{ is } A_1' \text{ and } \ldots \text{ and } x_k \text{ is } A_k' \ ; t \text{ is } C_{true}'$$
$$\Rightarrow conclusion\text{: } y \text{ is } B'; \ t \text{ is } C_{true}''.$$

The partially true proposition is represented by the statement of type-2 fuzzy sets,

$$rule\text{: if } x_1 \text{ is } \tilde{A}_1 \text{ and } \ldots \text{ and } x_k \text{ is } \tilde{A}_k \text{ then } y \text{ is } \tilde{B}$$
$$fact\text{: if } x_1 \text{ is } \tilde{A}_1' \text{ and } \ldots \text{ and } x_k \text{ is } \tilde{A}_k'$$
$$\Rightarrow conclusion\text{: } y \text{ is } \tilde{B}',$$

where $i = 1, \ldots, k$. Secondary grades represent as $f_{C_{true}}(u)$, $f_{C_{true}'}(u)$ and $f_{C_{true}''}(u)$, respectively. That is, these type-2 fuzzy sets are

$$\tilde{A}_i = \sum_{x_i \epsilon X_i} [\sum_{u \epsilon J_{x_i}} f_{C_{true}}(u)/u]/x_i,$$

$$\tilde{A}_i' = \sum_{x_i \epsilon X_i} [\sum_{u \epsilon J_{x_i}} f_{C_{true}'}(u)/u]/x_i,$$

$$\tilde{B} = \sum_{y \epsilon Y} [\sum_{w \epsilon J_y} f_{C_{true}}(w)/w]/y,$$

$$\tilde{B}' = \sum_{y \epsilon Y} [\sum_{w \epsilon J_y} f_{C_{true}''}(w)/w]/y.$$

The case of truth qualification proposition is shown in the following example.

Example 1: Herein, an example of the proposition from (Raha and Ray 1999), "It is almost fairly_true that people will feel not so uncomfortable, when it is true that humidity is moderate," is referenced from the general knowledge that "It is fairly_true when humidity is high and people feel uncomfortable". Accordingly, the conclusion of the form is derived with the proposed method. Here, the following is assumed.

$\tilde{A}$=(humidity is) high,

$\tilde{B}$=(human's tolerance is) uncomfortable,

C_{true}=*farily_true*,

$\widetilde{A}'$=(humidity is) moderate,

C'_{true}=*true*.

Consequently, the propositions are listed as follows.

rule: if humidity is *high* then tolerance is *uncomfortable*, the truth is *fairly_true*.

fact: if humidity is *moderate*, the truth is *true*.

Then, the purpose is to represent the inexact concepts in the propositions as type-2 fuzzy sets based on an appropriate universe of discourses. Let the universes of discourse be denoted as follows,

$$\text{Percentile humidity} \in [0,1],$$

$$\text{Percentile tolerance index} \in [0,1],$$

$$\text{Truth value} \in [0,1].$$

Thus, at any particular time, the humidity of air is normalized in the choice of universe. Similarity, tolerance index '1.0' means "feeling comfortable"; anything less than '1.0' means "partially comfortable", and '0.0' means "absolutely uncomfortable". Hence, the definitions of type-2 fuzzy sets are listed as follows,

$$high = \frac{0.3/0.25}{0.25} + \frac{0.6/0.5}{0.25} + \frac{0.8/0.75}{0.75} + \frac{1/1}{1},$$

$$uncomfortable = \frac{0.9/0.8}{0.0} + \frac{0.7/0.65}{0.125} + \frac{0.6/0.55}{0.25} + \frac{0.4/0.35}{0.5} + \frac{0.3/0.2}{0.75} + \frac{0.2/0.1}{1.0},$$

$$moderate = \frac{0.65/0.5}{0.25} + \frac{0.9/0.75}{0.5} + \frac{1/1}{0.75} + \frac{0.9/0.75}{1}.$$

Accordingly, the similarity is given by

$$\tilde{S}(high, moderate) = \frac{1}{4}\left(\frac{0.3\cdot 0.5 + 0.65\cdot 0.5}{0.5^2 + 0.65^2} + \frac{0.6\cdot 0.5 + 0.5\cdot 0.9}{0.6^2 + 0.9^2} + \frac{0.8\cdot 0.5 + 0.5\cdot 0.1}{0.8^2 + 1^2} + \frac{1\cdot 0.5 + 0.9\cdot 0.5}{1^2 + 0.9^2}\right) = 0.61,$$

and, the fuzzy relation $\tilde{Q}$ computed as follows:

$$\tilde{Q} = \frac{\frac{0.3}{0.5}}{(0.25,0.0)} + \frac{\frac{0.3}{0.5}}{(0.25,0.125)} + \frac{\frac{0.3}{0.5}}{(0.25,0.25)} + \frac{\frac{0.3}{0.5}}{(0.25,0.5)} + \frac{\frac{0.3}{0.5}}{(0.25,0.75)} + \frac{\frac{0.2}{0.1}}{(0.25,1.0)} + \frac{\frac{0.6}{0.5}}{(0.5,0.0)} + \frac{\frac{0.6}{0.5}}{(0.5,0.125)} + \frac{\frac{0.6}{0.5}}{(0.5,0.25)} + \frac{\frac{0.4}{0.35}}{(0.5,0.5)} + \frac{\frac{0.3}{0.2}}{(0.5,0.75)}$$

$$+ \frac{\frac{0.2}{0.1}}{(0.5,1.0)} + \frac{\frac{0.8}{0.75}}{(0.75,0.0)} + \frac{\frac{0.7}{0.65}}{(0.75,0.125)} + \frac{\frac{0.6}{0.55}}{(0.75,0.25)} + \frac{\frac{0.4}{0.35}}{(0.75,0.5)}$$

$$+ \frac{\frac{0.3}{0.2}}{(0.75,0.75)} + \frac{\frac{0.2}{0.1}}{(0.75,1.0)} + \frac{\frac{0.9}{0.8}}{(1.0,0.0)} + \frac{\frac{0.7}{0.65}}{(1.0,0.125)} + \frac{\frac{0.6}{0.55}}{(1.0,0.25)}$$

$$+ \frac{\frac{0.4}{0.35}}{(1.0,0.5)} + \frac{\frac{0.3}{0.2}}{(1.0,0.75)} + \frac{\frac{0.2}{0.1}}{(1.0,1.0)}.$$

Furthermore, the relation $\tilde{Q}'$ is adjusted by

$$\tilde{Q}' = \frac{\frac{0.49}{0.82}}{(0.25,0.0)} + \frac{\frac{0.49}{0.82}}{(0.25,0.125)} + \frac{\frac{0.49}{0.82}}{(0.25,0.25)} + \frac{\frac{0.49}{0.57}}{(0.25,0.5)} + \frac{\frac{0.49}{0.33}}{(0.25,0.75)} + \frac{\frac{0.33}{0.16}}{(0.25,1.0)} + \frac{\frac{0.98}{0.82}}{(0.5,0.0)} + \frac{\frac{0.98}{0.82}}{(0.5,0.125)} + \frac{\frac{0.98}{0.82}}{(0.5,0.25)} + \frac{\frac{0.66}{0.57}}{(0.5,0.5)} + \frac{\frac{0.49}{0.33}}{(0.5,0.75)}$$

$$+ \frac{\frac{0.33}{0.16}}{(0.5,1.0)} + \frac{\frac{1}{1}}{(0.75,0.0)} + \frac{\frac{1}{1}}{(0.75,0.125)} + \frac{\frac{0.98}{0.9}}{(0.75,0.25)} + \frac{\frac{0.66}{0.57}}{(0.75,0.5)}$$

$$+ \frac{\frac{0.49}{0.33}}{(0.75,0.75)} + \frac{\frac{0.33}{0.16}}{(0.75,1.0)} + \frac{\frac{1}{1}}{(1.0,0.0)} + \frac{\frac{1}{1}}{(1.0,0.125)} + \frac{\frac{0.98}{0.9}}{(1.0,0.25)}$$

$$+ \frac{\frac{0.66}{0.57}}{(1.0,0.5)} + \frac{\frac{0.49}{0.33}}{(1.0,0.75)} + \frac{\frac{0.33}{0.16}}{(1.0,1.0)}.$$

Consequently, $\tilde{Q}'$ is projected to obtain the conclusion according to

$$\tilde{B}' = \frac{\left(\frac{0.49}{0.82}\right) \bar{\cup} \left(\frac{0.98}{0.82}\right) \bar{\cup} \left(\frac{1}{1}\right) \bar{\cup} \left(\frac{1}{1}\right)}{0.0} + \frac{\left(\frac{0.49}{0.82}\right) \bar{\cup} \left(\frac{0.98}{0.82}\right) \bar{\cup} \left(\frac{1}{1}\right) \bar{\cup} \left(\frac{1}{1}\right)}{0.125}$$

$$+ \frac{\left(\frac{0.49}{0.82}\right) \bar{\cup} \left(\frac{0.98}{0.82}\right) \bar{\cup} \left(\frac{0.98}{0.9}\right) \bar{\cup} \left(\frac{0.98}{0.9}\right)}{0.25} + \frac{\left(\frac{0.49}{0.57}\right) \bar{\cup} \left(\frac{0.66}{0.57}\right) \bar{\cup} \left(\frac{0.66}{0.57}\right) \bar{\cup} \left(\frac{0.66}{0.57}\right)}{0.5}$$

$$+ \frac{\left(\frac{0.49}{0.33}\right) \bar{\cup} \left(\frac{0.49}{0.33}\right) \bar{\cup} \left(\frac{0.49}{0.33}\right) \bar{\cup} \left(\frac{0.49}{0.33}\right)}{0.75} + \frac{\left(\frac{0.33}{0.16}\right) \bar{\cup} \left(\frac{0.33}{0.16}\right) \bar{\cup} \left(\frac{0.33}{0.16}\right) \bar{\cup} \left(\frac{0.33}{0.16}\right)}{1.0}$$

$$= \frac{\left(\frac{1}{1}\right)}{0.0} + \frac{\left(\frac{1}{1}\right)}{0.125} + \frac{\left(\frac{0.98}{0.9}\right)}{0.25} + \frac{\left(\frac{0.66}{0.57}\right)}{0.5} + \frac{\left(\frac{0.49}{0.33}\right)}{0.75} + \frac{\left(\frac{0.33}{0.16}\right)}{0.125}.$$

Hence, the conclusion describes the tolerance when the humidity is moderate and the truth condition is true. According to the derivation, when the humidity is moderate, the people will feel less uncomfortable due to the secondary grades rising after the tolerance index "0.25". Moreover, this case was also applied by the ordinary fuzzy implication in Pappis and Karacapilidis (1993), wherein the results were obtained as

$$B' = \frac{1}{0.0} + \frac{1}{0.125} + \frac{0.775}{0.25} + \frac{0.543}{0.5} + \frac{0.31}{0.75} + \frac{0.07}{1.0},$$

and the truth function was given as

$$truth = \frac{0.25}{0.75} + \frac{0.5}{0.8} + \frac{0.75}{0.85} + \frac{1}{0.9} + \frac{0.95}{0.95} + \frac{0.9}{1.0}.$$

The corresponding results show that the general fuzzy implication is not sufficient to handle the fuzzy proposition with the truth function, because the truth function is independent of the fuzzy processing, and the results of the truth function are hard to associate to the original proposition. Conversely, the partial truth statements are associated with the proposition in our proposed method; the set-theoretic considerations can be used to derive partial true knowledge.

4 The Relationship between Intuitionistic Fuzzy Sets and Type-2 Fuzzy Sets

Atanassov (1996) associated a mapping from $IFSs(X)$ to $FS_2(X)$, where the set of all intuitionistic fuzzy sets and type-2 fuzzy sets in X are representing as $IFS(X)$ and $FS_2(X)$. That study also defined the following operator:

$$\breve{A} = \{< x, \mu_A(x_i), v_A(x_i) > | x \in X\} \rightarrow$$
$$f_\alpha(\breve{A}) = \{< x, \mu_A(x_i) + \alpha\pi_A(x_i), 1 - \mu_A(x_i) - \alpha\pi_A(x_i) > | x \in X\},$$

where $f_\alpha: IFSs(X) \rightarrow FSs(X)$ (Atanassov 1986). The operator f_α coincides with the operator D_α given in Atanassov. However, the limitations of above the equation are listed as follows,

(a). The equation considers only the membership degree and omits the imperfect information (non-membership degree).

(b). The operator cannot handle the reverse switching. i.e. from fuzzy sets to intuitionistic fuzzy sets.

(c). Considering the complementation of FSs such as that of Sugeno (1977) or Yager (1979), the sum of membership and non-membership from one, the result is a negative number. Therefore, the elementary intuitionism condition given by Atanassov is not satisfied.

In addition for each element $x \in X$, the type-2 fuzzy set can model hesitation and additional uncertainties by using additional degrees. For instance, the polarizing concepts, i.e., more/less, optimistic/pessimistic, membership/non-membership, can be inferred by the secondary grades. Restated, the secondary grades are defined to determine the magnitudes, allowing us to weight the degree of intuitionism of an intuitionistic fuzzy set. Thus, to eliminate the above limitations, the relationship is introduced as follows.

Definition 3. Let $\breve{A} \in IFSs(X)$ define as follows,

$$\breve{A} = \sum_{i=1}^{n} [x_i, \mu_A(x_i), v_A(x_i)],$$

where $x_i \in X$. Then, the association from $IFSs(X)$ to $FS_2(X)$ is defined as

$$\tilde{A} = \sum_{i=1}^{n} [1/(\mu_A(x_i) + p\pi_A(x_i)) + 0/(1 - v_A(x_i) + p\pi_A(x_i))] /x_i, \quad (3)$$

The secondary grade in (3) represents the indeterminacy index of membership/non-membership degree, which models the un-hesitancy of determining the extent to which an object satisfies a specific property. Restated, 1/0 in secondary grades implies the total certain/uncertain with respect to membership/non-membership herein. Additionally, according to P1 (properties of IFSs), the definition of non-membership degree of an element x is $v_{\breve{A}}(x_i) \geq v_{\breve{B}}(x_i) \Leftrightarrow \breve{A} \leq \breve{B}$. This notion contradicts a natural generalization of a standard fuzzy set of the containment statement of Zadeh, $v_{\breve{A}}(x_i) \leq v_{\breve{B}}(x_i) \Leftrightarrow \breve{A} \leq \breve{B}$. Thus, the non-membership value in (3) is obtained by subtracting the $v_{\breve{B}}(x_i)$ from one.

Accordingly, the following proposition is proven to validate the relationship between T2FS and IFS. To simply the proof, intuitionistic fuzzy set refers to the extension of the fuzzy sets. Namely, the membership and non-membership degrees are added an equal to one.

Proposition 1. Let $\breve{A}, \breve{B} \in IFSs(X)$, denote that the switch from $IFSs(X)$ to $FS_2(X)$ are validated, then $\breve{A} \leq \breve{B}$ if and only if $\tilde{A} \leq \tilde{B}$.
Proof. Assume that two IFSs $\breve{A}$ and $\breve{B}$ are defined as follows:

$$\breve{A} = \sum_{i=1}^{n} [x_i, \mu_{\breve{A}}(x_i), v_{\breve{A}}(x_i)],$$

and

$$\breve{B} = \sum_{i=1}^{n} [x_i, \mu_{\breve{B}}(x_i), v_{\breve{B}}(x_i)],$$

where $\forall x_i \in X$.
Thus, two corresponding type-2 fuzzy sets are defined as:

$$\tilde{A} = \sum_{i=1}^{n} [1/(\mu_{\breve{A}}(x_i) + p\pi_{\breve{A}}(x_i)) + 0/(1 - v_{\breve{A}}(x_i) + p\pi_{\breve{A}}(x_i))] /x_i ,$$

and

$$\tilde{B} = \sum_{i=1}^{n} [1/(\mu_{\breve{B}}(x_i) + p\pi_{\breve{B}}(x_i)) + 0/(1 - v_{\breve{B}}(x_i) + p\pi_{\breve{B}}(x_i))] /x_i.$$

Hence, assume that $\breve{A} \leq \breve{B}$, then

$$\mu_A(x_i) \leq \mu_B(x_i) \text{ and } v_A(x_i) \geq v_B(x_i)$$
$$\Rightarrow \mu_A(x_i) - v_A(x_i) \leq \mu_B(x_i) - v_B(x_i).$$

On the contrary, assume that $\tilde{A} \leq \tilde{B}$, that is

$$\mu_{\breve{A}}(x_i) + p\pi_{\breve{A}}(x_i) \leq \mu_{\breve{B}}(x_i) + p\pi_{\breve{B}}(x_i),$$

and

$$1 - v_{\breve{A}}(x_i) + p\pi_{\breve{A}}(x_i) \leq 1 - v_{\breve{B}}(x_i) + p\pi_{\breve{B}}(x_i),$$
$$\Rightarrow \mu_{\breve{A}}(x_i) - v_{\breve{A}}(x_i) \leq \mu_{\breve{B}}(x_i) - v_{\breve{B}}(x_i).$$

Hence, the proposition is proven. □

Furthermore, it is known that if A is a fuzzy set on a referential X and $c: [0,1] \rightarrow [0,1]$ is a fuzzy complement, the set

$$A = \sum_{i=1}^{n} [x_i, \mu_A(x_i), c(\mu_A(x_i))] \quad , \tag{4}$$

defined as an intuitionistic fuzzy set (De et al. 2001). However, according to our previous statement, non-membership value is not a natural generalization of a standard FS. Besides, if we take Sugeno's negation (Sugeno and Terano 1977)

$$c_\lambda(x) = \frac{1 - x}{1 + \lambda x}, \text{with } -1 < \lambda < 0,$$

or Yager's negation (1979)

$$c_\lambda(x) = (1 - x^\omega)^{1/\omega}, \text{with } 1 < \omega,$$

as a fuzzy complement, (3) is not an intuitionistic fuzzy set, because $\mu_A(x_i) + c(\mu_A(x_i)) > 1$ and therefore, $\pi_A(x_i) < 0$. Hence, for the purpose to clear state the relation of type-2 fuzzy sets and intuitionistic fuzzy sets, the reverse relationship is defined as follows:

Definition 4. Let $\tilde{A} \in FS_2(X)$ define as follows,

$$\tilde{A} = \sum_{i=1}^{n} [1/\mu_1(x_i) + 0/\mu_2(x_i)] / x_i,$$

where $x_i \in X$. Then, the one way transforms from type-2 fuzzy set to intuitionistic fuzzy set is defined as:

$$\breve{A} = \sum_{i=1}^{n} [x_i, \mu_1(x_i) - p\pi_{\breve{A}}(x_i), 1 - \mu_2(x_i) - p\pi_{\breve{A}}(x_i)], \tag{5}$$

where

$$\pi_{\breve{A}}(x_i) = \begin{cases} \dfrac{\mu_1(x_i) - \mu_2(x_i)}{2p - 1}, & if\ \mu_1(x_i) > \mu_2(x_i), \\ \dfrac{\mu_2(x_i) - \mu_1(x_i)}{1 - 2p}, & if\ \mu_1(x_i) \leq \mu_2(x_i), \end{cases}$$

and $p \in [0,1]$.

Proposition 2. According to (5), then membership degree, non-membership degree and hesitation part are summed to one.
Proof. We intend to obtain

$$\mu_1(x_i) - p\pi_{\tilde{A}}(x_i) + 1 - \mu_2(x_i) - p\pi_{\tilde{A}}(x_i) + \pi_{\tilde{A}}(x_i) = 1.$$

$$\text{Thus,} \because \mu_1(x_i) - \mu_2(x_i) + 1 + (1-2p)\pi_{\tilde{A}}(x_i)$$
$$= \mu_1(x_i) - \mu_2(x_i) + 1 + (1-2p)\frac{(\mu_1(x_i) - \mu_2(x_i))}{2p-1}.$$

Accordingly, if $\mu_1(x_i) > \mu_2(x_i)$, and then the above equation is summarized as

$$\mu_1(x_i) - \mu_2(x_i) + 1 + (1-2p)\frac{\mu_1(x_i) - \mu_2(x_i)}{2p-1}$$
$$\Rightarrow \mu_1(x_i) - \mu_2(x_i) + 1 - (\mu_1(x_i) - \mu_2(x_i)) = 1.$$

$\mu_1(x_i) \leq \mu_2(x_i)$ denotes the same. Thus, the proposition has been proved. □

Proposition 3. Let $\tilde{A}$ is defined as follows:

$$\tilde{A} = \sum_{i=1}^{n} [1/\mu_{\tilde{A}}(x_i) + 0/c(\mu_{\tilde{A}}(x_i))]/x_i,$$

and the fuzzy complement is Sugeno's negation. Assume that the conversion from $FS_2(X)$ to $IFSs(X)$ are validated, then

$$\mu_A(x_i) - p\pi_{\tilde{A}}(x_i) + (1 - c(\mu_A(x_i))) - p\pi_{\tilde{A}}(x_i) \leq 1, \quad (6)$$

when $-1 < \lambda < 0$.

Proof. Accordingly, to the above equation, it means that we need to approve.

$$\mu_A(x_i) - p\pi_{\tilde{A}}(x_i) - c(\mu_A(x_i)) - p\pi_{\tilde{A}}(x_i) \leq 0.$$

$$\because p \in [0,1], \text{and } -p\pi_{\tilde{A}}(x_i) \text{ is always negative}$$
$$\therefore \text{we only need to approve } \mu_A(x_i) - c(\mu_A(x_i)) \leq 0,$$

According to Sugeno and Terano (1977), $-1 < \lambda < 0$ is derived to obtain the bigger output then input values in Sugeno's class. Thus, the above equation is always true when $-1 < \lambda < 0$.

Conversely, assume that $\mu_A(x_i) + (1 - c(\mu_A(x_i))) \leq 1$, then we intend to apply the Sugeno's negation, $\mu_A(x_i) + (1 - \frac{1-\mu_A(x_i)}{1+\lambda\mu_A(x_i)}) \leq 1$.

$$\Rightarrow \frac{\mu_A(x_i) + \lambda(\mu_A(x_i))^2 - 1 + \mu_A(x_i)}{1 + \lambda\mu_A(x_i)} \leq 0.$$

$$\because 1 + \lambda\mu_A(x_i) \geq 0, \text{then}$$

$$\therefore 2\mu_A(x_i) + \lambda(\mu_A(x_i))^2 - 1 \leq 0. \quad (7)$$

If we can limit λ in $-1 < \lambda < 0$, then we can approve the Expression(7).

Thus, the proposition is proven. □

5 Application to Medical Diagnosis

A medical knowledge base focuses on how to properly diagnose D for a patient T with given values of symptoms S. Therefore, this section introduces a method for handling medical diagnostic problems based on the type-2 similarity. In a given pathology, assume that S denotes a set of symptoms, D denotes a set of diagnoses, and T denotes a set of patients. Analogous to De *et al.* notation of "Intuitionistic Medical Knowledge" (De et al. 2001), the authors defined "Type-2 Similarity Medical Knowledge" as a reasoning process from the set of symptoms S to the set of diagnoses D.

Consider four patients Al, Bob, Joe and Ted, i.e. T={Al, Bob, Joe, Ted}. Their symptoms are high temperature, headache, stomach pain, cough and chest pain, i.e. S={Temperature, Headache, Stomach pain, Cough, Chest pain}. The set of diagnosis is defined, i.e. D={Viral Fever, Malaria, Typhoid, Stomach problem, Heart problem}. Tables 2 and 3 summarize the intuitionistic fuzzy relations $T \rightarrow S$ and $S \rightarrow D$.

Hence, attempts to calculate for each patient t_j of his symptoms from a set of symptoms s_i characteristic of each diagnosis d_k. The reasoning process is as follows. (I) switch the acquired medical knowledge base from $IFS(X)$ to $FS_2(X)$, (II) to calculate the similarity of symptoms s_i between each patient t_j and each diagnosis d_k, where $i = 1, \cdots, 5, j = 1, \cdots, 4$ and $k = 1, \cdots, 5$, (III) to determine higher similarities, implying a proper diagnosis. The relationships of $T \rightarrow S$, $D \rightarrow S$, that is, the mapping from $IFS(\mathrm{X})$ to $FS_2(\mathrm{X})$ is switched and listed as follows (for the sake of simplicity, take *Al* and *Temperature* for example):

$$Al = \frac{\frac{1}{0.8+p\cdot 0.1}+\frac{0}{0.9+p\cdot 0.1}}{Temperature} + \frac{\frac{1}{0.6+p\cdot 0.3}+\frac{0}{0.9+p\cdot 0.3}}{Headache} + \frac{\frac{1}{0.2}+\frac{0}{0.2}}{Stomach\,pain} + \frac{\frac{1}{0.6+p\cdot 0.3}+\frac{0}{0.9+p\cdot 0.3}}{Cough} + \frac{\frac{1}{0.1+p\cdot 0.3}+\frac{0}{0.4+p\cdot 0.3}}{Chest\,pain}$$

and

$$Viral\,fever = \frac{\frac{1}{0.4+p\cdot 0.6}+\frac{0}{1+p\cdot 0.6}}{Temperature} + \frac{\frac{1}{0.3+p\cdot 0.2}+\frac{0}{0.5+p\cdot 0.2}}{Headache} + \frac{\frac{1}{0.1+p\cdot 0.2}+\frac{0}{0.3+p\cdot 0.2}}{Stomach\,pain} + \frac{\frac{1}{0.4+p\cdot 0.3}+\frac{0}{0.7+p\cdot 0.3}}{Cough} + \frac{\frac{1}{0.1+p\cdot 0.2}+\frac{0}{0.3+p\cdot 0.2}}{Chest\,problem}.$$

Notably, $p \in [0,1]$. Tables 4 and 5 list the similarity measure (2) for each patient from the considered set of possible diagnoses for $p = 0, 1$, respectively. In Table 6, Szmidt and Kacprzyk diagnosed by estimating the distances of three parameters: membership function, non-membership and the hesitation margin for all patient symptoms. That study also developed a geometrical interpretation to evaluate the similarity between IFSs, as well as properly diagnose in Table 7. De *et al.* defined the "Intuitionistic Medical Knowledge" to discuss how symptoms and diagnosis are related. Table 8 summarizes those results. The formal fuzzy similarity

$$S(A,B) = \frac{\sum_n \{1 - |\mu_A(x) - \mu_B(x)|\}}{n}$$

is applied in the medical diagnosis shown in Table 9 (Raha et al. 2002).

Table 10 displays all of the results from the above-mentioned methods. Medical software or computer systems assist physicians in patient care and facilitate the diagnosis of complex medical conditions. Determining which method can facilitate an exact diagnosis is extremely difficult. According to Table 10, Bob obviously suffers from stomach problems (all methods agree) and Joe is inflicted with typhoid (five out of the six methods agree). Above results demonstrate that the proposed theoretical method can facilitate the diagnosis. As for diagnoses of viral fever and malaria, our results indicate that these two diagnoses have difficulty in accuracy (nearly half of the methods approve one or the other diagnosis); in addition, these two symptoms are involved with each other. The proposed method differs from other methods in this respect. Moreover, the proposed method can be used as a facilitator. Distinct diagnosis results can be obtained by type-2 fuzzy sets, including a high effectiveness in handling imprecise and imperfect information than intuitionistic fuzzy sets can.

Table 1 Symmetric discrimination measures ('*' marks as the recognizing result)

	P_1	P_2	P_3
Dengfeng's Method [26]	0.78	0.8	0.85*
Mitchell's Method [27]	0.54	0.54	0.61*
Vlachos's Method [28]	0.4492	0.3487	0.2480*
Proposed Method	2.0169	2.0585	2.1253*

Table 2 Symptoms characteristic for the patients considered

	Temperature	*Headache*	*Stomach Pain*	*Cough*	*Chest pain*
Al	(0.8,0.1)	(0.6,0.1)	(0.2,0.8)	(0.6,0.1)	(0.1,0.6)
Bob	(0.0,0.8)	(0.4,0.4)	(0.6,0.1)	(0.1,0.7)	(0.1,0.8)
Joe	(0.8,0.1)	(0.8,0.1)	(0.0,0.6)	(0.2,0.7)	(0.0,0.5)
Ted	(0.6,0.1)	(0.5,0.4)	(0.3,0.4)	(0.7,0.2)	(0.3,0.4)

Table 3 Symptoms characteristic for the diagnoses considered

	Viral fever	*Malaria*	*Typhoid*	*Stomach problem*	*Chest problem*
Temperature	(0.4,0.0)	(0.7,0.0)	(0.3,0.3)	(0.1,0.7)	(0.1,0.8)
Headache	(0.3,0.5)	(0.2,0.6)	(0.6,0.1)	(0.2,0.4)	(0.0,0.8)
Stomach pain	(0.1,0.7)	(0.0,0.9)	(0.2,0.7)	(0.8,0.0)	(0.2,0.8)
Cough	(0.4,0.3)	(0.7,0.0)	(0.2,0.6)	(0.2,0.7)	(0.2,0.8)
Chest pain	(0.1,0.7)	(0.1,0.8)	(0.1,0.9)	(0.2,0.7)	(0.8,0.1)

Table 4 Result is measured by type-2 similarity of $p = 0$ ('*' marks as the diagnosis result)

	Viral fever	*Malaria*	*Typhoid*	*Stomach problem*	*Chest problem*
Al	0.68*	0.59	0.47	0.41	0.38
Bob	0.47	0.41	0.51	0.75*	0.42
Joe	0.59	0.44	0.62*	0.50	0.36
Ted	0.65*	0.56	0.51	0.48	0.32

Table 5 Result is measured by type-2 similarity of $p = 1$ ('*' marks as the diagnosis result)

	Viral fever	*Malaria*	*Typhoid*	*Stomach problem*	*Chest problem*
Al	0.71*	0.69	0.46	0.47	0.46
Bob	0.54	0.51	0.53	0.83*	0.44
Joe	0.63	0.48	0.65	0.66*	0.45
Ted	0.77*	0.62	0.55	0.56	0.4

Table 6 Result is measured by Szmidt et al. in (Szmidt et al., 2001) ('*' marks as the diagnosis result)

	Viral fever	*Malaria*	*Typhoid*	*Stomach problem*	*Chest problem*
Al	0.29	0.25*	0.32	0.53	0.58
Bob	0.43	0.56	0.33	0.14*	0.46
Joe	0.36	0.41	0.32*	0.52	0.57
Ted	0.25*	0.29	0.35	0.43	0.5

Table 7 Result is measured by Szmidt et al. in (Szmidt et al., 2004). ('*' marks as the diagnosis result)

	Viral fever	*Malaria*	*Typhoid*	*Stomach problem*	*Chest problem*
Al	0.75*	1.19	1.31	3.27	∞
Bob	2.1	3.73	1.1	0.35*	∞
Joe	0.87	1.52	0.46*	2.61	∞
Ted	0.95	0.77*	1.67	∞	2.56

Table 8 Results measured by De et al in (De et al., 2001). ('*' marks as the diagnosis result)

	Viral fever	*Malaria*	*Typhoid*	*Stomach problem*	*Chest problem*
Al	0.35	0.68*	0.57	0.04	0.08
Bob	0.2	0.08	0.32	0.57*	0.04
Joe	0.35	0.68*	0.57	0.04	0.05
Ted	0.32	0.68*	0.44	0.18	0.18

Table 9 Results measured by the formal fuzzy similarity. ('*' marks as the diagnosis result)

	Viral fever	*Malaria*	*Typhoid*	*Stomach problem*	*Chest problem*
Al	0.8	0.84*	0.82	0.56	0.52
Bob	0.74	0.58	0.8	0.86*	0.66
Joe	0.74	0.74	0.8*	0.54	0.5
Ted	0.78	0.82*	0.76	0.62	0.58

Table 10 All the considered results

	$p = 0$	$p = 1$	Szmidt in (2001)	Szmidt in (2004)	De in (2001)	Fuzzy similarity
Al	*Viral fever*	*Viral fever*	*Malaria*	*Viral fever*	*Malaria*	*Malaria*
Bob	*Stomach problem*	*Stomach problem*	*Stomach problem*	*Stomach problem*	*Stomach problem*	*Stomach problem*
Joe	*Typhoid*	*Typhoid*	*Malaria*	*Typhoid*	*Typhoid*	*Typhoid*
Ted	*Viral fever*	*Viral fever*	*Malaria*	*Malaria*	*Malaria*	*Malaria*

6 Conclusions

This study represents a new direction in approximate reasoning based on vague knowledge, which is associated with partial or incomplete truth values. The proposed theoretical method can handle vague quantities by converting this partial truth-value into a precisely quantified statement based on the type-2 fuzzy inference system. The membership functions of type-2 fuzzy sets have more parameters than those of type-1 fuzzy sets, providing a greater degree of design freedom. Therefore, type-2 fuzzy sets may outperform type-1 fuzzy sets, particularly in uncertain environments. Moreover, the proposed method can perform reasoning with incomplete knowledge; helping to yield meaningful resolutions using fuzzy sentential logic, and systematically compute uncertainty. Furthermore, a mutual switch between type-2 fuzzy sets and intuitionistic fuzzy sets was defined, and a medical diagnosis was generalized through switching and reasoning according to type-2 similarity. Consequently, easy comprehension and axiomatic definitions are provided during the switching process. Importantly, the proposed method makes it possible to extend the usage of type-2 fuzzy sets and renews the relationship between type-2 fuzzy sets and intuitionistic fuzzy sets.

Acknowledgements

The author would like to thank the National Science Council of the Republic of China, Taiwan for partially supporting this research under Contract No. NSC 98-2218-E-129 -001, NSC 99-2221-E-129 -014.

References

Atanassov, K.: Intuitionistic fuzzy sets. Fuzzy Sets and Systems 20, 87–96 (1986)

Baldwin, J.F.: A new approach to approximate reasoning using a fuzzy logic. Fuzzy Sets and Systems 2, 309–325 (1979)

Bellman, R.E., Zadeh, L.A.: Local and fuzzy logics. In: Epstein, G. (ed.) Modern Uses of Multiple-valued Logic, The Netherlands (1977)

Chaira, T.: Intuitionistic fuzzy segmentation of medical images. IEEE Transactions on Biomedical Engineering 57, 1430–1436 (2010)

Chen, S.M.: Measures of similarity between vague sets. Fuzzy Sets and Systems 74, 217–223 (1995)

Chen, S.M.: Similarity measures between vague sets and between elements. IEEE Trans. Syst. Man Cybernetic 27, 153–158 (1997)

De, S.K., Biswas, A., Roy, R.: Some operations on intuitionistic fuzzy sets. Fuzzy Sets and Systems 114, 477–484 (2001)

Dengfeg, L., Chuntian, C.: New similarity measure of intuitionistic fuzzy sets and application to pattern recognitions. Pattern Recognition Letters 23, 221–225 (2002)

Dubois, D., Prade, H.: Fuzzy sets and systems: theory and application. Academic Press INC, London (1980)

Fan, L., Zhangyan, X.: Similarity measures between vague sets. J. Software 12, 922–927 (2001)

Fu, X., Shen, Q.: Fuzzy compositional modeling. IEEE Transactions on Fuzzy Systems 18, 823–840 (2010)

Hong, D.H., Kim, C.: A note on similarity measures between vague sets and between elements. Information Sciences 115, 83–96 (1999)

Hung, W.L., Yang, M.S.: Similarity measures of intuitionistic fuzzy sets based on hausdorff distance. Pattern Recognition Letters 25, 1603–1611 (2004)

Li, H.X., Zhang, L., Cai, K.Y., Chen, G.: An improved robust fuzzy –PID controller with optimal fuzzy reasoning. IEEE Transactions on Systems, Man, and Cybernetics Part B: Cybernetics 33, 1283–1294 (2005)

Li, Y., Zhongxian, C., Degin, Y.: Similarity measures between vague sets and vague entropy. J. Computer Science 29, 129–132 (2002)

Mendel, J.M.: Uncertainty, fuzzy logic, and signal processing. Signal Processing 80, 913–933 (2000)

Mendel, J.M., John, R.I., Liu, F.: Interval type-2 logic systems made simple. IEEE Transactions on Fuzzy System 14, 808–821 (2006)

Mitchell, H.B.: On the dengfeng-chuntian similarity measure and its application to pattern recognition. Pattern Recognition Letters 24, 3101–3104 (2003)

Mushrif, M.M., Ray, A.K.: A-IFS histone based multithresholding algorithm for color image segmentation. IEEE Signal Processing Letters 16, 168–171 (2009)

Pappis, C.P., Karacapilidis, N.I.: A comparative assessment of measures of similarity of fuzzy values. Fuzzy Sets and Systems 56, 171–174 (1993)

Petry, F.E., Yager, R.R.: A framework for use of Imprecise Categorization in Developing Intelligent Systems. IEEE Transactions on Fuzzy Systems 18, 348–361 (2010)

Raha, S., Pal, N.R., Ray, K.S.: Similarity-based approximate reasoning: methodology and application. IEEE Transactions on Systems, Man, and Cybernetics PART A: Systems and Humans 32, 41–47 (2002)

Raha, S., Ray, K.S.: Reasoning with vague default. Fuzzy Sets and Systems 91, 327–338 (1997)

Raha, S., Ray, K.S.: Reasoning with vague truth. Fuzzy Sets and Systems 105, 385–399 (1999)

Raha, S., Ray, K.S.: Approximate reasoning with time. Fuzzy Sets and Systems 107, 59–79 (2000)

Sledge, I.J., Bezdek, J.C., Havens, T.C., Keller, J.M.: Relational generalizations of Cluster validity indices. IEEE Transactions on Fuzzy Systems 18, 771–786 (2010)

Song, J.: Real 2D presentation of spatial solids and real presentation rate with fuzzy pattern recognition I-concept and principles. In: 9th International Conference on CAID/CD, vol. 22, pp. 529–534 (2008)

Sugeno, M., Terano, T.: A model of learning based on fuzzy information. Cybernetes 6, 157–166 (1977)

Tang, Y.: A collective decision model involving vague concepts and linguistic expressions. IEEE Transactions on Systems, Man, and Cybernetics Part B: Cybernetics 38, 421–428 (2008)

Tianjiang, W., Zhengding, L., Fan, L.: Bidirectional approximate reasoning based on weighted similarity measures of vague sets. J. Computer Science 24, 96–100 (2002)

Xu, T.: Intuitionistic fuzzy aggregation operators. IEEE Transactions on Fuzzy Systems 15, 1179–1187 (2007)

Yager, R.R.: On the measure of fuzziness and negation. Part II: Lattices, Information, and Control 44, 236–260 (1979)

Zadeh, L.A.: A theory of approximate reasoning. Machine Intelligence 9, 149–194 (1979)

Zhizhen, L., Pengfei, S.: Similarity measures on intuitionistic fuzzy sets. Pattern Recognition Letters 24, 2687–2693 (2003)

Decision Making with Second Order Information Granules

R.A. Aliev, W. Pedrycz, O.H. Huseynov, and L.M. Zeinalova

Abstract. Decision-making under uncertainty has evolved into a mature field. However, in most parts of the existing decision theory, one assumes decision makers have complete decision-relevant information. The standard framework is not capable to deal with partial or fuzzy information, whereas, in reality, decision-relevant information about outcomes, probabilities, preferences etc is inherently imprecise and as such described in natural language (NL). Nowadays, there is no decision theory with second-order uncertainty in existence albeit real-world uncertainties fall into this category. This applies, in particular, to imprecise probabilities expressed by terms such as *likely, unlikely, probable, usually* etc. We call such imprecise evaluations second-order information granules.

In this study, we develop a decision theory with second-order information granules. The first direction we consider is decision making with fuzzy probabilities. The proposed theory differs from the existing ones as one that accumulates non-expected utility paradigm with NL-described decision-relevant information. Linguistic preference relations and fuzzy utility functions are used instead of their classical counterparts as forming a more adequate description of human preferences expressed under fuzzy probabilities. Fuzzy probability distribution is incorporated into the suggested fuzzy utility model by means of a fuzzy number-valued fuzzy measure instead of a real-valued non-additive probability. We provide representation theorems for a fuzzy utility function described by a fuzzy number-valued Choquet integral with a fuzzy number-valued integrand and a fuzzy number-valued fuzzy measure. The proposed theory is intended to solve decision problems when the environment of fuzzy states and fuzzy outcomes are

R.A. Aliev · O.H. Huseynov · L.M. Zeinalova
Department of Computer-Aided Control Systems, Azerbaijan State Oil Academy, Azerbaijan, Baku, Azadlyg ave. 20, AZ1010
e-mail: raliev@asoa.edu.az, oleg_huseynov@yahoo.com, Zeynalova-69@mail.ru

W. Pedrycz
Department of Electrical & Computer Engineering University of Alberta, Edmonton AB T6R 2G7, Canada, and
System Research Institute, Polish Academy of Sciences Warsaw, Poland
e-mail: pedrycz@ee.ualberta.ca

W. Pedrycz and S.-M. Chen (Eds.): Granular Computing and Intell. Sys., ISRL 13, pp. 117–153.
springerlink.com

characterized by fuzzy probabilities. As the second direction in this realm we consider hierarchical imprecise probability models. Such models allow us to take into account imprecision and imperfection of our knowledge, expressed by interval values of probabilities of states of nature and degrees of confidence associated with such values. A decision-making process analysis and a choice of the most preferable alternative subject to variation of intervals at the lower and upper levels of models and the types of distribution on the sets of random values of probabilities of states of nature is also of significant interest.

We apply the developed theories and methodologies to solving real-world economic decision-making problems. The obtained results show validity of the proposed approaches.

Keywords: Decision Making, Imprecise Probabilities, Fuzzy Utility, Fuzzy Function, Fuzzy Measure, Linguistic Preference, Second-Order Uncertainty, Choquet Integral.

1 Introduction

Decision theory is composed of a large variety of mathematical methods for modeling human behavior in presence of uncertainty (Gilboa 2009). Two main parts of the classical theory of choice are the theory of expected utility (EU) of von Neumann and Morgenstern (von Neumann and Morgenstern 1944) and the theory of subjective expected utility (SEU) of Savage (Savage 1954, Wakker and Zank 1999) in which it is assumed that individuals are motivated by material incentives (Akerlof and Shiller 2009) and can assign objective or consistent subjective probabilities to all outcomes. These theories are well-composed and have strong analytical power. However, they define human behavior as "ideal", i.e., inanimate, and do not correspond to the computational abilities of humans. Works of Allais (Allais and Hagen 1979, Allais 1953), Markowitz (Markowitz 1952), Ellsberg (Ellsberg 1961) and the ensuing works of psychologists and economists collected a substantial evidence that individuals systematically violate the basic assumptions of the EU such as independency, probabilistic beliefs, descriptive and procedural invariance etc.

Prospect theory (PT) and Cumulative Prospect theory (CPT) (Kahneman and Tversky 1979, Tversky and Kahneman 1992) are of the most famous theories on the new view at utility concept. These theories exhibit a significant success as they include psychological aspects that form human behavior. The Maximin expected utility model (MMEU) developed to resolve Allais and Ellsberg paradoxes assumes that individuals derive probabilities based on their experience and consider them varying within intervals. Fuzzy integrals (Grabish 1995, Yoneda et al. 1993, Grabish et al. 2000, Grabish 1996, Gajdos et al. 2004, Sims and Zhenyuan 1990, Zhang 1992, Zhong 1990) and Choquet expected utility (CEU) (Chateauneuf et al. 2000, Chateauneuf and Eichberger 2003, Modave et al. 1997, Modave and Grabish 1998, Gilbert et al. 2004, Jeleva 2000, Mangelsdorff and Weber 1994, Narukawa and Murofushi 2004) are other well known and effective non-expected utility models based on the use of non-additive measures that are more suitable to model human behavior than additive measures. CEU reflects ambiguity aversion and various risk preferences. The Choquet integral (Choquet 1953, Grabish et al. 2008, Pedrycz and Gomide 2007) is also a very useful tool for solving multicriteria decision-making (Grabish and Labreuche 2008) problems, because there exists a formal parallelism between the latter and decision making under uncertainty (Modave et al. 1997).

Non-expected utility models such as the PT, CPT, rank-dependent expected utility, MMEU, CEU etc. are more suitable for modeling choice principles mainly because of the use of a non-linear transformation of probabilities and non-additive measures (Billot 1992, Jamison and Lodwick 2002, Scmeidler 1989, Chateauneuf 2000, Machina 2004, Kobberling and Wakker 2003, Wang and Li 1999, Denneberg 1994). Each of them comes with advantages and shortcomings and there is no general approach to realistic modeling of the utility function.

Most of the existing utility models assume the utility function and probability distribution are accurately known. But in reality, human preferences, events, probabilities are not exactly known – they are imprecise being described in natural language (NL), and therefore, should be considered as information granules. In real world decision-making problems, we also often encounter uncertainty about uncertainty, or uncertainty^2 (Zadeh 2008a). Instances of uncertainty^2 are fuzzy sets of type 2, imprecise probabilities (Gajdos et al. 2004, Giraud 2005, Pedrycz and Peters 1998, Grabish and Roubens 2000, Yager 1999, Chen et al. 2000, Hable 2009a, Hable 2009b, Cooman and Walley 2002, Troffaes 2007) etc. There exist a huge number of methods able to handle first-order uncertainty, but there is no general theory of decision analysis under uncertainty^2. Most parts of the existing decision theory assume a decision maker (DM) has complete knowledge about the distribution on states of nature and does not offer methodology to deal with partial or fuzzy information. The standard probability theory is more than an adequate tool to deal with physical, inanimate systems where human judgment, perceptions, and emotions do not play a role, but is not designed to deal with perceptions which play a prominent role for humanistic systems. Perceptions are intrinsically imprecise (Zadeh 2008a, Zadeh 1997) or, more specifically, perceptions are intrinsically *fuzzy*. In a perceptions-based approach, probabilities, events, utilities etc are imprecise and are described in the NL.

Natural and useful interpretations of imprecise probabilities as second-order information granules are interval probabilities, fuzzy (linguistic) probabilities (Augustin et al. 2009, Pedrycz and Gomide 2007, Yager 1999) etc. The reason to use fuzzy probabilities is that in most applications the objective probabilities are not exactly known, and we are not able to compute unique prior subjective probabilities. We have only a fuzzy constraint on probability distribution instead of appropriate probability distribution. We only know linguistic estimates of our probabilities such as “this is likely” or “that is unlikely”. Handling imprecise probabilities include such problems as describing imprecise priors and sampling distributions; proceeding from approximate priors and sampling distributions to approximate posteriors and posterior related quantities; making decisions with imprecise probabilities etc. Today, computation with imprecise probabilities (Yager 1999, Augustin et al. 2009, Eichberger et al. 2007, Jaffray 1999, Ekenberg and Thorbiornson 2001, Utkin and Augustin 2003, Utkin 2005, Zadeh 2008a, Alo et al. 2002, Walley 1991, Walley 1997, Gil and Jain 1992) is an active field of research. However, much of the literature is concerned with elicitation rather than computation and imprecise probabilities are considered in isolation (Zadeh 2008a), whereas in reality, they generally occur in an environment of fuzzy events, fuzzy relations and fuzzy constraints. An important aspect of human behavior that does not have its strong general theoretical reflection in the existing utility models is that in reality utilities very often are evaluated by humans using NL. So, utility functions are not accurately known and it is necessary to create strong theoretical basis for this. Fuzzy utility functions, fuzzy alternatives, imprecise probabilities, fuzzy preferences, linguistic preferences relations (LPR) are considered in (Alo et al. 2002, Gil and Jain 1992, Mathieu-Nicot 1986, Billot 1995, Borisov et al. 1989, Zadeh 1975, Wilde 2004, Cantarella and Fedele 2003, Aliev and Fazlollahi 2008). LPR defined on the base of fuzzy preference relation and the extension principle is considered in (Zadeh 1975). In (Borisov et al. 1989) axioms for LPR in terms of linguistic lotteries over fuzzy outcomes are given on the base of which fuzzy expected utility is defined. Paper (Billot 1995) presents conditions on representation of fuzzy preorder by a non-fuzzy (numerical) function. In the (Mathieu-Nicot 1986) existence and continuity of fuzzy EU under traditional conditions (reflexivity, transitivity, continuity etc) for the case fuzzy preference is given. The author proves theorems on existence of a fuzzy EU for the cases of probabilistic and possibilistic information on states of nature. However, information is considered as numerical (non-fuzzy). Of course, the possibilistic case, as the author mentions, is more adequate for real-world problems. Fuzzy preference relations, fuzzy random variables and expected fuzzy number-valued utility are considered in (Gil and Jain, 1992). In (Alo et al. 2002) fuzzy utility defined as a Choquet integral with respect to a numerical fuzzy measure obtained on the basis of the set of possible probability distributions and with a fuzzy integrand are considered. Various approaches to describing imprecision of utility function and probabilities are done in (Alo et al. 2002). But, unfortunately, an overall model for decision making under uncertainty[2] has not been suggested yet.

Adequate and intuitively meaningful models for describing information structures of decision-making problem which are the second-order imprecise hierarchical models. First a hierarchical model where probabilities were normally imprecise was proposed in (Good 1962). It is argued in (Ferson et al. 2002, de Cooman and Walley 2002) that more general hierarchical models are Bayesian models. Unfortunately, probabilities in these models are supposed to be exact that significantly limits the use of this type of models. Further the hierarchical models were modified and developed in (Walley 1997, de Cooman 1998, Nau 1992). In (de Cooman 1998) a hierarchical uncertainty model that presents vague probability assessment and inference on its base is considered. This model is based on Walley's theory of imprecise probabilities. In (de Cooman and Walley 2002, Utkin 2007) the various decision making criteria with imprecise probabilities are discussed. In (Augustin 2002, Aven 2003) the unified behavioral theory of lower prevision and some open problems in the theory are discussed. In (Aven 2003,

Baudrit and Dubois 2006) the problem of combination of experts` beliefs represented by imprecise probabilities is considered. The possibilistic imprecise second-order probability model is considered in (Berger 1985, Bernard 2001, Walley 1997). Modeling linguistic uncertainty in terms of upper probabilities which are given in behavioral interpretation is considered (Walley and de Cooman 2001). A quasi-Bayesian model of subjective uncertainty where beliefs are represented by lower and upper probability is investigated in (Bernard 2002).

Shortcomings of the mentioned models are that they require the usage of exact probability distribution at the second level of a hierarchical model. It limits a practical use of these models. Further development of a hierarchical model is related to the formation of hierarchical models that have interval probabilities at the second level. More comprehensive review and some new results are shown in (Utkin 2007). More universal and popular model for interval probabilities processing is Walley's theory of imprecise probabilities and a principle of natural extension. Main drawback of this approach is impossibility to use it when experts' beliefs are contradictory. Other drawback is related to a huge number of optimization problems.

Use of the mentioned above hierarchical imprecise models is analyzed in (Utkin 2007). We have to mention that the decision procedures in (Utkin 2007) are applied to simple classical decision making problems. The models in (Levi 1974, Walley 1991) and others which use such type of up-to-date decision making problems are open problems.

So, we need a new utility theory which would take into account impreciseness of utility functions and probability distribution and would be based on the use of non-additive measures. What can be used here is fuzzy logic (Zadeh 2008b, Aliev 2008, Aliev et al. 2004, Aliev R.A. and Aliev R.R. 2001) and fuzzy mathematics (Aliev R.A. and Aliev R.R. 2001, Mordeson and Nair 2001) that are capable to interpret impreciseness of utility evaluations adequately and mathematically-strictly. The same concerns empirical probability evaluations under conditions of absence of experimental data.

Taking into account the above mentioned facts, we propose fundamentals of Theory of Decision Making with Second Order Information Granules. The first direction of the proposed theory is a theory of decision making with fuzzy probabilities. This theory differs from the existing methods as one that accumulates non-expected utility paradigm with NL-described decision-relevant information. Fuzzy probabilities, LPR and fuzzy utility functions are used instead of their classical analogs as more adequate description of decision-relevant information. We present representation theorems for a fuzzy utility function described by a fuzzy number-valued Choquet integral with a fuzzy number-valued integrand and a fuzzy number-valued fuzzy measure. The proposed theory is intended to solve decision problems when environment of fuzzy states and fuzzy outcomes are characterized by fuzzy probabilities.

The second avenue of the proposed theory is devoted to hierarchical imprecise probability models. Such models take into account an imprecision and imperfection of our knowledge, expressed by interval values of probabilities of states of nature and degrees of trust to such values. By completing a review of state-of- the art of decision making problems with hierarchical probability models, we consider the decision making problem with the second-order uncertainty where beliefs at the first and the second levels of hierarchical probability models are represented by interval probabilities. Imprecise hierarchical models of Walley may be viewed as a special case of the considered model. A decision-making process analysis and a choice of the most preferable alternative subject to variation of intervals at the lower and upper levels of models and the types of distribution of probabilities of states of nature is of most interest.

We apply the introduced theories and methodologies to solving real-world decision making problems. The obtained results demonstrate the validity of the proposed approaches.

2 Decision Making with Fuzzy Probabilities

2.1 Preliminaries

Let E^n be the space of fuzzy subsets of R^n which are normal, convex, and upper semicontinuous with compact support. Denote by $E^1_{[a,b]}$ the corresponding space of fuzzy sets of $[a,b]\subset R$.

Definition 1 ***Fuzzy Haussdorf distance*****(Aliev and Pedrycz 2009, Aliev 2008).** Let $\tilde{A}$ and $\tilde{B}$ are fuzzy sets, namely $\tilde{A},\tilde{B}\in E^n$. Then the fuzzy Hausdorff distance d_{fH} between $\tilde{A}$ and $\tilde{B}$ is defined as

$$\tilde{d}_{fH}(\tilde{A},\tilde{B}) = \bigcup_{r\in[0,1]} r\left[d_H(A^{r=1},B^{r=1}),\sup_{r\leq\bar{r}\leq 1} d_H(A^{\bar{r}},B^{\bar{r}})\right]$$

Fuzzy functions. By a fuzzy function we mean a function, whose values are fuzzy numbers (for more details see (Bede and Gal 2005, Diamond and Kloeden 1994, Lakshmikantham and Mohapatra 2003). Let $\tilde{f}$ be a fuzzy function, $\mu_{\tilde{f}(x)}$ denotes the membership function of the fuzzy number $\tilde{f}(x)$, and for $0<r\leq 1$, $\tilde{f}_2^r(x)$ stands for $\sup\{z\in dom(\mu_{\tilde{f}(x)}) | \mu_{\tilde{f}(x)}(z)\geq r\}$ and $\tilde{f}_1^r(x)$ denotes $\inf\{z\in dom(\mu_{\tilde{f}(x)}) | \mu_{\tilde{f}(x)}(z)\geq r\}$. Functions $\tilde{f}_1^r(x)$ and $\tilde{f}_2^r(x)$ are level functions of $\tilde{f}$.

Definition 2 (Zhang 1992). Let $\tilde{a}$ be a fuzzy number. For every positive real number M, there exists a $r_0\in(0,1]$ such that $M<a_2^{r_0}$ or $a_1^{r_0}<-M$. Then $\tilde{a}$ is called fuzzy infinity, written as $\tilde{\infty}$.

Definition 3 (Zhang 1992, Zhong 1990). A subclass $\tilde{F}$ of $\tilde{\mathrm{F}}(\Omega)$ is called a fuzzy σ-algebra if it has the following properties:

(1) $\varnothing, \Omega \in \tilde{F}$

(2) if $\tilde{B} \in \tilde{F}$, then $\tilde{B}^c \in \tilde{F}$

(3) if $\{\tilde{B}_n\} \subset \tilde{F}$, then $\bigcup_{n=1}^{\infty} \tilde{B}_n \in \tilde{F}$

Definition 4 (Zhang 1992). A fuzzy number-valued fuzzy measure ((*z*) fuzzy measure) on $\tilde{\mathrm{F}}$ is a fuzzy number-valued fuzzy set function $\tilde{\eta}: \tilde{\mathrm{F}} \to E^1$ with the properties:

(1) $\tilde{\eta}(\varnothing) = 0$;

(2) if $\tilde{B} \subset \tilde{C}$ then $\tilde{\eta}(\tilde{B}) \le \tilde{\eta}(\tilde{C})$;

(3) if $\tilde{B}_1 \subset \tilde{B}_2 \subset ..., \tilde{B}_n \subset ... \in \tilde{F}$, then $\tilde{\eta}(\bigcup_{n=1}^{\infty} \tilde{B}_n) = \lim_{n\to\infty} \tilde{\eta}(\tilde{B}_n)$;

(4) if $\tilde{B}_1 \supset \tilde{B}_2 \supset ..., \tilde{B}_n \in \tilde{F}$, and there exists n_0 such that $\tilde{\eta}(\tilde{B}_{n_0}) \ne \tilde{\infty}$ then $\tilde{\eta}(\bigcap_{n=1}^{\infty} \tilde{B}_n) = \lim_{n\to\infty} \tilde{\eta}(\tilde{B}_n)$.

Here limits are taken in terms of the $\tilde{d}_{fH}$ distance.

Definition 5 (Yang et al. 2005). Let $\tilde{f}: \Omega \to E^n$ be a fuzzy measurable fuzzy-valued function on Ω and $\tilde{\eta}$ be a fuzzy-number-valued fuzzy measure on $\tilde{F}$. The Choquet integral of $\tilde{f}$ with respect to $\tilde{\eta}$ is defined by

$$\int \tilde{f} d\tilde{\eta} = \bigcup_{r \in [0,1]} r \int \bar{f}_r d\tilde{\eta}$$

Definition 6 (Borisov et al. 1989). Let $x = \{x_1, ..., x_n\}$ be a discrete random variable. The linguistic probability assessment over x is the assignment of a collection of n linguistic constraints $P(x_i)\ is\ \tilde{A}_i, \tilde{A}_i \in E^1_{[0,1]}$.

Definition 7 (Borisov et al. 1989). ***Fuzzy set-valued random variable***. Let the discrete variable $\tilde{x}$ takes a value from the set $\{\tilde{x}_1, ..., \tilde{x}_n\}$ of possible linguistic values, each of which is a fuzzy variable $\langle x_i, U_x, \tilde{x}_i \rangle$ described by a fuzzy set $\tilde{x}_i = \int_{U_x} \mu_{\tilde{x}_i}(x)/x$. Let a probability that $\tilde{x}$ will take a linguistic value $\tilde{x}_i$ be characterized by linguistic probability $\tilde{P}_i \in \tilde{P}^l, \tilde{P}^l = \{\tilde{P} \mid \tilde{P} \in E^1_{[0,1]}\}$. The variable $\tilde{x}$ is then called fuzzy set-valued random variable.

Definition 8 ***Linguistic lottery*** **(Borisov et al. 1989).** Linguistic lottery is a fuzzy set-valued random variable with known linguistic probability distribution and is represented by a vector $\tilde{L}=\left(\tilde{P}_1,\tilde{x}_1;...;\tilde{P}_i,\tilde{x}_i;...;\tilde{P}_n,\tilde{x}_n\right)$

Definition 9. ***Lower prevision*** **(Walley, 1991, Aliev R.A. and Aliev R.R 2001, Aliev et al. 2004, Aliev et al. 1999).** Let S be a set of states of nature. Lower prevision of $X \subset S$ is the lower envelope of a closed convex set P of linear previsions $P(\cdot)$

$$\underline{P}(X)=\inf\{P(X): P\in \mathrm{P}, X\subset S\}$$

Definition 10. ***Second-order hierarchical model*** **(Cooman and Walley 2002, Pedrycz and Gomide 2007, Walley 1991, Utkin 2007).** The second-order hierarchical uncertainty models describe uncertainty of a random quantity by means of two levels. Suppose that there is a set of expert judgments related to some measures of states of nature $s_i\,(i=\overline{1,n})$. There are intervals of probabilities on state s_i $[\underline{a}_i,\overline{a}_i]$ at the first level. Let the creditability of interval probabilities at the first level be characterized by interval of probabilities $[\underline{b}_i,\overline{b}_i]$.The second-order uncertainty $[\underline{b}_i,\overline{b}_i]$ forms an imprecise probability, described by a set N of distributions on the set M of all distributions on S.

Generally, imprecise information in the hierarchical model can be expressed as follows:

$$P\{\underline{a}_i \le f_i \le \overline{a}_i\}\in[\underline{b}_i,\overline{b}_i]$$

2.2 Problem Statement

The axiomatization of decision making problem we adopt in our investigation is based on those used by Anscombe and Aumann (Anscombe and Aumann, 1963) and by Schmeidler (Schmeidler 1986, 1989). But the model we suggest is based on imperfect information framework, where as the Schmeidler's model (Schmeidler 1989), which is a more developed than the model in (Anscombe and Aumann, 1963), is constructed for perfect information framework. The essential aspects of our model are as follows: 1) Spaces of fuzzy sets (Diamond and Kloeden 1994, Lakshmikantham and Mohapatra 2003) instead of the classical framework are used for modelling states of nature, outcomes, and actions 2) Fuzzy probabilities (Walley 1991, Zadeh 2008a) are considered instead of numerical probabilities 3) LPR (Zadeh 1975, Borisov et al. 1989) is used instead of classical preference relation 4) Fuzzy functions (Diamond and Kloeden 1994, Lakshmikantham and Mohapatra 2003, Borisov et al. 1989, Bede and Gal 2005, Aliev and Pedrycz 2009) instead of real-valued functions are used to model utility function 5) Fuzzy number-valued fuzzy measure (Zhang 1992) is used instead of a real-valued nonadditive probability. These aspects form fundamentally a new statement of the problem - the problem of decision making with second-order information granules.

It is obvious that under uncertainty humans evaluate alternatives or choices linguistically using such evaluations as "much worse", "a little better", "much better", "almost equivalent" etc. In contrast to the classical preference relation, LPR consistently expresses "degree of preference" allowing the analysis of preferences under uncertainty. In our approach, a utility function is described as fuzzy-number-valued (fuzzy, for short) Choquet integral (Yang et al. 2005) with respect to fuzzy number-valued fuzzy measure generated by fuzzy probabilities to better reflect uncertainty we mostly encounter in real world.

In what follows, we formulate in general a problem of decision making with fuzzy probabilities.

Let $S \subset E^n$ be a set of fuzzy states of the nature, $X \subset E^n$ be a set of fuzzy outcomes, Y be a set of fuzzy probability distributions (Borisov et al. 1989, Yager 1999) over X with finite supports, i.e. Y is a set of fuzzy number-valued functions (Diamond and Kloeden 1994, Lakshmikantham and Mohapatra 2003, Bede and Gal 2005): $Y = \left\{ \tilde{y} \middle| \tilde{y} : X \to E^1_{[0,1]} \right\}$. For notational simplicity, we can identify X with the subset $\left\{ \tilde{y} \in Y \middle| \tilde{y}(\tilde{x}) = 1 \, for\ some\ \tilde{x} \in X \right\}$ of Y. Denote by $\tilde{F}_S$ a σ-algebra of subsets of S. Denote by A_0 the set of all $\tilde{F}_S$-measurable fuzzy finite valued step functions (Mordeson and Nair 2001) from S to Y and denote by A_c the constant fuzzy functions in A_0. Let A be a convex subset (Nanda 1991) of Y^S which includes A_c. Y can be considered as a subset of some linear space, and Y^S can then be considered as a subspace of the linear space of all fuzzy functions from S to the first linear space. Let us now pointwise define convex combinations (Nanda 1991) in Y: for $\tilde{y}$ and $\tilde{z}$ in $Y \subset E^1_{[0,1]}$, and $\lambda \in [0,1]$, $\lambda\tilde{y} + (1-\lambda)\tilde{z} = \tilde{r}$, where $\tilde{r}(x) = \lambda\tilde{y}(\tilde{x}) + (1-\lambda)\tilde{z}(\tilde{x})$. Convex combinations in A are also defined pointwise, i.e., for $\tilde{f}$ and $\tilde{g}$ in A $\lambda\tilde{f} + (1-\lambda)\tilde{g} = \tilde{h}$ where $\lambda\tilde{f}(\tilde{s}) + (1-\lambda)\tilde{g}(\tilde{s}) = \tilde{h}(\tilde{s})$ on S.

To model LPR it is adequate to introduce a linguistic variable "degree of preference" (Zadeh 1975, Borisov et al. 1989) with term-set $\mathrm{T} = (T_1, ..., T_n)$. Terms can be labeled as "equivalence", "little preference", "high preference", and each can be described by a fuzzy number defined over some scale, for example [0,1] or [0,10] etc. The fact that preference of $\tilde{f}$ against $\tilde{g}$ is described by some $T_i \in \mathrm{T}$ is written as $\tilde{f}\, T_i\, \tilde{g}$. Let us denote LPR as $\succsim_l$ and for simplicity below we write $\tilde{f} \succsim_l \tilde{g}$ or $\tilde{f} \succ_l \tilde{g}$ instead of $\tilde{f}\, T_i\, \tilde{g}$.

Definition 11. Two acts $\tilde{f}$ and $\tilde{g}$ in Y^S are said to be comonotonic if for no $\tilde{s}$ and $\tilde{t}$ in S, $\tilde{f}(\tilde{s}) \succ_l \tilde{f}(\tilde{t})$ and $\tilde{g}(\tilde{t}) \succ_l \tilde{g}(\tilde{s})$ hold.

Two real-valued functions a and b are comonotonic iff $(a(\tilde{s})-a(\tilde{t}))(b(\tilde{s})-b(\tilde{t})) \geq 0$ for all $\tilde{s},\tilde{t} \in S$. For a fuzzy function $\tilde{a}: S \to E^1$ denote by $a^r, r \in [0,1]$ its r-cut and note that $a^r = \left[a_1^r, a_2^r\right]$, where $a_1^r, a_2^r : S \to R$.

Two fuzzy functions $\tilde{a},\tilde{b}: S \to E^1$ are said to be comonotonic iff real-valued functions $a_1^r, b_1^r : S \to R$ are comonotonic (Schmeidler 1989) and $a_2^r, b_2^r : S \to R$ are comonotonic (Schmeidler 1989). A constant act $\tilde{f}$ i.e., $\tilde{f} = \tilde{y}^S$, $\tilde{y} \in Y$, and any act $\tilde{g}$ are comonotonic. An act $\tilde{f}$ whose statewise lotteries $\{\tilde{f}(\tilde{s})\}$ are mutually indifferent, i.e., $\tilde{f}(\tilde{s})_l \sim \tilde{y}$ for all $\tilde{s} \in S$, and any act $\tilde{g}$ are comonotonic.

In the suggested framework we extend classical neo-Bayesian nomenclature. More specifically X is a set of fuzzy outcomes; Y is a set of linguistic lotteries; A is a set of fuzzy acts; S is a set of fuzzy states of nature; $\tilde{F}_S$ is a set of fuzzy events.

A problem of decision-making with fuzzy probabilities can be formalized as a 4-tuple $\left(S, Y, A, \succsim_l\right)$. Below we give a series of axioms of the LPR $\succsim_l$ over A underlying the proposed utility model.

(i). ***Weak-order:*** (a) Completeness of LPR. Any two alternatives are comparable with respect to LPR: for all $\tilde{f}$ and $\tilde{g}$ in A: $\tilde{f} \succsim_l \tilde{g}$ or $\tilde{g} \succsim_l \tilde{f}$. This means that for all $\tilde{f}$ and $\tilde{g}$ there exists such $T_i \in \mathrm{T}$ that $\tilde{f} T_i \tilde{g}$ or $\tilde{g} T_i \tilde{f}$ (b) Transitivity. For all $\tilde{f}, \tilde{g}$ and $\tilde{h}$ in A: If $\tilde{f} \succsim_l \tilde{g}$ and $\tilde{g} \succsim_l \tilde{h}$ then $\tilde{f} \succsim_l \tilde{h}$. This means that if there exist such $T_i \in \mathrm{T}$ and $T_j \in \mathrm{T}$ that $\tilde{f} T_i \tilde{g}$ and $\tilde{g} T_j \tilde{h}$, then there exists such $T_k \in \mathrm{T}$ that $\tilde{f} T_k \tilde{h}$. Transitivity of LPR is defined on the base of extension principle and fuzzy preference relation (Zadeh 1975, Borisov et al. 1989). This axiom states that any two alternatives are comparable and assumes one of the fundamental properties of preferences (transitivity) for the case of imperfect information framework.

(ii). ***Comonotonic Independence***: For all pairwise comonotonic acts $\tilde{f}, \tilde{g}$ and $\tilde{h}$ in A if $\tilde{f} \succsim_l \tilde{g}$, then $\alpha\tilde{f} + (1-\alpha)\tilde{h} \succsim_l \alpha\tilde{g} + (1-\alpha)\tilde{h}$ for all $\alpha \in (0,1)$. This means that if there exists such $T_i \in \mathrm{T}$ that $\tilde{f} T_i \tilde{g}$ then there exists such $T_k \in \mathrm{T}$ that $\alpha \tilde{f} + (1-\alpha)\tilde{h} T_k \alpha \tilde{g} + (1-\alpha)\tilde{h}$, with $\tilde{f}, \tilde{g}$ and $\tilde{h}$ pairwise comonotonic. The axiom extends the property of comonotonic independence in case of imperfect information.

(iii). ***Continuity:*** For all $\tilde{f}, \tilde{g}$ and $\tilde{h}$ in A : if $\tilde{f} \succ_l \tilde{g}$ and $\tilde{g} \succ_l \tilde{h}$ then there are α and β in $(0,1)$ such that $\alpha \tilde{f} + (1-\alpha)\tilde{h} \succ_l \tilde{g} \succ_l \beta \tilde{f} + (1-\beta)\tilde{h}$. This means that if there exist such $T_i \in \mathrm{T}$ and $T_j \in \mathrm{T}$ that $\tilde{f} T_i \tilde{g}$ and $\tilde{g} T_j \tilde{h}$ then there exist such $T_k \in \mathrm{T}$ and $T_m \in \mathrm{T}$ that define preference of $\alpha \tilde{f} + (1-\alpha)\tilde{h} T_k \tilde{g} T_m \beta \tilde{g} + (1-\beta)\tilde{h}$. The axiom is an extension of the classical continuity axiom for imperfect information framework.

(iv). ***Monotonicity.*** For all $\tilde{f}$ and $\tilde{g}$ in A: If $\tilde{f}(\tilde{s}) \succsim_l \tilde{g}(\tilde{s})$ on S then $\tilde{f} \succsim_l \tilde{g}$. This means that if for any $\tilde{s} \in S$ there exists such $T \in \mathrm{T}$ that $\tilde{f}(\tilde{s}) T \, \tilde{g}(\tilde{s})$ then there exists $T_i \in \mathrm{T}$ such that $\tilde{f} T_i \tilde{g}$. The axiom is an extension of the classical monotonicity axiom to imperfect information framework.

(v). ***Nondegeneracy:*** Not for all $\tilde{f}, \tilde{g} \in A, \tilde{f} \succsim_l \tilde{g}$.

LPR $\succsim_l$ on A induces LPR denoted also by $\succsim_l$ on Y : $\tilde{y} \succsim_l \tilde{z}$ iff $\tilde{y}^S \succsim_l \tilde{z}^S$ where $\tilde{y}^S$ denotes the constant function $\tilde{y}$ on S. The presented axioms are formulated in accordance with imperfect information framework and the essence of human preferences. This formulation mandates a fuzzy representation of utility function. To determine the best action one could use a fuzzy utility function $\tilde{U}$ such that

$$\forall \tilde{f}, \tilde{g} \in A, \tilde{f} \succsim_l \tilde{g} \Leftrightarrow \tilde{U}(\tilde{f}) \geq \tilde{U}(\tilde{g})$$

We assume the existence of a fuzzy measure on S that better reflect human behavior because of non-additivity property. A fuzzy measure $\tilde{\eta}$ is to be obtained from linguistic probability distribution over S expressed as $\tilde{P}^l = \tilde{P}_1 / \tilde{s}_1 + \tilde{P}_2 / \tilde{s}_2 + \tilde{P}_3 / \tilde{s}_3 =$ *small/small* + *high/medium* + *small/large*, with the understanding that a term such as *high/medium* means that the probability that $\tilde{s}_2 \in S$ is *medium* is *high*. In our approach, computation with fuzzy probabilities plays a pivotal role.

2.3 Fuzzy Valued Utility Function

Analyzing the existing literature on decision analysis, we have arrived at the conclusion that among various non-expected approaches the models based Choquet integral, are more universal and effective (Hey et al. 2007). An extensive experimental investigation conducted in (Hey et al. 2007) shows that CEU takes the first place among SEU, CPT, MMEU, Maximax and Minimum Regret models as the model that better reflects human choices. However, most of the existing works are devoted to decision analysis under first-order uncertainty and they use real-valued utility functions and real-valued fuzzy measures that are not sufficiently adequate to modeling human evaluations. In view of this, in the present work we consider decision problems characterized by fuzzy probabilities and fuzzy events. Now let us introduce the definition of fuzzy utility function defined over an arbitrary set Z of fuzzy alternatives.

Definition 12. Fuzzy-number-valued function $\tilde{U}(\cdot): \mathrm{Z} \to E^1$, is a utility function if it represents linguistic preferences $\succsim_l$ such that for any pair of alternatives $\tilde{\mathrm{z}},\tilde{\mathrm{z}}' \in \mathrm{Z}$ $\tilde{\mathrm{z}} \succsim_l \tilde{\mathrm{z}}'$ holds if and only if $\tilde{U}(\tilde{\mathrm{z}}) \geq \tilde{U}(\tilde{\mathrm{z}}')$. The degree of preference is defined by the distance between $\tilde{U}(\tilde{\mathrm{z}})$ and $\tilde{U}(\tilde{\mathrm{z}}')$.

Here we consider a set Z of alternatives as a set A of actions $\tilde{f}: S \to Y$ and a value of utility function $\tilde{U}$ for action $\tilde{f}$ is determined as a fuzzy Choquet integral over S:

$$\tilde{U}(\tilde{f}) = \int_S \tilde{u}(\tilde{f}(\tilde{s}))d\tilde{\eta}_{\tilde{P}^l} \quad (1)$$

If A is a finite set $A = \{\tilde{f}_1,\ldots,\tilde{f}_m\}$ then after determining utility values for every alternative, the best alternative can be found using some fuzzy ranking method.

Below we present theorems on existence of a fuzzy Choquet-integral-based utility function representing weak order relation defined over the set A of fuzzy alternatives under conditions of linguistic probability distribution $\tilde{P}^l$ over a set S.

THEOREM 1. *Assume that LPR $\succsim_l$ on $A = A_0$ satisfies (i) weak order, (ii) continuity, (iii) comonotonic independence, (iv) monotonicity, and (v) nondegeneracy. Then, there exists a unique fuzzy number-valued fuzzy measure $\tilde{\eta}$ on $\tilde{F}_S$ and an affine fuzzy number-valued function $\tilde{u}$ on Y such that for all $\tilde{f}$ and $\tilde{g}$ in A:*

$$\tilde{f} \succsim_l \tilde{g} \text{ iff } \int_S \tilde{u}(\tilde{f}(\tilde{s}))d\tilde{\eta} \geq \int_S \tilde{u}(\tilde{g}(\tilde{s}))d\tilde{\eta},$$

where $\tilde{u}$ *is unique up to positive linear transformations.*
The proof is given in (Zadeh et al. 2009).

THEOREM 2. *For a nonconstant affine fuzzy number-valued function* $\tilde{u}$ *on* Y *and a fuzzy number-valued fuzzy measure* $\tilde{\eta}$ *on* $\tilde{F}_S$ *a fuzzy Choquet integral* $\tilde{U}(f) = \int_S \tilde{u}(\tilde{f}(\tilde{s}))d\tilde{\eta}$ *induces such LPR on* A_0 *that satisfies (i)-(v) shown above. Additionally,* $\tilde{u}$ *is unique up to positive linear transformations.*

The proof is given in (Zadeh et al. 2009).

If S is a finite set $S = \{\tilde{s}_1, ..., \tilde{s}_n\}$, (1) can be written down as follows:

$$\tilde{U}(\tilde{f}) = \sum_{i=1}^{n} \left(\tilde{u}(\tilde{f}(\tilde{s}_{(i)}) -_h \tilde{u}(\tilde{f}(\tilde{s}_{(i+1)})\right)(\tilde{\eta}_{\tilde{P}^l}(\tilde{B}_{(i)})) \tag{2}$$

Subscript $(\cdot)$ shows that the indices are permuted in order to have $\tilde{u}(\tilde{f}(\tilde{s}_{(1)})) \geq ... \geq \tilde{u}(\tilde{f}(\tilde{s}_{(n)}))$ using some fuzzy ranking method, $\tilde{u}(\tilde{f}(\tilde{s}_{(n+1)}) = 0$, and $\tilde{B}_{(i)} = \{\tilde{s}_{(1)}, ..., \tilde{s}_{(i)}\}$. The optimal action $\tilde{f}^*$ is found as an action with utility value $\tilde{U}(\tilde{f}^*) = \max_{\tilde{f} \in A} \left\{ \int_S \tilde{u}(\tilde{f}(\tilde{s}))d\tilde{\eta}_{\tilde{P}^l} \right\}$.

To calculate a value of a fuzzy Choquet integral for a given action $\tilde{f}$ one has to solve a problem of construction of a fuzzy measure over the set $S = \{\tilde{s}_1, ..., \tilde{s}_n\}$ of the fuzzy states of nature. In this research, it is assumed that only NL-described reasonable knowledge about probability distribution over S is available. It means that a state $\tilde{s}_i$ is assigned a linguistic probability $\tilde{P}_i$ that can be described by a fuzzy number defined over [0,1]. Initial data in the problem are represented by given linguistic probabilities for $n-1$ fuzzy states of nature whereas for one of the given fuzzy states the probability is unknown. First it is required to obtain unknown linguistic probability $\tilde{P}(\tilde{s}_j) = \tilde{P}_j$ After that we construct the desired fuzzy measure. In the framework of Computing with Words (Zadeh et al. 2009, Kacprzyk and Zadeh 1999a, 1999b, Mendel 2007, Zadeh 2001, Zadeh 1999, Zadeh 1996, Zadeh 2006, Zadeh 2008b) the problem of obtaining the unknown linguistic probability for state $\tilde{s}_j$ given linguistic probabilities of all other states is a problem of propagation of generalized constraints (Zadeh 1996, Zadeh 2006). Formally this problem is formulated as follows (Borisov et al. 1989, Zadeh 2006):

$$\text{Given } \tilde{P}(\tilde{s}_i) = \tilde{P}_i;\ \tilde{s}_i \in E^n,\ \tilde{P}_i \in E^1_{[0,1]},\ i \in \{1, ..., j-1, j+1, ..., n\} \tag{3}$$

$$\text{find unknown } \tilde{P}\left(\tilde{s}_j\right)=\tilde{P}_j,\ \tilde{P}_j \in E^1_{[0,1]} \tag{4}$$

This problem reduces to a variational problem of constructing the membership function $\mu_{\tilde{P}_j}(\cdot)$ of an unknown fuzzy probability $\tilde{P}_j$:

$$\mu_{\tilde{P}_j}(p_j)=\sup_\rho \min_{i=\{1,\dots,j-1,j+1,\dots,n\}}(\mu_{\tilde{P}_i}(\int_S \mu_{\tilde{s}_i}(s)\rho(s)ds)) \tag{5}$$

$$\text{subject to} \quad \int_S \mu_{\tilde{s}_j}(s)\rho(s)ds = p_j,\ \int_S \rho(s)ds = 1 \tag{6}$$

This problem is of high computational complexity for solving of which we suggested a novel neuro-fuzzy-evolutionary approach utilizing effective Soft Computing methodologies (Pedrycz 1995, Aliev and Aliev 2001, Aliev et al. 2004) such as Differential Evolution Optimization (DEO), artificial neural networks, fuzzy computations and their combinations. The suggested neuro-fuzzy-evolutionary approach is described in (Zadeh et al. 2009).

When $\tilde{P}_j$ is found, we have linguistic probability distribution $\tilde{P}^l$ over all states $\tilde{s}_i$:

$$\tilde{P}^l = \tilde{P}_1/\tilde{s}_1 + \tilde{P}_2/\tilde{s}_2 + \ldots + \tilde{P}_n/\tilde{s}_n$$

This linguistic probability distribution contains a fuzzy set $\tilde{P}^\rho$ of possible numeric probability distributions $\rho(\cdot)$ over S. If we have linguistic probability distribution the important problem that arises is the verification of its consistency, completeness and redundancy. Given consistent, complete and not redundant linguistic probability distribution $\tilde{P}^l$ we can obtain from it a fuzzy set $\tilde{P}^\rho$ of possible probability distributions $\rho(s)$. We will construct a fuzzy measure from $\tilde{P}^\rho$ as its lower prevision (Nguyen and Walker 2000) by taking into account a degree of correspondence of to $\tilde{P}^l$. We denote this fuzzy-number-valued fuzzy measure with $\tilde{\eta}_{\tilde{P}^l}$ because it is derived from the given linguistic probability distribution $\tilde{P}^l$. A degree of membership of an arbitrary probability distribution $\rho(s)$ to $\tilde{P}$ (a degree of correspondence of $\rho(s)$ to $\tilde{P}^l$) can be as

$$\pi_{\tilde{P}}(\rho(s)) = \min_{i=1,n}(\pi_{\tilde{P}_i}(p_i)) \ ,$$

where p_i is numeric probability of fuzzy state $\tilde{s}_i$ defined by $\rho(s)$. So, $\pi_{\tilde{P}_i}(p_i)=\mu_{\tilde{P}_i}\left(\int_S \rho(s)\mu_{\tilde{s}_i}(s)ds\right)$ is the membership degree of p_i to $\tilde{P}_i$. To derive a fuzzy-number-valued fuzzy measure $\tilde{\eta}_{\tilde{P}^l}$ we use the formulas:

$$\tilde{\eta}_{\tilde{P}^l}(\tilde{B}) = \bigcup_{r\in(0,1]} r\left[\tilde{\eta}^{\,r}_{\tilde{P}^l{}_1}(\tilde{B}), \tilde{\eta}^{\,r}_{\tilde{P}^l{}_2}(\tilde{B})\right] \tag{7}$$

$$\tilde{\eta}_{\tilde{P}^l{}_1}^{\,r}(\tilde{B}) = \inf\left\{ \int_S \rho(s)\mu_{\tilde{B}}(s)ds \,\middle|\, \rho(s) \in \tilde{P}^{\rho^r} \right\},$$

$$\tilde{\eta}_{\tilde{P}^l{}_2}^{\,r}(\tilde{B}) = \inf\left\{ \int_S \rho(s)\mu_{\tilde{B}}(s)ds \,\middle|\, \rho(s) \in \tilde{P}^{\rho^{r=1}} \right\}, \tag{8}$$

$$\tilde{P}^{\rho^r} = \left\{ \rho(s) \,\middle|\, \min_{i=1,n}(\pi_{\tilde{P}_i}(p_i)) \geq r \right\}, \tilde{B} \subset S$$

For $r = 0$, $\tilde{\eta}_{\tilde{P}^l}^{r=0}$ is considered as the support of $\tilde{\eta}_{\tilde{P}^l}$ and is defined as $\tilde{\eta}_{\tilde{P}^l}^{r=0} = cl\left(\bigcup_{r \in (0,1]} \tilde{\eta}_{\tilde{P}^l}^{r} \right)$.

The problem of decision analysis with fuzzy probabilities consists in determination of an optimal action $\tilde{f}^* \in A$ with $\tilde{U}(\tilde{f}^*) = \max_{\tilde{f} \in A}\left\{ \int_S \tilde{u}(\tilde{f}(\tilde{s}))d\tilde{\eta} \right\}$. For this purpose first it is needed to assign linguistic utility values $\tilde{u}(\tilde{f}_j(\tilde{s}_i))$ to every action $\tilde{f}_j \in A$ taken at a state $\tilde{s}_i \in S$. Then it is required to construct a fuzzy measure over S based on partial knowledge available. Here given known probabilities one has to find an unknown probability $P\left(\tilde{s}_j\right) = \tilde{P}_j, \tilde{P}_j \in E^1_{[0,1]}$ by solving the problem (5)-(6). As a result one would obtain a linguistic probability distribution $\tilde{P}^l$ expressed over all the states of nature. If some additional information about the probability over S is received (e.g. from indicator events), it is required to update $\tilde{P}^l$ on the base of this information using a fuzzy Bayes' formula. Then based on the latest $\tilde{P}^l$ it is needed to construct fuzzy measure $\tilde{\eta}_{\tilde{P}^l}$ by solving the problem expressed by (7)-(8). Next the problem of calculation of a fuzzy Choquet integral (Yang et al. 2005) for every action $\tilde{f}_j$ is solved. Here first it is required for an action $\tilde{f}_j$ to rearrange indices of states $\tilde{s}_i$ using fuzzy ranking and find such new indices (i) that $\tilde{u}(\tilde{f}_j(\tilde{s}_{(1)})) \geq \ldots \geq \tilde{u}(\tilde{f}_j(\tilde{s}_{(n)}))$ and calculate fuzzy values $\tilde{U}(\tilde{f}_j)$ of a fuzzy Choquet integral for every action $\tilde{f}_j$ by using (2). Finally, by using a suitable fuzzy ranking method, an optimal $\tilde{f}^* \in A$, that is $\tilde{f}^* \in A$ for which $\tilde{U}(\tilde{f}^*) = \max_{\tilde{f} \in A}\left\{ \int_S \tilde{u}(\tilde{f}(\tilde{s}))d\tilde{\eta}_{\tilde{P}^l} \right\}$, is determined.

3 Decision Making with Second-Order Imprecise Probability

The assessed intervals of probabilities are reflective of expert's or DM's experience. His/her assessment is also imprecise and can be described as an interval. A decision making problem with the second-order information granules, where the probabilities at the first and at the second levels are given as intervals is considered in this section.

Denote by $\underline{a}_i, \overline{a}_i$ the lower and upper bounds, respectively, of interval probability of a state of nature at the first level, i.e. a state s_i is assigned a probability P_i that can be described by a interval $[\underline{a}_i, \overline{a}_i]$. Denote by $\underline{b}_i, \overline{b}_i$ the lower and upper bounds respectively of interval probability describing expert's confidence at the second level, i.e. an expert's confidence can be described by interval $[\underline{b}_i, \overline{b}_i]$. Then we can express the imprecise hierarchical probability model as

$$\Pr\{P_i = [\underline{a}_i; \overline{a}_i]\} = [\underline{b}_i; \overline{b}_i], i = \overline{1,n}. \tag{9}$$

Probability distributions of the second level of hierarchical model are defined on a set of probability distributions of the first level. The suggested decision making methodology based on a given information structure (9) uses utility function for description of preferences. To represent the utility function we use the Choquet integral with non-additive measure:

$$U(h) = \int_S u(h(s))d\eta \tag{10}$$

where η is a nonadditive measure.

Then the decision making problem consists in determination of an optimal action $h^* \in A$ such that

$$U(h^*) = \max_{h \in A}\{\int_S u(h(s))d\eta\} \tag{11}$$

Here η - non-additive measure represented by lower prevision which is constructed from (9).

The suggested decision making method includes the following stages. At first stage it is needed to assign utility values for actions $h_i \in A$ taken at a state $s_i \in S$. The second stage consists in construction of lower prevision to calculate value of utility function for $h_i \in A$. In this study, the lower prevision is determined as

$$\eta(B) = \inf\{P(B) : P \in M, B \subset S\} \tag{12}$$

where M is defined taking into account the following constraints:

$$\underline{b} \leq \int_{\underline{a}}^{\overline{a}} f(p)dp \leq \overline{b} \quad (13)$$

$$\int_{0}^{1} f(p)dp = 1 \quad (14)$$

Here $f(p)$ is a density function. Relevant decision information on imprecise probability is represented by interval probabilities for n-1 states of nature. It is required to obtain unknown imprecise probability $P(s_j) = P_j$ given $P(s_i) = P_i, i \in \{1,..., j-1, j+1,...,n\}$. The lower and upper bounds of the j interval $P_j = [\underline{a}_j; \overline{a}_j]$ are determined as follows:

$$\underline{a}(s_j) = \min(1 - \sum\nolimits_{i=1, i \neq j}^{n} p(s_i)) \quad (15)$$

$$\overline{a}(s_j) = \max(1 - \sum\nolimits_{i=1, i \neq j}^{n} p(s_i)) \quad (16)$$

Here $p(s_i) \in [\underline{a}_i, \overline{a}_i]$ is a basic value of the probability at the first level of the hierarchical model.

Next the problem of calculation of a value of utility function for each $h_i \in A$ as a Choquet integral in accordance with (10) is performed. Finally, an optimal action $h^* \in A$ is obtained in accordance with (11).

For calculations in accordance with the presented methodology the next tools have been developed:

1) The solver for unknown probability calculation.

2) The module for lower prevision calculation in accordance with (12)-(14) on the base of non-linear programming approach.

3) The module for Choquet integral calculation.

4) The module for ranking values of utility function to determine preferences among alternatives.

We will consider application of the suggested approach to economic real-life problem.

4 Experiments

4.1 Decision Making on Oil Extraction at a Potentially Oil-Bearing Region

We consider problem of determination of an optimal decision when a DM is not credibly informed on an actual state of nature. The information on probability of states of the nature is available only for some of the states, and moreover, is imprecise being described in NL. This is conditioned by the fact that there are no statistical data in existence and information available is conveyed as experts' opinion commonly represented by linguistic statements. Utilities of various outcomes a DM will meet taking various actions at various states of nature are also evaluated in NL. The problem is a determination of an optimal action.

Assume that a manager of an oil-extracting company needs to make a decision on oil extraction at a potentially oil-bearing region. The manager's knowledge is described in NL and comes in the following form:

"probability of "occurrence of commercial oil deposits" is lower than medium"

The manager can make a decision based on this information, or at first having conducted seismic investigation. Concerning the seismic investigation used, its accuracy is such that it with the probability "*very high*" confirms occurrence of commercial oil deposits and with the probability "*high*" confirms their absence.

Let us now develop a formal description of the problem. The set of the fuzzy states of the nature is $S = \{\tilde{s}_1, \tilde{s}_2\}$, where $\tilde{s}_1$ – "occurrence of commercial oil deposits" and $\tilde{s}_2$ – "absence of commercial oil deposits". $\tilde{s}_1$ and $\tilde{s}_2$ are represented by triangular fuzzy numbers (TFN) $\tilde{s}_1 = (1;1;0)$, $\tilde{s}_2 = (0;1;1)$. The linguistic probability distribution $\tilde{P}^l$ over the states of nature that corresponds the manager's knowledge is $\tilde{P}^l = \tilde{P}_1 / \tilde{s}_1 + \tilde{P}_2 / \tilde{s}_2$. $\tilde{P}_1 = (0.3;0.4;0.5)$ represents linguistic term "lower than medium" and $\tilde{P}_2$ is unknown.

Taking into account the opportunities of the manager, we consider the following set of possible actions: $A = \{\tilde{f}_1, \tilde{f}_2, \tilde{f}_3, \tilde{f}_4, \tilde{f}_5, \tilde{f}_6\}$. The NL-based description of the actions $\tilde{f}_i$, $i = \overline{1,6}$ is given in Table 1.

Table 1 Possible actions of the manager

Notation	NL-based description
$\tilde{f}_1$	Conduct seismic investigation and extract oil if seismic investigation shows occurrence of commercial oil deposits
$\tilde{f}_2$	Conduct seismic investigation and do not extract oil if seismic investigation shows occurrence of commercial oil deposits
$\tilde{f}_3$	Conduct seismic investigation and extract oil if seismic investigation shows absence of commercial oil deposits
$\tilde{f}_4$	Conduct seismic investigation and do not extract oil if seismic investigation shows absence of commercial oil deposits
$\tilde{f}_5$	Extract oil without seismic investigation
$\tilde{f}_6$	Abandon seismic investigation and oil extraction

We have two types of events: geological events (states of nature) - "occurrence of commercial oil deposits" ($\tilde{s}_1$) and "absence of commercial oil deposits" ($\tilde{s}_2$) and two seismic events (results of seismic investigation) - "seismic investigation shows occurrence of commercial oil deposits" (b_1) and "seismic investigation shows absence of commercial oil deposits" (b_2). Possible combinations of geological and seismic events with their fuzzy probabilities are the following:

$b_1/\tilde{s}_1$ - there is oil and seismic investigation confirms their occurrence, $\tilde{P}(b_1/\tilde{s}_1)=(0.7;0.8;0.9)$

$b_2/\tilde{s}_1$ - there is oil but seismic investigation shows their absence, $\tilde{P}(b_2/\tilde{s}_1)$ is unknown;

$b_1/\tilde{s}_2$ - there is almost no oil but seismic investigation shows their occurrence, $\tilde{P}(b_1/\tilde{s}_2)$ is unknown;

$b_2/\tilde{s}_2$ - there is almost no oil and seismic investigation confirms their absence, $\tilde{P}(b_2/\tilde{s}_2)=(0.6;0.7;0.8)$

According to (5)-(6) we have obtained unknown conditional probabilities $\tilde{P}(b_2/\tilde{s}_1)=(0.1;0.2;0.3)$ and $\tilde{P}(b_1/\tilde{s}_2)=(0.2;0.3;0.4)$.

Seismic investigation allows updating the prior knowledge about actual state of nature to obtain more credible information. Given a result of seismic investigation, the manager can revise prior probabilities of the states of nature on the base of fuzzy probabilities $\tilde{P}(b_j/\tilde{s}_k), k=\overline{1,2}, j=\overline{1,2}$ of possible combinations $\tilde{s}_k/b_j$ of geological and seismic events. The combinations are shown in Table 2.

Table 2 Possible combinations of seismic and geological events

Seismic events	Geological events	Notation
Seismic investigation shows occurrence of oil	occurrence of commercial oil deposits	$\tilde{s}_1/b_1$
Seismic investigation shows absence of oil	occurrence of commercial oil deposits	$\tilde{s}_1/b_2$
Seismic investigation shows occurrence of oil	absence of commercial oil deposits	$\tilde{s}_2/b_1$
Seismic investigation shows absence of oil	absence of commercial oil deposits	$\tilde{s}_2/b_2$

To revise probability of $\tilde{s}_k$ given seismic investigation result b_j we obtain a fuzzy posterior probability $\tilde{P}(\tilde{s}_k/b_j)$ based on the fuzzy Bayes' formula (Buckley, 2006). $\tilde{P}(\tilde{s}_1/b_j)$, $j=\overline{1,2}$ are shown in Fig.1 and 2.

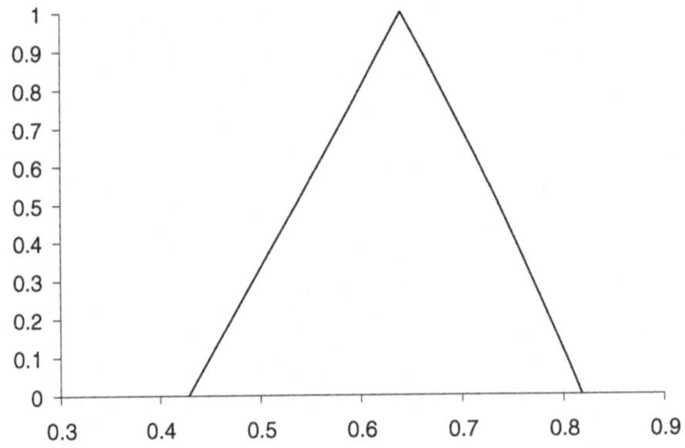

Fig. 1 Posterior probability $\tilde{P}(\tilde{s}_1 / b_1)$

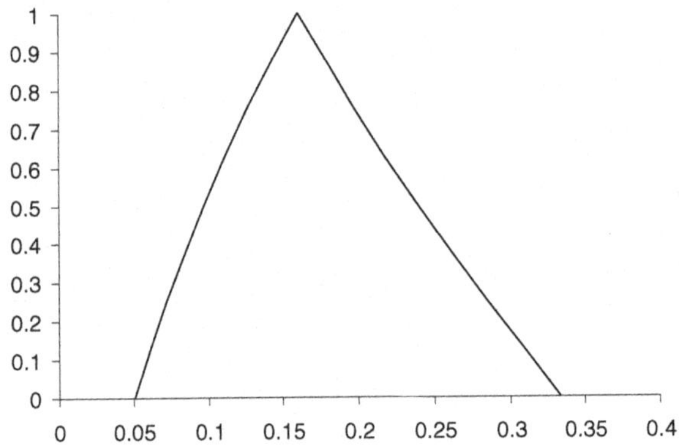

Fig. 2 Posterior probability $\tilde{P}(\tilde{s}_1 / b_2)$

Now we have revised (posterior) fuzzy probabilities $\tilde{P}(\tilde{s}_1 / b_1)$ and $\tilde{P}(\tilde{s}_1 / b_2)$ for the state $\tilde{s}_1$ obtained given possible seismic investigation results b_1, b_2, respectively. We denote $\tilde{P}(\tilde{s}_1 / b_1)$ and $\tilde{P}(\tilde{s}_1 / b_2)$ by $\tilde{P}_{1r}^{b_1}$ and $\tilde{P}_{1r}^{b_2}$, respectively. For this case we have obtained unknown probabilities $\tilde{P}_{2r}^{b_1}$ and $\tilde{P}_{2r}^{b_2}$ of absence of oil by using the neuro-fuzzy-evolutionary approach presented in (Musayev et al. 2009, Zadeh et al. 2009). Fuzzy probabilities $\tilde{P}_{1r}^{b_1}$, $\tilde{P}_{2r}^{b_1}$ and $\tilde{P}_{1r}^{b_2}$, $\tilde{P}_{2r}^{b_2}$ are shown in the Fig. 3 and 4 respectively.

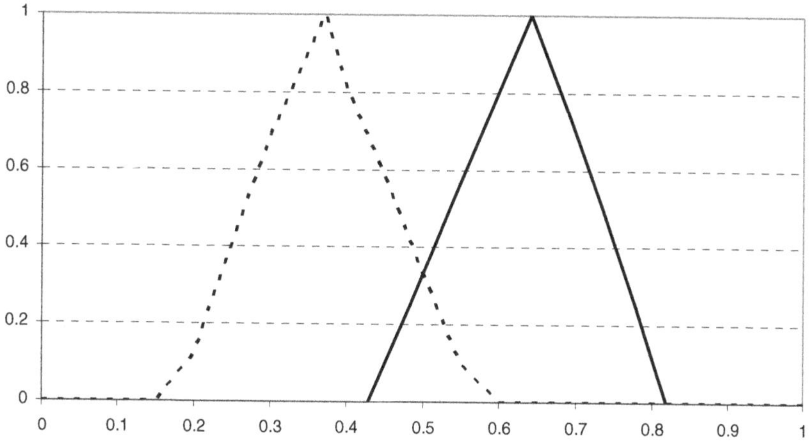

Fig. 3 Posterior probability $\tilde{P}_{1r}^{b_1}$ (solid curve) and $\tilde{P}_{2r}^{b_1}$ (dotted curve)

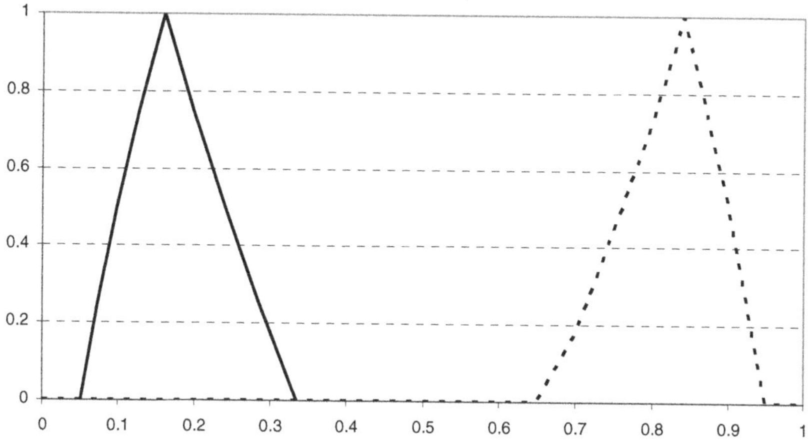

Fig. 4 Posterior probability $\tilde{P}_{1r}^{b_2}$ (solid curve) and $\tilde{P}_{2r}^{b_2}$ (dotted curve)

For actions $\tilde{f}_1, \tilde{f}_2, \tilde{f}_3, \tilde{f}_4$ depending on seismic investigations, the manager will use $\tilde{P}_{1r}^{b_1}$ and $\tilde{P}_{2r}^{b_1}$ or $\tilde{P}_{1r}^{b_2}$ and $\tilde{P}_{2r}^{b_2}$ instead of prior $\tilde{P}_1$. For action $\tilde{f}_5$ the unknown $\tilde{P}_2$ is obtained from the $\tilde{P}_1$ as triangular fuzzy number $(0.5;0.6;0.7)$ using neuro-fuzzy-evolutionary approach described in (Musayev et al. 2009, Zadeh et al. 2009).

In this way, we have all required probabilities of the states $\tilde{s}_1$ and $\tilde{s}_2$. Assume that because of incomplete and uncertain information about possible values of profit from oil sale and possible costs for seismic investigation and drilling of a well, the manager linguistically evaluates utilities for various actions taken at various states of the nature. Assume that the manager's linguistic utility evaluations are as shown in the Table 3.

Table 3 Linguistic utility evaluations

	$\tilde{f}_1$	$\tilde{f}_2$	$\tilde{f}_3$	$\tilde{f}_4$	$\tilde{f}_5$	$\tilde{f}_6$
$\tilde{s}_1$	Positive significant	Negative very high	High	Negative very high	Positive high	0
$\tilde{s}_2$	Negative low	Negative very low	Negative low	Negative very low	Negative insignificant	

Let us give the representation of linguistic utilities by TFN $\tilde{u}(\tilde{f}_i(\tilde{s}_k)) = \tilde{u}_{ik}$ defined on the scale$[-1,1]: \tilde{u}_{11} = (0.65;0.75;0.85)$, $\tilde{u}_{12} = (-0.11;-0.1;-0.09)$, $\tilde{u}_{21} = (-0.88;-0.85;-0.82)$, $\tilde{u}_{22} = (-0.07;-0.04;-0.01)$, $\tilde{u}_{31} = (0.65;0.75;0.85)$, $\tilde{u}_{32} = (-0.11;-0.1;-0.09)$, $\tilde{u}_{41} = (-0.88;-0.85;-0.82)$, $\tilde{u}_{42} = (-0.07;-0.04;-0.01)$, $\tilde{u}_{51} = (0.7;0.8;0.9)$, $\tilde{u}_{52} = (-0.08;-0.07;-0.06)$, $\tilde{u}_6 = 0$. In other words, we have fuzzy utility function $\tilde{U}(\tilde{f}(\tilde{s}))$ defined over the set of combinations $\left(\tilde{f}_i, \tilde{s}_k\right)$, $i = \overline{1,6}$; $k = \overline{1,2}$. To find the optimal action we first calculate for each $\tilde{f}_i$ its utility as a Choquet integral

$$\tilde{U}(\tilde{f}_i) = \int_S \tilde{u}(\tilde{f}_i(\tilde{s}))d\tilde{\eta}_{\tilde{P}^l},$$

where $\tilde{\eta}_{\tilde{P}^l}$ is a fuzzy measure obtained from the linguistic probability distribution as a solution to the problem (7)-(8) based on the neuro-fuzzy-evolutionary technique covered in (Musayev et al. 2009, Zadeh 2009). Depending upon actions, a fuzzy measure will be constructed by considering either prior or posterior probability distributions. For $\tilde{f}_1, \tilde{f}_2$, a fuzzy measure is constructed on the basis of $\tilde{P}_r^{b_1}$ and for $\tilde{f}_3, \tilde{f}_4$ a fuzzy measure is constructed based on $\tilde{P}_r^{b_2}$ (seismic investigation is used). For $\tilde{f}_5$ a fuzzy measure will be constructed on the basis of prior distribution. Utility of $\tilde{f}_6$ is obviously equal to zero. Fuzzy measures $\tilde{\eta}_1$ and $\tilde{\eta}_2$ (approximated to TFN) defined on the base of $\tilde{P}_r^{b_1}$ and $\tilde{P}_r^{b_2}$ respectively are shown in Table 4.

Table 4 Fuzzy number-valued measures obtained from the posterior probabilities

$\tilde{B} \subset S$	$\{\tilde{s}_1\}$	$\{\tilde{s}_2\}$	$\{\tilde{s}_1, \tilde{s}_2\}$
$\tilde{\eta}_1(\tilde{B})$	(0.43, 0.64, 0.64)	(0.18, 0.36, 0.36)	1
$\tilde{\eta}_2(\tilde{B})$	(0.05, 0.16, 0.16)	(0.67, 0.84, 0.84)	1

The fuzzy measure $\tilde{\eta}$ (approximated to TFN) defined on the set of all subsets of S obtained on the base of prior probability is shown in Table 5.

Table 5 Fuzzy measure obtained from the prior probabilities

$\tilde{B} \subset S$	$\{\tilde{s}_1\}$	$\{\tilde{s}_2\}$	$\{\tilde{s}_1, \tilde{s}_2\}$
$\tilde{\eta}(\tilde{B})$	(0.3, 0.4, 0.4)	(0.5, 0.6, 0.6)	1

As utilities for $\tilde{u}_{ik}$ are fuzzy numbers, the corresponding values of Choquet integrals will also be fuzzy. We calculate a fuzzy utility for every action $\tilde{f}_i$ as a fuzzy value of a Choquet integral. For $\tilde{f}_1$ we have

$$\tilde{U}(\tilde{f}_1) = \sum_{i=1}^{2} \left(\tilde{u}(\tilde{f}_1(\tilde{s}_{(i)})) -_h \tilde{u}(\tilde{f}_1(\tilde{s}_{(i+1)})) \right) \tilde{\eta}_{\tilde{P}^l} \left(\tilde{B}_{(i)} \right) =$$

$$= \left(\tilde{u}(\tilde{f}_1(\tilde{s}_{(1)})) -_h \tilde{u}(\tilde{f}_1(\tilde{s}_{(2)})) \right) \tilde{\eta}_1 \left(\{\tilde{s}_{(1)}\} \right) + \tilde{u}(\tilde{f}_1(\tilde{s}_{(2)})) \tilde{\eta}_1 \left(\{\tilde{s}_{(1)}, \tilde{s}_{(2)}\} \right)$$

As
$\tilde{u}(\tilde{f}_1(\tilde{s}_{(1)})) = \tilde{u}_{11} = (0.65; 0.75; 0.85)$, $\tilde{u}(\tilde{f}_1(\tilde{s}_2)) = \tilde{u}_{12} = (-0.11; -0.1; -0.09)$, we find $\tilde{u}_{11} \geq \tilde{u}_{12}$. Then $\tilde{u}(\tilde{f}_1(\tilde{s}_{(1)})) = \tilde{u}_{11}$, $\tilde{u}(\tilde{f}_1(\tilde{s}_{(2)})) = \tilde{u}_{12}$ and $\tilde{s}_{(1)} = \tilde{s}_1$, $\tilde{s}_{(2)} = \tilde{s}_2$. The Choquet integral for f_1 is equal to

$$\tilde{U}(\tilde{f}_1) = \left(\tilde{u}(\tilde{f}_1(\tilde{s}_1)) -_h \tilde{u}(\tilde{f}_1(\tilde{s}_2)) \right) \tilde{\eta}_1 \left(\{\tilde{s}_1\} \right) + \tilde{u}(\tilde{f}_1(\tilde{s}_2)) \tilde{\eta}_1 \left(\{\tilde{s}_1, \tilde{s}_2\} \right) =$$
$$= \left(\tilde{u}_{11} -_h \tilde{u}_{12} \right) \tilde{\eta}_1 \left(\{\tilde{s}_1\} \right) + \tilde{u}_{12} \tilde{\eta}_1 \left(\{\tilde{s}_1, \tilde{s}_2\} \right) =$$
$$= \left(\tilde{u}_{11} -_h \tilde{u}_{12} \right) \tilde{\eta}_1 \left(\{\tilde{s}_1\} \right) + \tilde{u}_{12} =$$
$$= \left((0.65; 0.75; 0.85) -_h (-0.11; -0.1; -0.09) \right) \cdot (0.43, 0.64, 0.64) +$$
$$+ (-0.11; -0.1; -0.09)$$

The obtained result approximated by TFN is $\tilde{U}(\tilde{f}_1)=(0.2168,0.444,0.5116)$. Based on this procedure, we have computed $\tilde{U}(\tilde{f}_i)$ the other actions $f_i, i=\overline{1,5}$ obtaining the following results:

$$\tilde{U}(\tilde{f}_2)=\sum_{i=1}^{2}\left(\tilde{u}(\tilde{f}_2(\tilde{s}_{(i)}))-_h \tilde{u}(\tilde{f}_2(\tilde{s}_{(i+1)}))\right)\tilde{\eta}_{\tilde{P}^l}\left(\tilde{B}_{(i)}\right)=$$
$$=\left(\tilde{u}(\tilde{f}_2(\tilde{s}_{(1)}))-_h \tilde{u}(\tilde{f}_2(\tilde{s}_{(2)}))\right)\tilde{\eta}_1\left(\{\tilde{s}_{(1)}\}\right)+\tilde{u}(\tilde{f}_2(\tilde{s}_{(2)}))\tilde{\eta}_1\left(\{\tilde{s}_{(1)},\tilde{s}_{(2)}\}\right)$$
$$=\left(\tilde{u}(\tilde{f}_2(\tilde{s}_2))-_h \tilde{u}(\tilde{f}_2(\tilde{s}_1))\right)\tilde{\eta}_1\left(\{\tilde{s}_2\}\right)+\tilde{u}(\tilde{f}_2(\tilde{s}_1))\tilde{\eta}_1\left(\{\tilde{s}_1,\tilde{s}_2\}\right)=$$
$$=\left(\tilde{u}_{22}-_h \tilde{u}_{21}\right)\tilde{\eta}_1\left(\{\tilde{s}_2\}\right)+\tilde{u}_{21}\tilde{\eta}_1\left(\{\tilde{s}_1,\tilde{s}_2\}\right)=$$

$$\left((-0.07;-0.04;-0.01)-_h(-0.88;-0.85;-0.82)\right)\cdot(0.67,\ 0.84,\ 0.84)+$$
$$+(-0.88;-0.85;-0.82)$$

$$\tilde{U}(\tilde{f}_3)=\sum_{i=1}^{2}\left(\tilde{u}(\tilde{f}_3(\tilde{s}_{(i)}))-_h \tilde{u}(\tilde{f}_3(\tilde{s}_{(i+1)}))\right)\tilde{\eta}_{\tilde{P}^l}\left(\tilde{B}_{(i)}\right)=$$
$$=\left(\tilde{u}(\tilde{f}_3(\tilde{s}_{(1)}))-_h \tilde{u}(\tilde{f}_3(\tilde{s}_{(2)}))\right)\tilde{\eta}_2\left(\{\tilde{s}_{(1)}\}\right)+\tilde{u}(\tilde{f}_3(\tilde{s}_{(2)}))\tilde{\eta}_2\left(\{\tilde{s}_{(1)},\tilde{s}_{(2)}\}\right)=$$
$$=\left(\tilde{u}(\tilde{f}_3(\tilde{s}_1))-_h \tilde{u}(\tilde{f}_3(\tilde{s}_2))\right)\tilde{\eta}_2\left(\{\tilde{s}_1\}\right)+\tilde{u}(\tilde{f}_3(\tilde{s}_2))\tilde{\eta}_2\left(\{\tilde{s}_1,\tilde{s}_2\}\right)=$$
$$=\left(\tilde{u}_{31}-_h \tilde{u}_{32}\right)\tilde{\eta}_2\left(\{\tilde{s}_1\}\right)+\tilde{u}_{32}\tilde{\eta}_2\left(\{\tilde{s}_1,\tilde{s}_2\}\right)=$$
$$=\left((0.65;0.75;0.85)-_h(-0.11;-0.1;-0.09)\right)\cdot(0.43,\ 0.64,\ 0.64)+$$
$$+(-0.11;-0.1;-0.09)$$

$$\tilde{U}(\tilde{f}_4)=\sum_{i=1}^{2}\left(\tilde{u}(\tilde{f}_4(\tilde{s}_{(i)}))-_h \tilde{u}(\tilde{f}_4(\tilde{s}_{(i+1)}))\right)\tilde{\eta}_{\tilde{P}^l}\left(\tilde{B}_{(i)}\right)=$$
$$=\left(\tilde{u}(\tilde{f}_4(\tilde{s}_{(1)}))-_h \tilde{u}(\tilde{f}_4(\tilde{s}_{(2)}))\right)\tilde{\eta}_2\left(\{\tilde{s}_{(1)}\}\right)+\tilde{u}(\tilde{f}_4(\tilde{s}_{(2)}))\tilde{\eta}_2\left(\{\tilde{s}_{(1)},\tilde{s}_{(2)}\}\right)=$$
$$=\left(\tilde{u}(\tilde{f}_4(\tilde{s}_2))-_h \tilde{u}(\tilde{f}_4(\tilde{s}_1))\right)\tilde{\eta}_2\left(\{\tilde{s}_2\}\right)+\tilde{u}(\tilde{f}_4(\tilde{s}_1))\tilde{\eta}_2\left(\{\tilde{s}_1,\tilde{s}_2\}\right)=$$
$$=\left(\tilde{u}_{42}-_h \tilde{u}_{41}\right)\tilde{\eta}_2\left(\{\tilde{s}_2\}\right)+\tilde{u}_{21}\tilde{\eta}_2\left(\{\tilde{s}_1,\tilde{s}_2\}\right)=$$
$$\left((-0.07;-0.04;-0.01)-_h(-0.88;-0.85;-0.82)\right)\cdot(0.67,\ 0.84,\ 0.84)+$$
$$+(-0.88;-0.85;-0.82)$$

$$\tilde{U}(\tilde{f}_5) = \sum_{i=1}^{2} \left(\tilde{u}(\tilde{f}_5(\tilde{s}_{(i)})) -_h \tilde{u}(\tilde{f}_5(\tilde{s}_{(i+1)})) \right) \tilde{\eta}_{\tilde{P}^l} \left(\tilde{B}_{(i)} \right) =$$

$$= \left(\tilde{u}(\tilde{f}_5(\tilde{s}_{(1)})) -_h \tilde{u}(\tilde{f}_5(\tilde{s}_{(2)})) \right) \tilde{\eta} \left(\{ \tilde{s}_{(1)} \} \right) + \tilde{u}(\tilde{f}_5(\tilde{s}_{(2)})) \tilde{\eta} \left(\{ \tilde{s}_{(1)}, \tilde{s}_{(2)} \} \right) =$$

$$= \left(\tilde{u}(\tilde{f}_5(\tilde{s}_1)) -_h \tilde{u}(\tilde{f}_5(\tilde{s}_2)) \right) \tilde{\eta} \left(\{ \tilde{s}_1 \} \right) + \tilde{u}(\tilde{f}_5(\tilde{s}_2)) \tilde{\eta} \left(\{ \tilde{s}_1, \tilde{s}_2 \} \right) =$$

$$= \left(\tilde{u}_{51} -_h \tilde{u}_{52} \right) \tilde{\eta} \left(\{ \tilde{s}_1 \} \right) + \tilde{u}_{52} \tilde{\eta} \left(\{ \tilde{s}_1, \tilde{s}_2 \} \right) =$$

$$= \left((0.7;0.8;0.9) -_h (-0.08;-0.07;-0.06) \right) \cdot (0.3, 0.4, 0.4) +$$

$$+(-0.08;-0.07;-0.06)$$

The obtained results approximated by TFNs are the following:

$\tilde{U}(\tilde{f}_2) = (\text{-}0.53,\text{-}0.33,\text{-}0.3)$;

$\tilde{U}(\tilde{f}_3) = (-0.072, 0.036, 0.0604)$;

$\tilde{U}(\tilde{f}_4) = (\text{-}0.8395,\text{-}0.7204,\text{-}0.6904)$;

$\tilde{U}(\tilde{f}_5) = (0.154, 0.278, 0.324)$.

These fuzzy numbers are also shown in Fig. 5

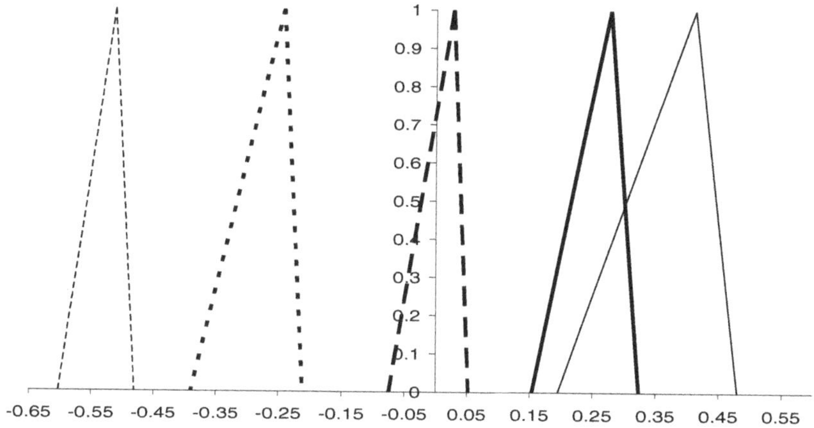

Fig. 5 Fuzzy values of Choquet integral for possible actions (for $\tilde{f}_1$ - thin solid line, for $\tilde{f}_2$ - thick dotted line, for $\tilde{f}_3$ - thick dashed line, for $\tilde{f}_4$ - thin dashed line, for $\tilde{f}_5$ - thick solid line)

As it can be seen that the highest fuzzy utilities are those of alternatives $\tilde{f}_1$ and $\tilde{f}_5$. Applying the fuzzy Jaccard compatibility-based ranking method (Setnes 1997) to compare $\tilde{U}(\tilde{f}_1)$ and $\tilde{U}(\tilde{f}_5)$ we arrive at the following results:

$\tilde{U}(\tilde{f}_1) \geq \tilde{U}(\tilde{f}_5)$ is satisfied with the degree 0.8748;

$\tilde{U}(\tilde{f}_5) \geq \tilde{U}(\tilde{f}_1)$ is satisfied with the degree 0.1635;

The best action is $\tilde{f}_1$ "Conduct seismic investigation and extract oil if seismic investigation shows occurrence of commercial oil deposits" with the highest fuzzy utility value $\tilde{U}(\tilde{f}_1) = (0.1953, 0.412, 0.4796)$.

4.2 Investment Problem

We consider the problem of investment under imprecise information about states of economy over a year. Suppose that we have three states of nature representing three possible states of economy during a year: economic growth (s_1), static economic situation (s_2), economic recession (s_3). The possible alternatives are the following: to buy bonds (h_1), to buy stocks of enterprise (h_2), to deposit money in a bank (h_3). The result of the each act depends on a state of economy that will actually take place. The utilities of the each act taken at various states of economy are given in Table 6.

Table 6 The utility values of actions under various states

	s_1	s_2	s_3
h_1	8	7	5
h_2	7	4	-1
h_3	7	8	5

Let the interval probabilities (first level-probabilities) for s_1 (economic growth) and s_2 (static economic situation) are $P(s_1) =$ [0.3;0.4] and $P(s_2) =$ [0.2;0.4] , respectively. The corresponding interval assessments of experts' confidence degree (second-level probabilities) are equal to [0.7; 0.9].

The problem is to determine an optimal action for investment. We need to find interval probability for s_3 (economic recession). For this we use (15), (16) and obtain the interval probability [0.2; 0.5]. So, an information structure for the given decision making problem may be described as: $\Pr\{P(s_1) = [0.3; 0.4]\} = [0.7; 0.9]$, $\Pr\{P(s_2) = [0.2; 0.4]\} = [0.7; 0.9]$, $\Pr\{P(s_2) = [0.2; 0.5]\} = [0.7; 0.9]$.

Given these data and following the proposed decision making method we have to calculate lower prevision measure. The results of these calculations are shown in Table 7.

Table 7 Lower prevision

Events	Lower prevision	Events	Lower prevision
$B_1=\{s_1\}$	$\eta_1=0.31$	$B_{12}=\{s_1,s_2\}$	$\eta_{12}=0.51$
$B_2=\{s_2\}$	$\eta_2=0.21$	$B_{23}=\{s_2,s_3\}$	$\eta_{23}=0.41$
$B_3=\{s_3\}$	$\eta_3=0.21$	$B_{13}=\{s_1,s_3\}$	$\eta_{13}=0.51$

Now we calculate the values of utility function described as Choquet integral for the each considered act in accordance with (10). The values of utility function for alternatives are the following: $U(h_1)=6.33$, $U(h_2)=6.13$, $U(h_3)=6.23$. So, we have $h_1 \succ h_3 \succ h_2$.

We have also investigated influence of interval probabilities at the first and second levels of hierarchical model to values of utility function for each act. The results are given in Tables 8, 9, respectively.

Table 8 Values of the utility function for the acts under variation of interval probabilities at the first level

CASES	(h_1)		(h_3)		(h_2)
$P(s_1)=[0.6;0.7]$ $P(s_2)=[0.1;0.2]$ $P(s_3)=[0.1;0.3]$	$U(h_1)$=7.03	$\succ$	$U(h_3)$=6.53	$\succ$	$U(h_2)$=6.13
$P(s_1)=[0.2;0.3]$ $P(s_2)=[0.3;0.5]$ $P(s_3)=[0.2;0.5]$	$U(h_1)$=6.23	$\prec$	$U(h_3)$=6.33	$\succ$	$U(h_2)$=5.23
$P(s_1)=[0.1;0.2]$ $P(s_2)=[0.8;0.9]$ $P(s_3)=[0.0;0.1]$	$U(h_1)$=6.93	$\prec$	$U(h_3)$=7.63	$\succ$	$U(h_2)$=4.33
$P(s_1)=[0.1;0.3]$ $P(s_2)=[0.5;0.6]$ $P(s_3)=[0.1;0.4]$	$U(h_1)$ =6.33	$\prec$	$U(h_3)$=6.73	$\succ$	$U(h_2)$=4.63
$P(s_1)=[0.9;0.95]$ $P(s_2)=[0.0;0.02]$ $P(s_3)=[0.03;0.1]$	$U(h_1)$ =7.73	$\succ$	$U(h_3)$=6.83	$\succ$	$U(h_2)$=6.73

As we can see from the Table 8, variations of interval probabilities at the first level of the hierarchical model change preferences between actions h_1 and h_3.

Table 9 Values of the utility function for the acts under variation of interval probabilities at the second level

CASES	(h_1)		(h_3)		(h_2)
$\Pr = [0.1;0.3]$	$U(h_1)$ =6.3	$\prec$	$U(h_3)$=6.356	$\succ$	$U(h_2)$ =5.5
$\Pr = [0.2;0.5]$	$U(h_1)$ =6.3	$\succ$	$U(h_3)$ =6.2	$\succ$	$U(h_2)$ =5.5
$\Pr = [0.4;0.6]$	$U(h_1)$ =6.3	$\prec$	$U(h_3)$=6.81	$\succ$	$U(h_2)$ =5.5
$\Pr = [0.6;0.8]$	$U(h_1)$=6.31	$\succ$	$U(h_3)$ =6.22	$\succ$	$U(h_2)$=5.51
$\Pr = [0.9;0.95]$	$U(h_1)$=6.36	$\succ$	$U(h_3)$=6.352	$\succ$	$U(h_2)$=5.57

As can be seen, the variations of interval probabilities at the second level of the hierarchical model also change preferences between actions h_1 and h_3.

The graphical representation of influence of changes of interval probabilities to values of utility function for acts is given in Fig. 6, 7 respectively.

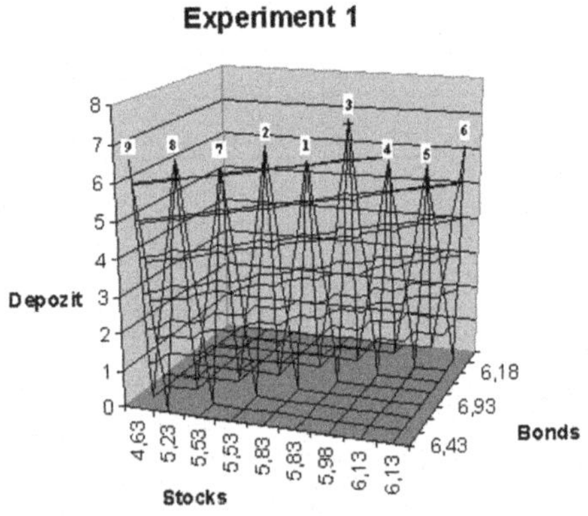

Fig. 6 The plot of surface for experiment 1

1- $P(s_1) = [0.4;0.6], P(s_2) = [0.1;0.2], P(s_3) = [0.2;0.5]$
$U(h_1) = 6.43,\ \ U(h_2) = 5.83,\ \ U(h_3) = 6.13;$

2- $P(s_1) = [0.5;0.7], P(s_2) = [0.2;0.3], P(s_3) = [0.0;0.3]$
$U(h_1) = 6.93,\ \ U(h_2) = 5.53,\ \ U(h_3) = 6.63;$

3- $P(s_1)=[0.6;0.7], P(s_2)=[0.3;0.4], P(s_3)=[0.0;0.1]$
$U(h_1)=7.43,\ \ U(h_2)=5.83,\ \ U(h_3)=7.13;$

4- $P(s_1)=[0.35;0.55], P(s_2)=[0.05;0.15], P(s_3)=[0.3;0.6]$
$U(h_1)=6.18,\ \ U(h_2)=5.98,\ \ U(h_3)=5.88;$

5- $P(s_1)=[0.3;0.5], P(s_2)=[0.0;0.1], P(s_3)=[0.4;0.7]$
$U(h_1)=5.93,\ \ U(h_2)=6.13,\ \ U(h_3)=5.64.$

6- $P(s_1)=[0.5;0.7], P(s_2)=[0.0;0.1], P(s_3)=[0.2;0.5]$
$U(h_1)=6.53,\ \ U(h_2)=6.13,\ \ U(h_3)=6.04.$

7- $P(s_1)=[0.3;0.5], P(s_2)=[0.2;0.3], P(s_3)=[0.2;0.5]$
$U(h_1)=6.33,\ \ U(h_2)=5.53,\ \ U(h_3)=6.23.$

8- $P(s_1)=[0.3;0.5], P(s_2)=[0.3;0.4], P(s_3)=[0.1;0.4]$
$U(h_1)=6.53,\ \ U(h_2)=5.23,\ \ U(h_3)=6.53.$

9- $P(s_1)=[0.2;0.5], P(s_2)=[0.4;0.5], P(s_3)=[0.0;0.4]$
$U(h_1)=6.43,\ \ U(h_2)=4.63,\ \ U(h_3)=6.63.$

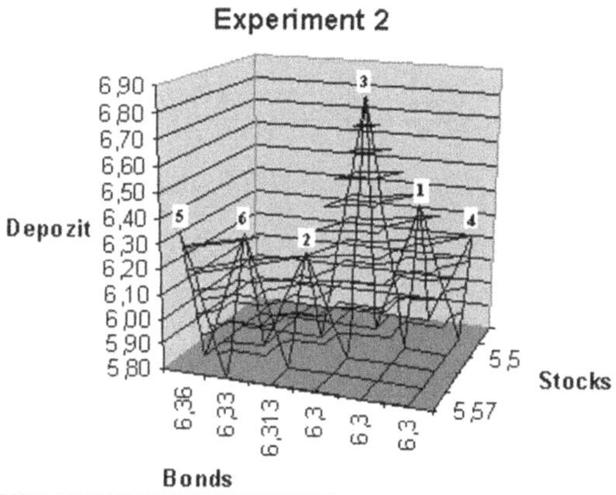

Fig. 7 The plot of surface for experiment 2

1 - $\Pr=[0.9;0.95]$
$U(h_1)=6.36,\ \ U(h_2)=5.57,\ \ U(h_3)=6.352;$

2 - $\Pr=[0.7;0.9]$
$U(h_1)=6.33,\ \ U(h_2)=5.53,\ \ U(h_3)=6.32;$

$\Pr = [0.6;0.8]$

3 - $U(h_1) = 6.31,\ U(h_2) = 5.51,\ U(h_3) = 6.22;$

$\Pr = [0.4;0.6]$

4 - $U(h_1) = 6.3,\ U(h_2) = 5.5,\ U(h_3) = 6.81;$

$\Pr = [0.2;0.5]$

5 - $U(h_1) = 6.3,\ U(h_2) = 5.5,\ U(h_3) = 6.2;$

$\Pr = [0.1;0.3]$

6 - $U(h_1) = 6.3,\ U(h_2) = 5.5,\ U(h_3) = 6.356;$

This graphical visualization offers us an ability to observe how actual economic situation may change the decisions made on investment.

To check robustness of the suggested approach, we applied small changes to interval probabilities of states s_1, s_2, s_3 and calculated for these changes the values of the utility function. The results are summarized in Tables 10, 11.

Let the interval assessments for states "Growth", "Static economic situation" are equal to: 1)[0.6; 0.7], [0.1; 0.2]; 2)[0.1; 0.3], [0.5; 0.6]; 3)[0.9; 0.95],[0.0; 0.02]. Then the results of preferences are shown in Table 10 (see the results of experiment 1).

Table 10 The table of preferences for checking of robustness

CASES	(h_1)		(h_3)		(h_2)
$P(s_1)=[0.6;0.7]$ $P(s_2)=[0.1;0.2]$ $P(s_3)=[0.1;0.3]$	$U(h_1)$=7.03	$\succ$	$U(h_3)$=6.53	$\succ$	$U(h_2)$=6.13
$P(s_1)=[0.1;0.3]$ $P(s_2)=[0.5;0.6]$ $P(s_3)=[0.1;0.4]$	$U(h_1)$=6.33	$\prec$	$U(h_3)$=6.73	$\succ$	$U(h_2)$=4.63
$P(s_1)=[0.9;0.95]$ $P(s_2)=[0.0;0.02]$ $P(s_3)=[0.03;0.1]$	$U(h_1)$=7.73	$\succ$	$U(h_3)$=6.83	$\succ$	$U(h_2)$=6.73

We change these interval assessments slightly. Let the interval assessments for events "Growth", "Static economic situation" are equal to:

1)[0.59; 0.69], [0.09; 0.19];2)[0.09; 0.29], [0.49; 0.59];3)[0.89; 0.94], [0;0.01].

Then the table of preferences will be as shown in the Table 11

Table 11 The table of preferences for checking of robustness

CASES	(h_1)		(h_3)		(h_2)
$P(s_1)=[0.59;0.69]$ $P(s_2)=[0.09;0.19]$ $P(s_3)=[0.12;0.32]$	$U(h_1)$=6.98	$\succ$	$U(h_3)$=6.48	$\succ$	$U(h_2)$=5.86
$P(s_1)=[0.09;0.29]$ $P(s_2)=[0.49;0.59]$ $P(s_3)=[0.12;0.42]$	$U(h_1)$=6.28	$\prec$	$U(h_3)$=6.68	$\succ$	$U(h_2)$=4.66
$P(s_1)=[0.89;0.94]$ $P(s_2)=[0.0;0.01]$ $P(s_3)=[0.05;0.11]$	$U(h_1)$ =7.7	$\succ$	$U(h_3)$=6.85	$\succ$	$U(h_2)$=6.826

As we can see from the tables 10 and 11 the preferences are stable.

5 Conclusions

The widely used principle of maximization of expected utility exhibits serious shortcomings. Experimental evidence has repeatedly shown that people violate the axioms of von Neumann-Morgenstern-Savage preferences in a systematic manner. It resulted in emergence of numerous non-expected theories to alleviate this discrepancy, from weighted utility to rank-dependent utility. Utility functions and non-additive measures used in non-expected utility models to model human preferences are mainly considered as real-valued functions despite of the fact that in reality, human preferences are imprecise and therefore are commonly described in NL. On the other hand, what is not available is a methodology for dealing with second-order uncertainty, that is, uncertainty about uncertainty, or uncertainty^2.

In this paper, we have developed the theory and methodology of decision making with second-order information granules. First, the part of our theory is devoted to decision making with fuzzy probabilities. We presented the characterization theorems about existence of fuzzy utility function representing weak order relation defined over the set of fuzzy alternatives. The proposed methodology based on these theorems is intended to help in solving problems when the environment of fuzzy events and fuzzy states are characterized by fuzzy probabilities. The proposed theory of decision analysis with fuzzy probabilities includes non-expected fuzzy-number-valued utility function represented by a fuzzy Choquet integral with a fuzzy-number-valued fuzzy measure generated by fuzzy probabilities.

The other direction of our theory is based on the use of the second-order imprecise hierarchical model. We have considered decision making method that is based on a two-level hierarchical probability model where probabilities both on the first and the second levels are interval probabilities. The proposed methodology based on this model is intended to solve a decision making problem when environment of states is characterized by second-order uncertainty, namely, by interval imprecise probabilities. The proposed theory of decision analysis with imprecise probabilities includes non-expected utility function represented by Choquet integral with lower prevision measure generated by imprecise probabilities both at the first and second levels. This approach to decision making forms a general framework that coincides with human-oriented assessment of imperfect information.

The developed theory and the methodology of decision making with fuzzy probabilities is applied to solve a problem of decision making in oil extraction from a potentially oil-bearing region. The suggested second-order imprecise probability model is also applied to solving investment problem. We have investigated robustness of the second-order hierarchical imprecise probability model and an impact of imprecise probabilities to decisions obtained. The obtained results showed the validity of the proposed theory.

Acknowledgments. Authors would like to deeply thank Prof. L.A. Zadeh for his outstanding idea to develop a theory of decision making with imprecise probabilities and are grateful to him for his valuable comments and suggestions.

References

Ahn, B.S.: Multiattribute Decision aid with Extended ISMAUT. IEEE Trans-actions on Systems, Man and Cybernetics, Part A: Systems and Humans 36(3), 507–520 (2006)

Akerlof, G.A., Shiller, R.J.: Animal Spirits, How Human Psychology Drives the Economy, and Why it Matters for Global Capitalism. Princeton University Press, Princeton (2009)

Aliev, R.A.: Modeling and stability analysis in fuzzy economics. Applied and Computational Mathematics 7(1), 31–53 (2008)

Aliev, R.A.: Decision And Stability Analysis in Fuzzy Economics. In: North American Fuzzy Information Processing Society Conference (NAFIPS 2009), Cincinnati, Ohio, USA, pp. 1–2 (2009)

Aliev, R.A., Aliev, R.R.: Soft Computing and its applications. World Scien-tific, Singapore (2001)

Aliev, R.A., Bonfig, K.W., Aliew, F.T.: Soft Computing. Technik Verlag, Ber-lin (2004)

Aliev, R.A., Fazlollahi, B.: Decision making and stability analysis in fuzzy economics. In: Aliev, R.A., Bonfig, K.W., Jamshidi, M., Pedrycz, W., Turksen, I.B. (eds.) Proceeding of the Seventh International Conference on Applications of Fuzzy Systems and Soft Computing (ICAFS 2008), Helsinki, pp. 9–44 (2008)

Aliev, R.A., Fazlollahi, B., Aliev, R.R.: Soft Computing and its Applications in Business and Economics. Springer, Heidelberg (2004)

Aliev, R.A., Jafarov, S.M., Gardashova, L.A., Zeinalova, L.M.: Principles of decision making and control under uncertainty. Nargiz, Baku (1999)

Aliev, R.A., Pedrycz, W.: Fundamentals of a fuzzy-logic-based generalized theory of stability. IEEE Transactions on Systems, Man, and Cybernetics, Part B: Cybernetics 39(4), 971–988 (2009)

Allais, M.: Le Comportement de l'Homme Rationnel devant le Risque, Criti-que des Postulats et Axiomes de l'Ecole Americaine. Econometrica 21(4), 503–546 (1953)

Allais, M., Hagen, O.: Expected Utility Hypotheses and the Allais Paradox. D. Reidel Publishing Co., Dordrecht (1979)

Alo, R., Korvin, A., Modave, F.: Fuzzy functions to select an optimal action in decision theory. In: Proceedings of the North American Fuzzy Information Proc-essing Society (NAFIPS), New Orleans, pp. 348–353 (2002)

Anscombe, F.J., Aumann, R.J.: A definition of subjective probability. The An-nals of Mathematical Statistics 34(1), 199–205 (1963)

Augustin, T.: Expected utility within a generalized concept of probability – a comprehensive framework for decision-making under ambiguity. Statistical Pa-pers 43, 5–22 (2002)

Augustin, T., Miranda, E., Vejnarova, J.: Imprecise probability models and their applications. International Journal of Approximate Reasoning 50(4), 581–582 (2009)

Aven, T.: Foundations of Risk Analysis: A Knowledge and Decision - Ori-ented Perspective. Wiley, England (2003)

Baudrit, C., Dubois, D.: Practical representations of incomplete probabilistic knowledge. Computational Statistics and Data Analysis 51, 86–108 (2006)

Bede, B., Gal, S.: Generalizations of the differentiability of fuzzy-number val-ued functions with applications to fuzzy differential equations. Fuzzy Sets and Systems 151(3), 581–599 (2005)

Berger, J.O.: Statistical Decision Theory and Bayesian Analysis. Springer, New York (1985)

Bernard, J.-M.: Non-parametric inference about an unknown mean using the imprecise Dirichlet model. In: Proceedings of the 2nd Int. Symposium on Impre-cise Probabilities and Their Applications (ISIPTA 2001), Ithaca, USA, pp. 40–50 (2001)

Bernard, J.-M.: Implicative analysis for multivariate binary data using an im-precise Dirichlet model. Journal of Statistical Planning and Inference 105(1), 83–104 (2002)

Bernard, J.-M.: Implicative analysis for multivariate binary data using an im-precise Dirichlet model. Journal of Statistical Planning and Inference 105(1), 83–104 (2002)

Billot, A.: From fuzzy set theory to non-additive probabilities: how have economists reacted. Fuzzy Sets and Systems 49(1), 75–90 (1992)

Billot, A.: An existence theorem for fuzzy utility functions: a new elementary proof. Fuzzy Sets and Systems 74(2), 271–276 (1995)

Borisov, A.N., Alekseyev, A.V., Merkuryeva, G.V., Slyadz, N.N., Gluschkov, V.I.: Fuzzy information processing in decision making systems. Radio i Svyaz (1989) (in Russian)

Buckley, J.J.: Fuzzy Probability and Statistics. Studies in Fuzziness and Soft Computing, vol. 196. Springer, Berlin (2006)

Cantarella, G.E., Fedele, V.: Fuzzy Utility Theory for Analyzing Discrete Choice Behaviour. In: Proceedings of the Fourth International Symposium on Uncertainty Modeling and Analysis (ISUMA 2003), pp. 148–154. College Park, Maryland (2003)

Chateauneuf, A., Dana, R.-A., Tallon, J.M.: Optimal risk-sharing rules and equilibria with Choquet-expected-utility. Journal of Mathematical Economics 34(2), 191–214 (2000)

Chateauneuf, A., Eichberger, J.: A Simple Axiomatization and Constructive Representation Proof for Choquet Expected Utility. Economic Theory 22(4), 907–915 (2003)

Chen, T.-Y., Wang, J.-C., Tzeng, G.-H.: Identification of General Fuzzy Measures By Genetic Algorithms Based on Partial Information. IEEE Transactions on Systems, Man, And Cybernetics – Part B: Cybernetics 30(4), 517–528 (2000)

Choquet, G.: Theory of capacities. Annales de l'Institut Fourier 5, 131–295 (1953)

Cooman, G., Walley, P.: A possibilistic hierarchical model for behaviour un-der uncertainty: Theory and Decision, vol. 52(4), pp. 327–374 (2002)

De Cooman, G.: Possibilistic previsions. In: Proceedings of IP-MU 1998, pp. 2–9. Editions EDK, Paris (1998)

Denneberg, D.: Non-additive Measure and Integral. Kluwer Academic Pub-lisher, Boston (1994)

Diamond, P., Kloeden, P.: Metric spaces of fuzzy sets. Theory and applica-tions. World Scientific, Singapore (1994)

Eichberger, J., Grant, S., Kelsey, D.: Updating Choquet beliefs. Journal of Mathematical Economics 43(7-8), 888–899 (2007)

Ekenberg, L., Thorbiornson, J.: Second–order decision analysis. International Journal of Uncertainty, Fuzziness and Knowledge-Based Systems 9(1), 13–37 (2001)

Ellsberg, D.: Risk, Ambiquity and the Savage Axioms. Quarterly Journal of Economics 75, 643–669 (1961)

Ferson, S., Ginsburg, L., Kreinovich, V., et al.: Uncertainty in risk analysis: Towards a general second-order approach combining interval, probabilistic, and fuzzy techniques. In: Proceedings of FUZZ-IEEE 2002, Hawaii, vol. 2, pp. 1342–1347 (2002)

Gajdos, T., Tallon, J.-M., Vergnaud, J.-C.: Decision making with imprecise probabilistic information. Journal of Mathematical Economics 40(6), 647–681 (2004)

Gil, M.A., Jain, P.: Comparison of Experiments in Statistical Decision Prob-lems with Fuzzy Utilities. IEEE Transactions On Systems, Man, And Cyber-neticts 22(4), 662–670 (1992)

Gilbert, W., Bassett, J., Koenker, R., Kordas, G.: Pessimistic Portfolio Alloca-tion and Choquet Expected Utility. Journal of Financial Econometrics 2(4), 477–492 (2004)

Gilboa, I.: Expected utility with purely subjective non additive probabilities. Journal of Mathematical Economics 16(1), 65–88 (1987)

Gilboa, I.: Theory of Decision under Uncertainty. Econometric Society Monographs, Cambridge (2009)

Giraud, R.: Objective imprecise probabilistic information, second order beliefs and ambiquity aversion: an axiomatization. In: Proceedings of ISIPTA-2005, Pittsburg, USA, pp. 183–192 (2005)

Good, I.J.: Subjective probability at the measure of non-measurable set. In: Nagel, E., Suppes, P., Tarski, A. (eds.) Logic, Methodology and Philosophy of Science, pp. 319–329. Stanford University Press, Stanford (1983)

Grabisch, M.: The application of Fuzzy integrals in multicriteria decision making. European Journal of Operational Research 89(3), 445–456 (1996)

Grabisch, M.: Alternative representations of discrete fuzzy measures for deci-sion making. International Journal of Uncertainty, Fuzziness and Knowledge Based Systems 5(5), 587–607 (1997)

Grabisch, M., Murofushi, T., Sugeno, M.: Fuzzy Measures and Integrals: The-ory and Applications. Physica-Verlag, Berlin (2000)

Grabisch, M., Roubens, M.: Application of the Choquet Integral in Multicrite-ria Decision Making. In: Grabisch, M., Murofushi, T., Sugeno, M. (eds.) Fuzzy Measures and Integrals: Theory and Applications, pp. 348–374. Physica-Verlag, Berlin (2000)

Grabisch, M.: Fuzzy integral in multicriteria decision making. Fuzzy Sets and Systems, Special issue on fuzzy information processing 69(3), 279–298 (1995)

Grabish, M., Kojadinovic, I., Meyer, P.: A review of methods for capacity identification in Choquet integral based multi-attribute utility theory. European Journal of Operational Research 186(2), 766–785 (2008)

Grabish, M., Labreuche, C.: A decade of application of the Choquet and Sugeno integrals in multi-criteria decision aid. A Quarterly Journal Of Operations Research 6(1), 1–44 (2008)

Hable, R.: Data-based decisions under imprecise probability and least favor-able models. International Journal of Approximate Reasoning 50(4), 642–654 (2009a)

Hable, R.: Finite approximations of data-based decision problems under im-precise probabilities. International Journal of Approximate Reasoning 50(7), 1115–1128 (2009b)

Helzner, J.: On the Application of Multi-attribute Utility Theory to Models of Choice. Theory and Decision 66(4), 301–315 (2009)

Hey, J.D., Lotito, G., Maffioletti, A.: Choquet OK? Discussion Papers of De-partment of Economics, University of York, December 7 (2010), `http://www.york.ac.uk/depts/econ/documents/dp/0712.pdf` (accessed September 22, 2010)

Huede, F.L., Grabisch, M., Labreuche, C., Saveant, P.: MCS-A new algorithm for multicriteria optimisation in constraint programming. Annals of Operations Research 147(1), 143–174 (2006)

Jaffray, J.Y.: Rational decision making with imprecise probabilities. In: Cooman, G., Cozman, F.G., Moral, S., Walley, P. (eds.) Proceedings of the First International Symposium on Imprecise Probabilities and their Applications (ISIPTA), Ghent, Belgium, pp. 324–332 (1999)

Jamison, K.D., Lodwick, W.A.: The construction of consistent possibility and necessity measures. Fuzzy Sets and Systems 132(1), 1–10 (2002)

Jeleva, M.: Background Risk, Demand for Insurance, and Choquet Expected Utility Preferences. The Geneva Papers on Risk and Insurance 25(1), 7–28 (2000)

Kacprzyk, J., Zadeh, L.A.: Computing with Words in Information. In: Intelligent Systems Part 1. Foundations, Physica-Verlag, Heidelberg (1999a)

Kacprzyk, J., Zadeh, L.A.: Computing with Words in Information. In: Intelligent Systems Part 2, vol. 609. Physica-Verlag, Heidelberg (1999b)

Kahneman, D., Tversky, A.: Prospect theory: an analysis of decision under un-certainty. Econometrica 47(2), 263–291 (1979)

Kobberling, V., Wakker, P.P.: Preference Foundations for Nonexpected Util-ity: A Generalized and Simplified Technique. Mathematics of Operations Re-seach 28(3), 395–423 (2003)

Lakshmikantham, V., Mohapatra, R.: Theory of fuzzy differential equations and inclusions. Taylor & Francis, London (2003)

Levi, I.: On indeterminate probabilities. Journal of Philosophy 71, 391–418 (1974)

Lukacs, G.: Towards an interactive query environment for a multi-attribute decision model with imprecise data and vague query conditions. In: Proceedings of the Eleventh International Workshop on Database and Expert Systems Appli-cations, pp. 708–712. IEEE Computer Society, London (2000)

Machina, M.J.: Non-Expected Utility Theory. In: Teugels, J., Sundt, B. (eds.) Encyclopedia of Actuarial Science, vol. 2, pp. 1173–1179. John Wiley and Sons, Chichester (2004)

Malak, R.J., Aughenbaugh, J.J., Paredis, C.J.J.: Multi-attribute utility analysis in set-based conceptual design. Computer-Aided Design 41(3), 214–227 (2009)

Mangelsdorff, L., Weber, M.: Testing Choquet expected utility. Journal of Economic Behavior and Organization 25, 437–457 (1994)

Marichal, J.L.: An axiomatic approach of the discrete Choquet integral as a tool to aggregate interacting criteria. IEEE Transactions on Fuzzy Systems 8(6), 800–807 (2000)

Marichal, J.L., Meyer, P., Roubens, M.: Sorting Multiattribute Alternatives: The Tomaso method. Computers and Operations Research 32(4), 861–877 (2005)

Marichal, J.L., Roubens, M.: Dependence between criteria and multiple crite-ria decision aid. In: 2nd Int. Workshop on Preferences and Decisions, Trento, Italy, pp. 69–75 (1998)

Markowitz, H.: The Utility of Wealth. Journal of Political Economy 60, 151–158 (1952)

Mathieu-Nicot, B.: Fuzzy Expected Utility. Fuzzy Sets and Systems 20(2), 163–173 (1986)

Mendel, J.M.: Computing with words and its relationships with fuzzistics. In-formation Sciences 179(8), 988–1006 (2007)

Modave, F., Grabisch, M.: Preference representation by the Choquet integral: The commensurability hypothesis. In: Proceedings of the 7th international con-ference on information processing and management of uncertainty in knowledge-based systems (IPMU), EDK Editions, Paris, France, pp. 164–171 (1998)

Modave, F., Grabisch, M., Dubois, D., Prade, H.: A Choquet Integral Repre-sentation in Multicriteria Decision Making. In: Association for the Advancement of Artificial Intelligence (AAAI) Fall Symposium, Boston (1997)

Mordeson, J.N., Nair, P.S.: Fuzzy mathematics: an introduction for engineers and scientists. Springer, Heidelberg (2001)

Murofushi, T.: Semiatoms in Choquet Integral Models of Multiattribute Deci-sion Making. Journal of Advanced Computational Intelligence and Intelligent In-formatics 9(5), 477–483 (2005)

Nanda, S.: Fuzzy linear spaces over valued fields. Fuzzy Sets and Systems 42(3), 351–354 (1991)

Narukawa, Y., Murofushi, T.: Logic of Programs 1981. LNCS, vol. 131, pp. 183–193 (2004)

Nau, R.F.: Indeterminate probabilities on finite sets. The Annals of Statistics 20, 1737–1767 (1992)

Nguyen, H.T., Walker, E.A.: A first course in Fuzzy Logic. CRC Press, Boca Raton (2000)

Pedrycz, W.: Fuzzy sets and neurocomputation: knowledge-based networks. In: Bezdek, J. (ed.) Proceedings of SPIE Applications of Fuzzy Logic Technology II, Orlando, Florida, USA, vol. 2493, pp. 2–20 (1995)

Pedrycz, W., Gomide, F.: Fuzzy Systems Engineering. In: Toward Human-Centric Computing. John Wiley & Sons, Hoboken (2007)

Pedrycz, W., Peters, J.F.: Computational Intelligence in Software Engineering. In: Advances in Fuzzy Systems, Applications and Theory, vol. 16. World Scientific, Singapoure (1998)

Quiggin, J.: A theory of anticipated utility. Journal of Economic Behavioral and Organization 3(4), 323–343 (1982)

Savage, L.J.: The Foundations of Statistics. Wiley, New York (1954); 2nd ed. Dover, New York (1972)

Schmeidler, D.: Integral representation without additivity. Proceedings of the American Mathematical Society 97(2), 255–261 (1986)

Schmeidler, D.: Subjective probability and expected utility without additivity. Econometrica 57(3), 571–587 (1989)

Setnes, M.: Compatibility-Based Ranking of Fuzzy numbers. In: Annual Meeting of the North American Fuzzy Information Processing Society (NAFIPS 1997), Syracuse, New York, pp. 305–310 (1997)

Sims, J.R., Zhenyuan, W.: Fuzzy measures and fuzzy integrals. International Journal of General Systems 17(2-3), 157–189 (1990)

Troffaes, M.C.M.: Decision making under uncertainty using imprecise prob-abilities. International Journal of Approximate Reasoning 45(1), 17–29 (2007)

Troffaes, M.C.M., De Cooman, G.: Uncertainty and conflict: A behavioral approach to the aggregation of expert opinions. In: Vejnarova, J. (ed.) Proceedings of the 6th Workshop on Uncertainty VSE, pp. 263–277. Oeconomica Publishers (2003)

Tversky, A., Kahneman, D.: Advances in Prospect theory: Cumulative Repre-sentation of Uncertainty. Journal of Risk and Uncertainty 5(4), 297–323 (1992)

Utkin, L.V.: Imprecise second-order hierarchical uncertainty model. Interna-tional Journal of Uncertainty, Fuzziness and Knowledge-Based System 13(2), 177–193 (2005)

Utkin, L.V.: Analysis of risk and decision making under incomplete information. Nauka, Saint Petersburg (2007)

Utkin, L.V., Augustin, T.: Decision making with imprecise second-order prob-abilities. In: Bernard, J.M., Seidenfeld, T., Zaffalon, M. (eds.) Proceedings of the Third International Symposium on Imprecise Probabilities and their Applications (ISIPTA 2003), Lugano, Switzerland, pp. 545–559. Carleton Scientific, Waterloo (2003)

Von Neumann, J., Morgenstern, O.: Theory of Games and Economic Behav-iour. Princeton University Press, Princeton (1944)

Wakker, P.P., Zank, H.: State Dependent Expected Utility for Savage's State Space. Mathematics of Operations Reseach 24(1), 8–34 (1999)

Walley, P.: Statistical Reasoning with Imprecise Probabilities. In: Monographs on Statistics and Applied Probability. Chapman and Hall, London (1991)
Walley, P.: Statistical inferences based on a second-order possibility distribu-tion. International Journal of General Systems 26(4), 337–383 (1997)
Walley, P., De Cooman, G.: A behavioral model for linguistic uncertainty. In-formation Sciences 134(1), 1–37 (2001)
Wang, G., Li, X.: On the convergence of the fuzzy valued functional defined by μ-integrable fuzzy valued functions. Fuzzy Sets and Systems 107(2), 219–226 (1999)
Wang, Z., Leung, K.-S., Wang, J.: A genetic algorithm for determining nonad-ditive set functions information fusion. Fuzzy Sets and Systems 102(3), 463–469 (1999)
Wilde, P.D.: Fuzzy Utility and Equilibria. IEEE Transactions on Systems, Man, and Cybernetics-Part B: Cybernetics 34(4), 1774–1785 (2004)
Yaari, M.E.: The dual theory of choice under risk. Econometrica 55(1), 95–115 (1987)
Yager, R.R.: Decision Making with Fuzzy Probability Assessments. IEEE Transactions on Fuzzy Systems 7(4), 462–467 (1999)
Yager, R.R.: A general approach to uncertainty representation using fuzzy measures. In: Proceedings of the Fourteenth International Florida Artificial Intel-ligence Research Society Conference, pp. 619–623. AAAI Press, Menlo Park (2001)
Yang, R., Wang, Z., Heng, P.-A., Leung, K.-S.: Fuzzy numbers and fuzzifica-tion of the Choquet integral. Fuzzy Sets and Systems 153(1), 95–113 (2005)
Yoneda, M., Fukami, S., Grabisch, M.: Interactive Determination of a Utility Function Represented as a Fuzzy Integral. Information Sciences 71(1-2), 43–64 (1993)
Zadeh, L.A.: The concept of a linguistic variable and its application to ap-proximate reasoning. Part I Information Sciences, 8, 199-249; Part II Information and Sciences 8, 301–357; Part III Information and Sciences, 9, 43–80 (1975)
Zadeh, L.A.: Fuzzy logic = Computing with Words. IEEE Transactions on Fuzzy Systems 4(2), 103–111 (1996)
Zadeh, L.A.: Toward a theory of fuzzy Information Granulation and its central-ity in human reasoning and fuzzy logic. Fuzzy Sets and Systems 90(2), 111–127 (1997)
Zadeh, L.A.: From computing with numbers to computing with words – from manipulation of measurements to manipulation with perceptions. IEEE Transac-tions on Circuits and Systems – I: Fundamental Theory and its Applications 45(1), 105–119 (1999)
Zadeh, L.A.: A new direction in AI – toward a computational theory of per-ceptions. AI Magazine 22(1), 73–84 (2001)
Zadeh, L.A.: Generalized theory of uncertainty (GTU) – principal concepts and ideas. Computational Statistics and Data Analysis 51, 15–46 (2006)
Zadeh, L.A.: Computation with imprecise probabilities. In: Proceedings of the Seventh International Conference on Applications of Fuzzy Systems and Soft Computing (ICAFS 2008), Helsinki, pp. 1–3 (2008a)
Zadeh, L.A.: Is there a need for fuzzy logic? Information Sciences, vol. 178(13), pp. 2751–2779 (2008b)
Zadeh, L.A., Aliev, R.A., Fazlollahi, B., Alizadeh, A.V., Guirimov, B.G., Huseynov, O.H.: Decision Theory with Imprecise Probabilities. Report on the contract "Application of Fuzzy Logic and Soft Computing to Communications, Planning and Management of Uncertainty". Berkeley, USA, Baku, Azerbaijan (2009), `http://www.raliev.com/report.pdf` (accessed September 21, 2010)
Zhang, G.-Q.: Fuzzy number-valued fuzzy measure and fuzzy number-valued fuzzy integral on the fuzzy set. Fuzzy Sets and Systems 49(3), 357–376 (1992)
Zhong, Q.: On fuzzy measure and fuzzy integral on fuzzy set. Fuzzy Sets and Systems 37(1), 77–92 (1990)

On the Usefulness of Fuzzy Rule Based Systems Based on Hierarchical Linguistic Fuzzy Partitions

Alberto Fernández, Victoria López, María José del Jesus, and Francisco Herrera

Summary. In the recent years, a high number of fuzzy rule learning algorithms have been developed with the aim of building the Knowledge Base of Linguistic Fuzzy Rule Based Systems. In this context, it emerges the necessity of managing a flexible structure of the Knowledge Base with the aim of modeling the problems with a higher precision. In this work, we present a short overview on the Hierarchical Fuzzy Rule Based Systems, which consists in a hierarchical extension of the Knowledge Base, preserving its structure and descriptive power and reinforcing the modeling of those problem subspaces with more difficulties by means of a hierarchical treatment (higher granularity) of the rules generated in these areas. Finally, this methodology includes a summarisation step by means of a genetic rule selection process in order to obtain a compact and accurate model. We will show the goodness of this methodology by means of a case of study in the framework of imbalanced data-sets in which we compare this learning scheme with some basic Fuzzy Rule Based Classification Systems and with the well-known C4.5 decision tree, using the proper statistical analysis as suggested in the specialised literature. Finally, we will develop a discussion on the usefulness of this methodology, analysing its advantages and proposing some new trends for future work on the topic in order to extract the highest potential of this technique for Fuzzy Rule Based Systems.

Keywords: Fuzzy Rule Based Classification Systems, Hierarchical Fuzzy Partitions, Hierarchical Systems of Linguistic Rules Learning Methodology, Granularity, Imbalanced Data-sets.

Alberto Fernández · María José del Jesus
Dept. of Computer Science, University of Jaén, Spain
e-mail: alberto.fernandez@ujaen.es, mjjesus@ujaen.es

Victoria López · Francisco Herrera
Dept. of Computer Science and A.I., University of Granada, Spain
e-mail: vlopez@decsai.ugr.es, herrera@decsai.ugr.es

W. Pedrycz and S.-M. Chen (Eds.): Granular Computing and Intell. Sys., ISRL 13, pp. 155–184.
springerlink.com

1 Introduction

Linguistic Fuzzy Rule Based Systems (FRBSs) (Yager and Filev, 1994) have demonstrated their ability for control problems (Palm et al, 1997), modeling (Pedrycz, 1996), classification or data mining (Kuncheva, 2000; Ishibuchi et al, 2004) in a huge number of applications. They provide an accurate model which is also easily interpretable by the end-user or expert by means of the use of linguistic labels. The main handicap in the application of linguistic systems is the hard restrictions on the fuzzy rule structure (Bastian, 1994), which may suppose a loss in accuracy when dealing with some complex systems, i.e. high dimensional problems, in the presence of noise or when the classes are overlapped (in classification tasks).

It is possible to make some considerations to face this drawback. Many different possibilities to improve the linguistic fuzzy modeling have been considered in the specialised literature. All of these approaches share the common idea of improving the way in which the linguistic fuzzy model performs the interpolative reasoning by inducing a better cooperation among the rules in the Knowledge Base (KB). This rule cooperation may be induced acting on three different model components:

- *Approaches acting on the Data Base (DB).* For example a priori granularity learning (Cordón et al, 2001b) or membership function tuning (Alcalá et al, 2007).
- *Approaches acting on the Rule Base (RB).* The most common approach is rule selection (Ishibuchi et al, 1995; Gacto et al, 2009) but also multiple rule consequent learning (Cordón and Herrera, 2000) could be considered.
- *Approaches acting on the whole KB.* This includes the KB derivation (Magdalena and Monasterio-Huelin, 1997) and a hierarchical linguistic rule learning (Ishibuchi et al, 1993; Cordón et al, 2002).

In this work we will focus on this last issue, studying the use of a hierarchical environment in order to improve the behaviour of linguistic FRBSs. This approach has been first proposed by Herrera and Martínez (Herrera and Martínez, 2001) in the field of Decision Making and later by Cordón et al. (Cordón et al, 2002) in the scenario of regression problems. The hierarchical model preserves the original descriptive power of FRBS and increases its accuracy by reinforcing those problem subspaces that are especially difficult by means of a hierarchical treatment of the rules generated in these areas producing a more general and well defined structure, the Hierarchical Knowledge Base (HKB).

Our aim is to provide a wide overview on the hierarchical methodology for linguistic fuzzy systems, describing the different approaches that have been developed on the topic including the basic hierarchical systems of linguistic rules learning methodology (HSLR-LM) (Cordón et al, 2002), the hybridization of weighted rule learning with the hierarchical approach (Alcalá et al, 2003) and the iterative scheme through different granularity levels of the HSLR-LM (Cordón et al, 2003). In order to show their usefulness, we will present a case of study on classification with imbalanced data-sets (He and Garcia, 2009; Sun et al, 2009), in which we have made use of the adaption of Hierarchical Fuzzy Rule Based Systems (HFRBSs) to this scenario (Fernández et al, 2009).

According to all these points, this work is organised as follows. First, Section 2 introduces the concept of hierarchal fuzzy partitions and the definition of the HKB. Next, Section 3 describes the learning methodology for HFRBSs and some extensions that have been developed this approach. In Section 4 we present the framework of imbalanced data-sets and the specific hierarchical fuzzy methodology that was designed for this scenario. Then, we provide a case of study for imbalanced data-sets in Section 5, showing some experimental results on this new topic. Finally, in Section 6 we will point out some concluding remarks about the study carried out and we will discuss some new challenges on the topic that can support further work from the basis previously presented.

2 Hierarchical Linguistic Fuzzy Partitions

As we have stated in the introduction of this work, the KB structure usually employed in the field of linguistic modeling has the drawback of its lack of accuracy when working with very complex systems. This fact is due to some problems related to the linguistic rule structure considered, which are a consequence of the inflexibility of the concept of linguistic variable (Zadeh, 1975). A summary of these problems may be found in (Bastian, 1994; Carse et al, 1996), and it is briefly enumerated as follows.

- There is a lack of flexibility in the FRBSs because of the rigid partitioning of the input and output spaces.
- When the system input variables are dependent themselves, it is very hard to fuzzy partition the input spaces.
- The homogenous partitioning of the input and output spaces when the input-output mapping varies in complexity within the space is inefficient and does not scale to high-dimensional spaces.
- The size of the RB directly depends on the number of variables and linguistic terms in the system. Obtaining an accurate FRBS requires a significant granularity amount, i.e., it needs of the creation of new linguistic terms. This granularity increase causes the number of rules to rise significantly, which may take the system to lose the capability of being interpretable for human beings.

At least two things could be done to solve many of these problems and to improve the model accuracy. On the one hand, we can use approximative fuzzy modeling, with the consequence of losing the model interpretability. On the other hand, we can refine a linguistic model trying not to change too much the meaning of the linguistic variables neither the descriptive power of the final FRBS generated.

Related to the previous issue, a crucial task for dealing with linguistic information is to determine the granularity of uncertainty, i.e., the cardinality of the fuzzy linguistic term set used to assess the linguistic variables. Depending on the uncertainty degree held by a source of information qualifying a phenomenon, the linguistic term set will have more or less terms (Bonissone and Decker, 1985; Herrera et al, 2000).

In order to overcome this drawback, Herrera and Martínez proposed in (Herrera and Martínez, 2001) the use of a set of multigranular linguistic contexts that they denoted as linguistic hierarchies term sets. A linguistic hierarchy is a set of levels, where each level is a linguistic term set with different granularity to the rest of levels of the hierarchy. The purpose of this extension is the flexibilisation of the KB to become an HKB. This is possible by the development of a new KB structure, where the linguistic variables of the linguistic rules could take values from fuzzy partitions with different granularity levels. An HKB is said to be composed of a set of layers ("levels" in the notation of Herrera and Martínez), and each layer is defined by its components in the following way:

$$layer(t,n(t)) = DB(t,n(t)) + RB(t,n(t)), \tag{1}$$

with $n(t)$ being the number of linguistic terms in the fuzzy partitions of layer t, $DB(t,n(t))$ being the DB which contains the linguistic partitions with granularity level $n(t)$ of layer t (t-linguistic partitions), and $RB(t,n(t))$ being the RB formed by those linguistic rules whose linguistic variables take values in $DB(t,n(t))$ (t-linguistic rules). For the sake of simplicity in the descriptions, the following notation equivalences are established:

$$DB(t,n(t)) \equiv DB^t \text{ and } RB(t,n(t)) \equiv RB^t \tag{2}$$

At this point, we should note that, At this point, we should note that, in this work, we are considering *linguistic partitions* with the same number of linguistic terms for all input variables, composed of symmetrical triangular-shaped and uniformly distributed membership functions (see Fig. 1). This type of membership functions is the most suitable for this environment easing the mapping between the different layers of the HKB. Furthermore, this environment can be extended to interval-valued fuzzy sets adding a degree of uncertainty in the definition of the support of each fuzzy term (Sanz et al, 2010).

Specifically, the number of linguistic terms in the t-linguistic partitions is defined in the following way:

$$n(t) = (n(1)-1)\cdot 2^{t-1} + 1, \tag{3}$$

with $n(1)$ being the granularity of the initial fuzzy partitions, linguistic hierarchy basic rules. This structure must satisfy the following rules:

1. To preserve all former modal points of the membership functions of each linguistic term from one level to the following one.
2. To make smooth transitions between successive levels. The aim is to build a new linguistic term set, new linguistic term will be added between each pair of terms belonging to the term set of the previous level. To carry out this insertion, we shall reduce the support of the linguistic labels in order to keep place for the new one located in the middle of them.

Fig.1 (left) graphically depicts the way in which a linguistic partition in DB^1 becomes a linguistic partition in DB^2. Each term of order k from DB^t, $S_k^{n(t)}$ ($S_k^{n(1)}$ in the figure), is mapped into the fuzzy set $S_{2\cdot k-1}^{2\cdot n(t)-1}$, preserving the former modal points, and a set of $n(t)-1$ new terms is created, each one between $S_k^{n(t)}$ and $S_{k+1}^{n(t)}$ ($k=1,\dots,n(t)-1$) (see Figure 1 right).

DB(t, n(t)) **DB(t+1, 2·n(t)-1)**

$S_{k-1}^{n(t)} \quad \rightarrow \quad S_{2k-3}^{2\cdot n(t)-1}$

$S_{2k-2}^{2\cdot n(t)-1}$

$S_k^{n(t)} \quad \rightarrow \quad S_{2k-1}^{2\cdot n(t)-1}$

$S_{2k}^{2\cdot n(t)-1}$

$S_{k+1}^{n(t)} \quad \rightarrow \quad S_{2k+1}^{2\cdot n(t)-1}$

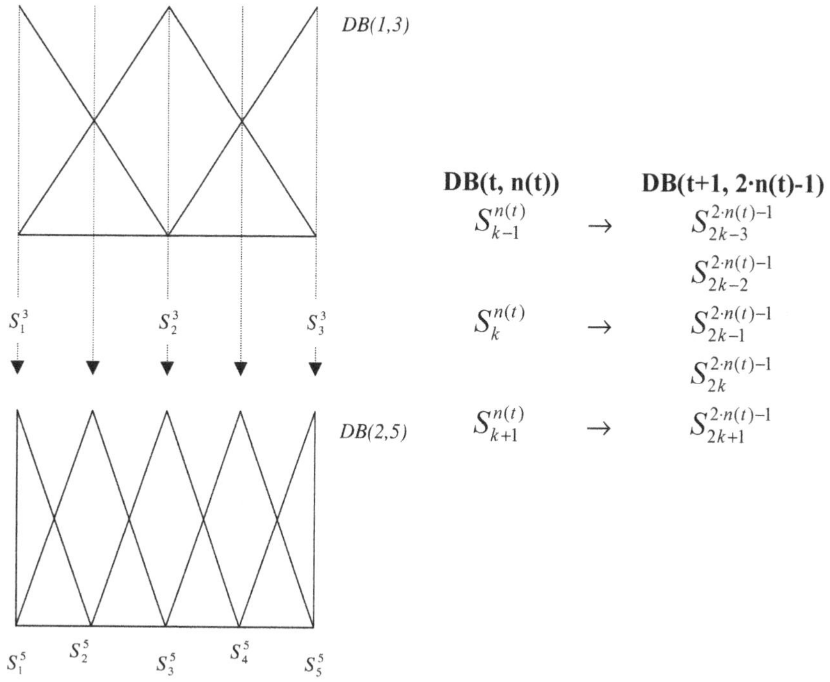

Fig. 1 Two layers of linguistic partitions which compose the HDB and mapping between terms from successive DBs.

The main purpose of developing a Hierarchical Rule Base (HRB) is to divide the problem space in a more accurate way. To do so, those linguistic rules from $RB(t,n(t))$ – RB^t – that classify a subspace with bad performance are expanded into a set of more specific linguistic rules, which become their image in $RB(t+1,2\cdot n(t)-1)$ – RB^{t+1} –. This set of rules classify the same subspace that the former one and replaces it. As a consequence of the previous definitions, we could now define the HKB as the union of every layer t:

$$HKB = \bigcup_t layer(t,n(t)) \qquad (4)$$

3 Hierarchical Fuzzy Rule Based Systems

In the previous section we have introduced the concept of hierarchical fuzzy partitions. As explained, this approach presents a more flexible KB structure that allows improving the accuracy of the FRBCSs without losing their interpretability: the HKB, which is composed of a Hierarchical Data Base (HDB) and an HRB.

In this section, we will first introduce the basic two-level HSLR-LM to generate an HFRBS (Cordón et al, 2002). Next, we describe the hybridization of weighted rule learning with the hierarchical approach which was developed with the aim of improving the system accuracy (Alcalá et al, 2003). Finally, an extension of the two-level learning method is presented as an iterative scheme through different granularity levels (Cordón et al, 2003). We must point out that in all these three cases, the scenario in which these approaches have been proposed is devoted to regression problems.

3.1 A Two-Level HSLR Learning Methodology

The first methodology to build an HFRBS was proposed by Cordón et al. (Cordón et al, 2001a) as a strategy to improve simple linguistic models preserving their structure and descriptive power, reinforcing only the modeling of those problem subspaces with more difficulties. For the sake of maintaining the interpretability of the final model, this basic HSLR was only based on two hierarchical levels, i.e., two layers.

In the following, the structure of the learning methodology and its most important components are described in detail. Specifically, the algorithm to obtain an HFRBS is based on two processes:

1. HKB Generation Process: An HRB is created from a simple RB obtained by an LRG-method.
2. HRB Genetic Selection Process: The best cooperative rules are selected by means of a Genetic Algorithm (GA).

To do so, it is needed to use an existing inductive LRG-method based on the existence of a set of input-output training data $X = \{x_1, \ldots, x_p, \ldots, x_m\}$ with $x_p = (x_{p1}, \ldots, x_{pn}, y_p)$, $p = 1,2,\ldots,m$ where x_{pi} is the ith attribute value ($i = 1,2,\ldots,n$) of the p-th training pattern y_p is the output value, and a previously defined DB^1. Usually, the LRG-method selected is aimed to obtain simple linguistic fuzzy models, such as the Wang and Mendel's algorithm (Wang and Mendel, 1992), or the Thrift's algorithm (Thrift, 1991). Two measures of error are used in the algorithm:

1. Global measure (used to evaluate the complete RB): The Mean Square Error (MSE) for a whole RB, calculated over X, is defined as:

$$MSE(X, RB) = \frac{\sum_{x_p \in X} (y_p - s(x_p))^2}{2 \cdot |X|}$$

with $s(x_p)$ being the output value obtained from the HLSR using the current RB when the input variable values are $x_p = (x_{p1}, \ldots, x_{pn}, y_p)$, and y_p is the known desired value.

2. Local measure (used to determine if an individual rule is expanded): The MSE for a simple rule[1], $R_i^{n(1)}$, calculated over X_i, is showed as follows:

$$MSE(X_i, RB_i^{n(1)}) = \frac{\sum_{x_p \in X_i} (y_p - s_i(x_p))^2}{2 \cdot |X_i|}$$

with X_i being a set of the examples matching the i-th rule antecedents to degree $\tau \in (0,1]$ and $s_i(x_p)$ being the output value from this rule.

Table 1 Two-level learning method

HIERARCHICAL KNOWLEDGE BASE
Step 0. $RB(1, n(1))$ *Generation Process*
Step 1. $RB(2, 2 \cdot n(1) - 1)$ *Generation Process*
Step 2. *Summarization Process*
HIERARCHICAL RULE BASE GENETIC SELECTION PROCESS
Step 3. *HRB Genetic Selection Process*

Now we will describe the HKB generation process (summarised in Table 1), which basically consists of the following steps:

Step 0. RB^1 **Generation.** Generate the rules from RB^1 by means of an existing LRG-method: $RB^1 = LRG - method(DB^1, X)$.

Step 1. RB^2 **Generation.** Generate RB^2 from RB^1, DB^1 and DB^2.

a) *Calculate the error of* RB^1 : $MSE(X, RB^1)$.

b) *Calculate the error of each 1-linguistic rule:* $MSE(X_i, RB_i^{n(1)})$.

c) *Select the 1-linguistic rules with bad performance which will be expanded (the expansion factor* α *may be adapted in order to have more or less expanded rules):*

If $MSE(X_i, R_i^{n(1)}) \geq \alpha \cdot MSE(X, RB^1)$ *Then* $R_i^{n(1)} \in RB_{bad}^1$

Else $R_i^{n(1)} \in RB_{good}^1$.

---

[1] Notice that other local error measures, such as the one showed in (Yen et al, 1998) could also be considered.

d) *Create* DB^2.

e) *For each bad performance 1-linguistic rule to be expanded,* $R_j^{n(1)} \in RB_{bad}^1$:

   i. *Select the 2-linguistic partitions terms* from DB^2 for each rule. For all linguistic terms considered in $R_j^{n(1)}$, i.e., $S_{jk}^{n(1)}$ defined in DB^1, select those terms $S_h^{2 \cdot n(1)-1}$ in DB^2 that significantly intersect them. We consider that two linguistic terms have a "*significant intersection*" between each other, if the maximum cross level between their fuzzy sets in a linguistic partition overcomes a predefined threshold δ :

$$I(S_{jk}^{n(1)}) = \{S_h^{2 \cdot n(1)-1} \in DB^2 / \max_{u \in U_k} \min\{\mu_{S_{jk}^{n(1)}}(u), \mu_{S_h^{2 \cdot n(1)-1}}(u)\} \geq \delta\} \quad (5)$$

   where $\delta \in [0,1]$.

   ii. *Combine the previously selected s sets* $I(S_{jk}^{n(1)})$ *by the following expression:*

$$I(R_j^{n(1)}) = I(S_{j1}^{n(1)}) \times \ldots \times I(S_{js}^{n(1)}) \quad (6)$$

   iii. *Extract 2-linguistic rules*, which are the expansion of the bad *1-linguistic rule* $R_j^{n(1)}$. This task is performed by the LRG-method, which takes $I(R_j^{n(1)})$ and the set of examples $X(R_j^{n(1)})$ as its parameters:

$$CLR(R_j^{n(1)}) = LRG - method(I(R_j^{n(1)}), X(R_j^{n(1)})) = \\ = \{R_{j_1}^{2 \cdot n(1)-1}, \ldots, R_{j_L}^{2 \cdot n(1)-1}\} \quad (7)$$

   with $CLR(R_j^{n(1)})$ being the image of the expanded linguistic rule $R_j^{n(1)}$, i.e., the candidates to be in the HRB from rule $R_j^{n(1)}$.

Step 2. **Summarization**. *Obtain a Joined set of Candidate Linguistic Rules* (JCLR), performing the union of the group of the new generated *2-linguistic rules* and the former good performance *1-linguistic rules*:

$$JCLR = RB_{good}^1 \cup (\cup_j CLR(R_j^{n(1)})), R_j^{n(1)} \in RB_{bad}^1 .$$

Example: In the following, we show an example of the whole expansion process. Let us consider $n(1) = 3$ and the following linguistic partitions:

$$DB_{x_1}(1,3) = DB_{x_2}(1,3) = DB_y(1,3) = \{S^3, M^3, L^3\},$$

$$DB_{x_1}(2,5) = DB_{x_2}(2,5) = DB_y(2,5) = \{VS^5, S^5, M^5, L^5, VL^5\},$$

where S stands for Small, M for Medium, L for large, and V for Very. Let us consider the following bad performance *1-linguistic rule* to be expanded (see Fig. 2):

$$R_i^3 : \text{IF } x_1 \text{ is } S_{i1}^3 \text{ and } x_2 \text{ is } S_{i2}^3 \text{ THEN } y \text{ is } B_i^3$$

where the linguistic terms are, $S_{i1}^3 = S^3$, $S_{i2}^3 = S^3$, $B_i^3 = S^3$, and the resulting sets I with $\delta = 0.5$ are:

$$I(S_{i1}^3) = \{VS^5, S^5\}, I(S_{i2}^3) = \{VS^5, S^5\}, I(B_i^3) = F(\cdot) \subseteq D_y(2,5)$$

$$I(R_i^3) = I(S_{i1}^3) \times I(S_{i2}^3) \times I(B_i^3).$$

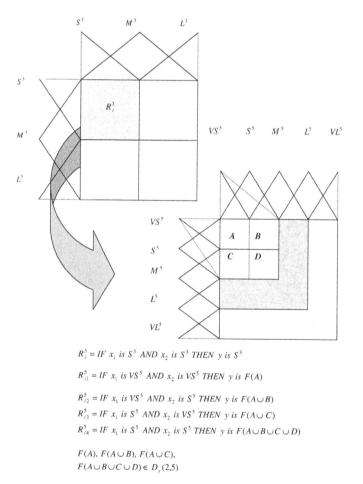

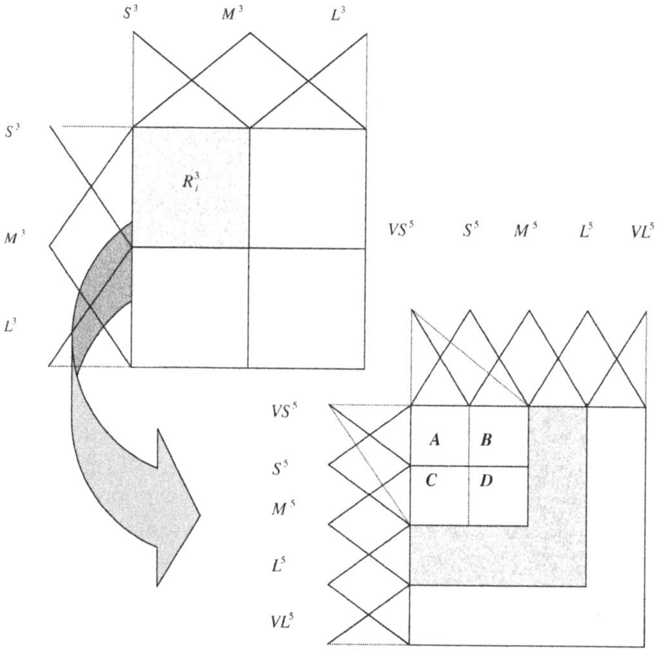

$R_i^3 = IF\ x_1\ is\ S^3\ AND\ x_2\ is\ S^3\ THEN\ y\ is\ S^3$

$R_{i1}^5 = IF\ x_1\ is\ VS^5\ AND\ x_2\ is\ VS^5\ THEN\ y\ is\ F(A)$

$R_{i2}^5 = IF\ x_1\ is\ VS^5\ AND\ x_2\ is\ S^5\ THEN\ y\ is\ F(A \cup B)$

$R_{i3}^5 = IF\ x_1\ is\ S^5\ AND\ x_2\ is\ VS^5\ THEN\ y\ is\ F(A \cup C)$

$R_{i4}^5 = IF\ x_1\ is\ S^5\ AND\ x_2\ is\ S^5\ THEN\ y\ is\ F(A \cup B \cup C \cup D)$

$F(A),\ F(A \cup B),\ F(A \cup C),$
$F(A \cup B \cup C \cup D) \in D_y(2,5)$

Fig. 2 Example of the HRB Generation Process.

Therefore, it is possible to obtain at most four *2-linguistic rules* generated by the LRG-method from the expanded R_i^3:

$$LRG(I(R_i^3), E_i) = \{R_{i1}^5, R_{i2}^5, R_{i3}^5, R_{i4}^5\}$$

This example is graphically showed in Fig. 2. In the same way, other bad performance neighbor rules could be expanded simultaneously.

Step 3. **HRB Selection.** *Simplify the set JCLR by removing the unnecessary rules from it and generating an HRB with good cooperation.* In JCLR –where rules of different hierarchical layers coexist–, it may happen that a complete set of *2-linguistic rules* which replaces an expanded *1-linguistic rule* does not produce good results. However, a subset of this set of *2-linguistic rules* may work properly. A genetic process is considered to put this task into effect, which is explained on detail in the next subsection.

$$HRB = Selection\,\mathrm{Pr}\,ocess(JCLR)$$

After applying this algorithm, the HKB is obtained as:

$$HKB = HDB + HRB$$

3.2 Linguistic Modeling with Hierarchical Systems of Weighted Linguistic Rules

In (Alcalá et al, 2003), Alcalá et al. proposed the hybridization of the hierarchical scheme with the use of rule weights by extending the two-level HSLR-LM proposed in (Cordón et al, 2002). The resulting Hierarchical System of Weighted Linguistic Rules (HSWLR), presents a model structure which is extended by permitting the use of weighted hierarchical linguistic rules. Besides, the summarization component –which has the aim of selecting the subset of rules best cooperating among the rules generated to obtain the final HKB– was modified by allowing it to jointly perform the rule selection and the rule weight derivation. A GA (Michalewicz, 1996) performing the rule selection together with the derivation of rule weights was developed for this task.

Hence, this extended methodology was intended as a meta-method over any other LRG-method, developed to improve simple linguistic fuzzy models by only reinforcing the modeling of those problem subspaces with more difficulties whereas the use of rule weights improved the way in which they interact. This extension of the learning methodology was named two-level HSWLR Learning Methodology (HSWLR-LM) and consists of two modifications:

- *Modification of the HRB structure and Inference System*, in order to consider the use of weights, obtaining Weighted HKB s (WHKB s).
- *Modification of the rule selection process (Step 3 of the two-level HSLR-LM algorithm)*, to consider the derivation of rule weights.

Weighted Hierarchical Knowledge Base

In this case, only the rule structure in the HRB has to be modified. The same structure of the weighted linguistic rules will be used to form the Weighted HRB (WHRB) and then the WHKB:

$$WHKB = HDB + WHRB$$

Therefore, the fuzzy reasoning process must be extended as in the case of weighted linguistic rules, considering the matching degree of the rules fired.

In this way, we can define the WHRB as a whole HRB together with their corresponding rule weights:

$$WHKB = \bigcup_t RB^t + \bigcup_t W^t .$$

with W^t being the set of weights associated to the rules from layer t. We should notice that these weights are obtained over the whole HRB (and not over the isolated layers) since they must consider the way in which all the rules interact, i.e., the weights considered in the different layers, W^t, are interdependent. Therefore, *they must be jointly derived once the whole HRB is available.*

Algorithm

The same operation mode of the two-level HSLR-LM algorithm will be followed to generate linguistic fuzzy models with this new structure, but including the weight learning. Again, the Wang and Mendel's algorithm (Wang and Mendel, 1992) was considered as LRG-method to obtain simple linguistic fuzzy models, although any other technique could be used. Therefore, the main steps of the extended algorithm are the following ones:

HIERARCHICAL KNOWLEDGE BASE GENERATION PROCESS

Step 0. $RB(1, n(1))$ *Generation Process*

Step 1. $RB(2, 2 \cdot n(1) - 1)$ *Generation Process*

Step 2. *Summarization Process* → **(Extract Repeated (JCLR) + Weights).**

HIERARCHICAL RULE BASE GENETIC SELECTION PROCESS

Step 3. Genetic Weight Derivation and Rule Selection Process

- Genetic selection of a subset of rules presenting good cooperation.
- Genetic derivation of the weights associated to these rules.

Fig. 3 presents the flowchart of this algorithm. Specifically, at Step 3 of the two-level HSWLR-LM, a GA with double coding scheme ($C = C_1 + C_2$) was employed for both *rule selection* and *weight derivation*.

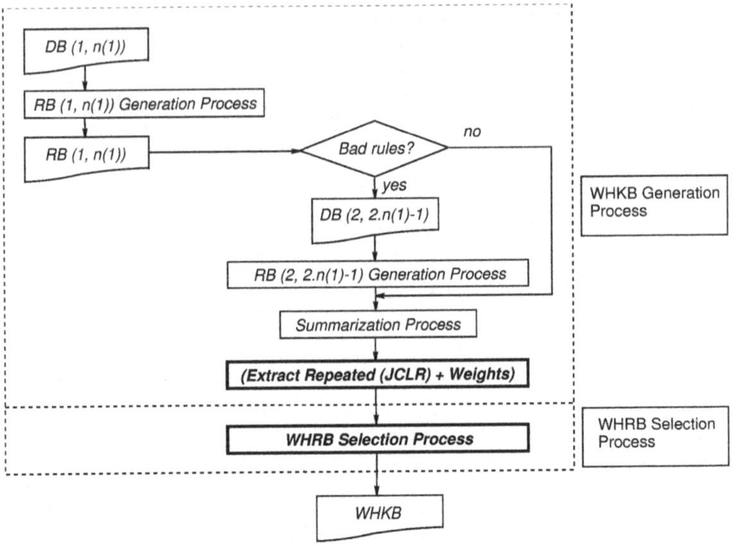

Fig. 3 HSWLR Learning Methodology.

3.3 *An Iterative Methodology for Hierarchical Linguistic Modeling*

In the beginning of this section we have introduced a basic approach to develop hierarchical models from a limited HKB focused on interpretability: the HSLRs of two levels (Cordón et al, 2002). In (Cordón et al, 2003), Cordón et al. extended the former model structure, i.e., the HKB, and proposed an iterative HSLR learning methodology to learn it from examples. As the name suggests, this methodology iteratively selects bad performance linguistic rules, which need more specificity, and expands them locally through different granularity levels. This fact produces a wide spectrum of solutions—from high interpretable to high accurate, and tradeoff solutions—and avoids typical drawbacks of prototype-based linguistic rule generation methods (LRG-methods).

The iterative HSLR-LM was developed as a parametrised methodology. The factor of expansion controls the level of bad performance that a rule should have to be expanded into more specific ones. Thus, a low factor implies a small expansion, a smaller number of rules, and a more interpretable model. In this sense, the basic approach (Cordón et al, 2002) is a special case which makes use of this parameter to obtain interpretable hierarchical models. Another parameter to be considered is the iteration of the algorithm. It is used to control the granularity level that more specific hierarchical rules, which replace those ones with bad performance, should have (see Fig. 4).

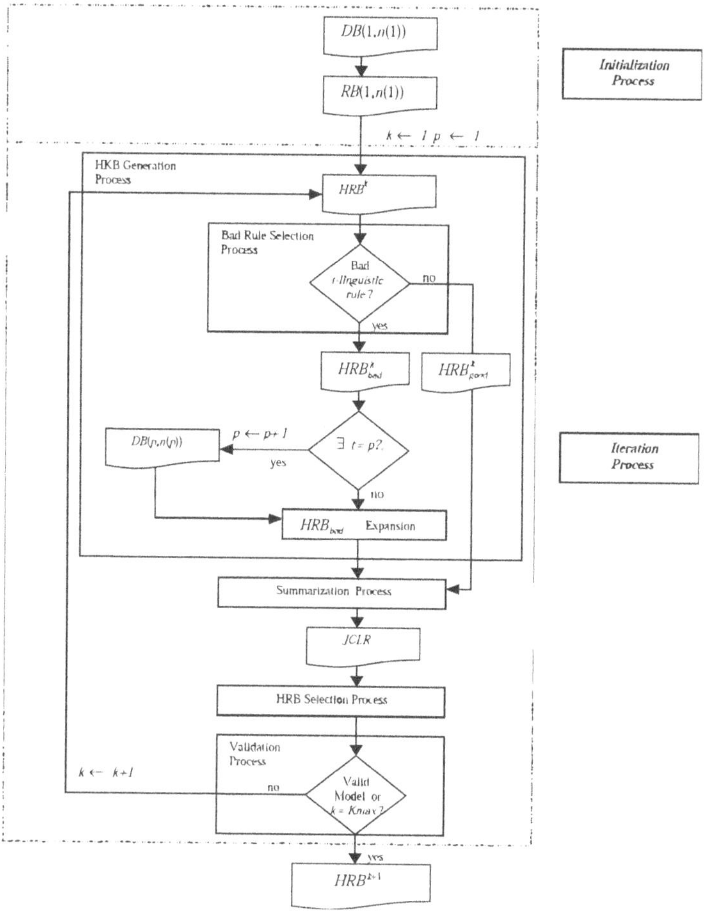

Fig. 4 Algorithm of the iterative HSLR-LM design process.

As the reader may have already noticed, the main change in the structure of the algorithm presented in the beginning of this section is the enabling of having several granularity levels and therefore to iterate steps 1 to 3 several times depending on a parameter k. According to this, it performs gradual and local-oriented refinements on problem subspaces that are badly modeled by previous models rather than in the whole problem domain. Furthermore, it integrates the improved local behavior with the whole model by summarization processes which ensure a good global performance.

4 On the Use of Hierarchical Fuzzy Rule Based Classification Systems on Imbalanced Data-Sets

In this section we will address a new and interesting problem, named as classification with imbalanced data-sets (He and Garcia, 2009; Sun et al, 2009) which consists in learning from highly skewed data, in general terms. Indeed, this problem has been recently identified as one of the hot topics in data mining (Yang and Wu, 2006) and therefore we must emphasise its significance and guide our efforts in dealing with this type of applications.

Regarding linguistic fuzzy systems, in a previous study of some of the authors we have shown the goodness of this type of models to deal with imbalanced data-sets (Fernández et al, 2008). In this section we will introduce a solution we have provided to improve the accuracy of Fuzzy Rule Based Classification Systems (FRBCS) in this framework (Fernández et al, 2009). Specifically we proposed a hierarchical environment, by means of the HSLR-LM described in the previous section, by increasing the granularity of the fuzzy partitions on the boundary areas between the classes, in order to obtain a better separability.

In order to do so, we will first introduce in this section the scenario of classification with imbalanced data-sets, that is, its main features and how to deal with this problem. Then, we will describe how to adapt the HSLR-LM to this specific framework.

4.1 An Introduction to Classification with Imbalance Data-Sets

Learning from imbalanced data is an important topic that has recently appeared in the Machine Learning community (Chawla et al, 2004; He and Garcia, 2009; Sun et al, 2009). This problem is very representative since it appears in a variety of real-world applications including, but not limited to, medical applications, finance, telecommunications, biology and so on.

In this framework, the class distribution is not uniform, resulting on a high number of examples belonging to one or more classes and a few from the rest. The minority classes are usually associated to the most interesting concepts from the point of view of learning and, due to that fact, the cost derived from a misclassification of one of the examples of these classes is higher than that of the majority classes. In this work we will focus on binary problems where there is just one positive and negative class.

Standard classifier algorithms usually have a bias towards the majority class, since the rules which predict the higher numbers of examples are positively weighted during the learning process in favour of the standard accuracy rate metric, which does not take into account the class distribution of the data. Consequently, the instances belonging to the minority class are misclassified more often than those belonging to the majority class.

Another important issue of this problem are the small disjuncts that can be found in the data set (Weiss and Provost, 2003) and the difficulty of most learning algorithms in detecting these areas (Orriols-Puig et al, 2009; Orriols-Puig and Bernadó-Mansilla, 2009). In fact, learning algorithms try to benefit those models with a higher degree of coverage and these small disjuncts imply the application of very specific models which are discarded in favour or more general ones.

Furthermore, another handicap of imbalanced data sets, which is related to the apparition of small disjuncts, is the overlapping between the examples of the positive and the negative class (García et al, 2008), in which the minority class examples can be simply treated as noise and ignored by the learning algorithm. These phenomena are depicted in Figure 5.a and 5.b respectively.

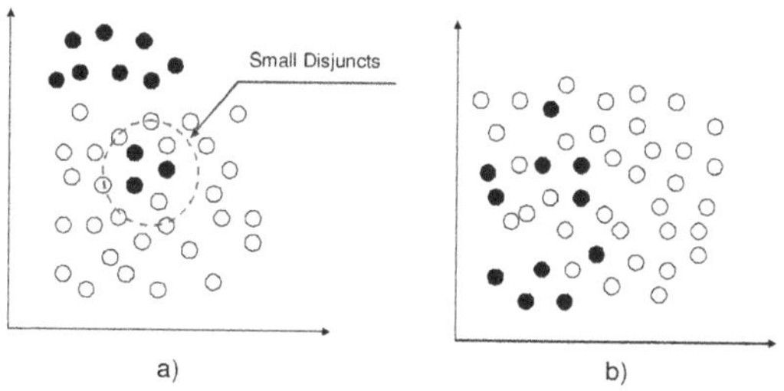

Fig. 5 Example of the imbalance between classes: a) small disjuncts b) overlapping between classes

A large number of approaches have previously been proposed for dealing with the class-imbalance problem. These approaches can be categorised into two groups: the internal approaches which create new algorithms or modify existing ones to take the class-imbalance problem into consideration (Barandela et al, 2003; Wu and Chang, 2005; Xu et al, 2007) and the external approaches which pre-process the data in order to diminish the effects of their class imbalance (Batista et al, 2004; Estabrooks et al, 2004). Furthermore, cost-sensitive learning solutions incorporating both the data and algorithmic level approaches assume higher misclassification costs with samples in the minority class and seek to minimise the high cost errors (Domingos, 1999; Zhou and Liu, 2006; Sun et al, 2007).

The great advantage of the external approaches is that they are more versatile, since their use is independent of the classifier selected. Furthermore, we may preprocess all data sets beforehand in order to use them to train different classifiers. In this manner, the computation time needed to prepare the data is lower.

Specifically, in the framework of fuzzy classification we analyzed the cooperation of some preprocessing methods with FRBCSs (Fernández et al, 2008), showing a good behaviour for the oversampling methods, especially in the case of the SMOTE methodology (Chawla et al, 2002).

4.2 *Adaptation of the Hierarchical Learning Process for Imbalanced Data*

As we stated previously, the framework of imbalanced data-sets requires some specific adaptations for the algorithms in order to obtain a good performance for both classes. In the remaining of this section we will first introduce the type of rules, rule weights and inference model used for standard classification tasks. Next, we will highlight the main changes carried out in the hierarchical learning process for its application to imbalanced problems.

Fuzzy Rule Based Classification Systems

Any classification problem consists of m training patterns $x_p = (x_{p1}, \ldots, x_{pn})$, $p = 1,2,\ldots,m$ from M classes where x_{pi} is the ith attribute value ($i = 1,2,\ldots,n$) of the p-th training pattern.

In this work we use fuzzy rules of the following form for our FRBCSs:

Rule R_j : If x_1 is A_{j1} and $\ldots$ and x_n is A_{jn} then Class = C_j with RW_j (8)

where R_j is the label of the jth rule, $x = (x_1, \ldots, x_n)$ is an n-dimensional pattern vector, A_{ji} is an antecedent fuzzy set (we use triangular membership functions), C_j is a class label, and RW_j is the rule weight.

In the specialised literature rule weights have been used in order to improve the performance of FRBCSs (Ishibuchi and Nakashima, 2001). For the framework of imbalanced data-sets and following the conclusions extracted in (Fernández et al, 2008), as heuristic method for the rule weight the Penalised Certainty Factor (Ishibuchi and Yamamoto, 2006) was employed:

$$RW_j = \frac{\sum_{x_p \in ClassC_j} \mu_{A_j}(x_p)}{\sum_{p=1}^{m} \mu_{A_j}(x_p)} - \frac{\sum_{x_p \notin ClassC_j} \mu_{A_j}(x_p)}{\sum_{p=1}^{m} \mu_{A_j}(x_p)} \tag{9}$$

The Fuzzy Reasoning Method of the winning rule (classical approach) (Cordón et al, 1999) was used for classifying new patterns by the RB. The single winner rule R_w is determined for a new pattern $x_p = (x_{p1}, \ldots, x_{pn})$ as

$$\mu_w(x_p) \cdot RW_w = \max\{\mu_j(x_p) \cdot RW_j\}; x_p \in X, j = 1 \ldots L \tag{10}$$

The new pattern x_p is classified as Class C_w , which is the consequent class of the winner rule R_w . If multiple fuzzy rules have the same maximum value but different consequent classes for the new pattern x_p in the previous equation, the classification of x_p is rejected. The classification is also rejected if no fuzzy rule is compatible with the new pattern x_p.

Two-Level Learning Method for Building HFRBCSs in Imbalanced Domains

The main scheme of the two-level HSLR-LM was maintained in this case, following exactly the same steps. In this case, as LRG-method the Chi et al. (Chi et al, 1996) approach must be considered for building a FRBCS.

The main change refers to the measures of error are used in the algorithm used both to evaluate the complete RB and to determine if an individual rule is expanded. Their expressions are defined below:

1. **Global measure.** We will employ the accuracy per class, computed as:

$$Acc_i(X_i, RB) = \frac{|\{x_p \in X_i / FRM(x_p, RB) = Class(x_p)\}|}{|X_i|} \tag{11}$$

where $|\cdot|$ is the number of patterns, with X_i being the subset of examples of the i-th class ($i \in 1 \ldots M$), $FRM(x_p, RB)$ is the output class computed following the fuzzy reasoning process using the current RB and $Class(x_p)$ is the class label for example x_p.

2. **Local measure.** The accuracy for a simple rule, $R_j^{n(1)}$, calculated over X, is showed as follows:

$$Acc(X, R_j^{n(1)}) = \frac{|X^+(R_j^{n(1)})|}{|X(R_j^{n(1)})|} \tag{12}$$

$$X^+(R_j^{n(1)}) = \{x_p \in X / \mu_{R_j^{n(1)}}(x_p) > 0.and.Class(x_p) = Class(R_j^{n(1)})\} \tag{13}$$

$$X(R_j^{n(1)}) = \{x_p \in X / \mu_{R_j^{n(1)}}(x_p) > 0\} \tag{14}$$

where $Class(\cdot)$ is a function that provides the class label for a pattern, or for a rule. We must note that $X^+(R_j^{n(1)})$ and $X(R_j^{n(1)})$ only include those examples that the rule actually classifies, because we are using as Fuzzy Reasoning Method the winning rule approach.

Additionally, another significant modification must be developed in the HRB genetic selection. Specifically, during the chromosome evaluation the fitness function must be in accordance with the framework of imbalanced data-sets and therefore the geometric mean of the true rates (Barandela et al, 2003) was used. This metric defined as:

$$GM = \sqrt{\frac{TP}{TP+FN} \cdot \frac{TN}{FP+TN}} \tag{15}$$

where TP and TN are the true rate for the positive and negative instances and FP and FN the rate for the false positives and negatives respectively. This metric attempts to maximise the accuracy of each one of the two classes with a good balance, being a performance metric that links both objectives.

5 Case of Study: Hierarchical Fuzzy Rule Based Classification Systems for Imbalanced Data-Sets

In the previous part of this work we have introduced the problem of imbalanced data-sets and the solution proposed to increase the performance of linguistic FRBCSs by means of the use of the HSLR-LM adapted for imbalanced problems (Fernández et al, 2009). Now, this section has the aim of presenting a case of study in which we show the goodness of the proposed methodology in contrast with other well-known fuzzy approaches and with C4.5 (Quinlan, 1993).

According to this, we will first introduce the configuration of the two-level learning method, determining all the parameters used in this experimental study and the selected benchmark data-sets. Next, we will present the statistical tests used in all our analysis. Finally, we will analyze the results of the HFRBCS when applied to imbalanced data-sets globally, and considering two different degrees of imbalance. This last part of the study is divided into two sections:

- A comparative study is carried out between the HFRBCS model and other fuzzy learning methodologies, including Chi et al.'s (Chi et al, 1996) and Ishibuchi et al.'s (Ishibuchi and Yamamoto, 2005) rule learning algorithms, and an approach proposed by Xu et al. for imbalanced data-sets, called E-Algorithm (Xu et al, 2007).
- The performance of the HFRBCSs is compared against the well-known C4.5 algorithm (Quinlan, 1993) as a well-known classifier that has been widely used for this type of problems (Batista et al, 2004; Estabrooks et al, 2004; Orriols-Puig et al, 2009; Su et al, 2006; Su and Hsiao, 2007; Sun et al, 2007).

5.1 Experimental Set-Up: Parameters and Data-Sets

In our former studies (Fernández et al, 2008) we selected as a good FRBCS model the use of the product T-norm as conjunction operator, together with the Penalised Certainty Factor (Ishibuchi and Yamamoto, 2005) approach for the rule weight and Fuzzy Reasoning Method of the winning rule. This configuration will be employed for all the FRBCSs used in this work, including Chi et al.'s method, Ishibuchi et al.'s approach and E-Algorithm.

After several trials, we selected the following values for the parameters in the learning method for building HFRBCSs:

- Rule Generation:
    - $\delta, n(t+1)$-linguistic partition} terms selector: 0.1
    - α, used to decide the expansion of the rule: 0.2
- GA Selection:
    - Number of evaluations: 10,000
    - Population length: 61

In order to reduce the effect of imbalance, we will employ the SMOTE preprocessing method (Chawla et al, 2002) for our experiments, consider only the

1-nearest neighbour to generate the synthetic samples (using the euclidean distance), and we balance the training data to the 50% class distribution. The E-Algorithm is always applied without preprocessing.

For Ishibuchi et al.'s rule generation method and E-Algorithm, only rules with three or less antecedent attributes are generated. Furthermore we have restricted the number of fuzzy rules in the RB to 30 for each class, using as selection measure the product of support and confidence. This configuration is the one indicated by the authors in (Ishibuchi and Yamamoto, 2005; Xu et al, 2007).

Table 2 Summary Description for imbalanced data-sets.

Data-set	**#Ex.**	**#Atts.**	**Class(min.,maj.)**	**%Class(min.; maj.)**	**IR**
Data-sets with Low Imbalance (IR 1.5 to 9)					
Glass1	214	9	(build-win-non_float-proc;remainder)	(35.51, 64.49)	1.82
Ecoli0vs1	220	7	(im;cp)	(35.00, 65.00)	1.86
Wisconsin	683	9	(malignant;benign)	(35.00, 65.00)	1.86
Pima	768	8	(tested-positive;tested-negative)	(34.84, 66.16)	1.90
Iris0	150	4	(Iris-Setosa;remainder)	(33.33, 66.67)	2.00
Glass0	214	9	(build-win-float-proc;remainder)	(32.71, 67.29)	2.06
Yeast1	1484	8	(nuc;remainder)	(28.91, 71.09)	2.46
Vehicle1	846	18	(Saab;remainder)	(28.37, 71.63)	2.52
Vehicle2	846	18	(Bus;remainder)	(28.37, 71.63)	2.52
Vehicle3	846	18	(Opel;remainder)	(28.37, 71.63)	2.52
Haberman	306	3	(Die;Survive)	(27.42, 73.58)	2.68
Glass0123vs456	214	9	(non-windowglass;remainder)	(23.83, 76.17)	3.19
Vehicle0	846	18	(Van;remainder)	(23.64, 76.36)	3.23
Ecoli1	336	7	(im;remainder)	(22.92, 77.08)	3.36
New-thyroid2	215	5	(hypo;remainder)	(16.89, 83.11)	4.92
New-thyroid1	215	5	(hyper;remainder)	(16.28, 83.72)	5.14
Ecoli2	336	7	(pp;remainder)	(15.48, 84.52)	5.46
Segment0	2308	19	(brickface;remainder)	(14.26, 85.74)	6.01
Glass6	214	9	(headlamps;remainder)	(13.55, 86.45)	6.38
Yeast3	1484	8	(me3;remainder)	(10.98, 89.02)	8.11
Ecoli3	336	7	(imU;remainder)	(10.88, 89.12)	8.19
Page-blocks0	5472	10	(remainder;text)	(10.23, 89.77)	8.77
Data-sets with High Imbalance (IR higher than 9)					
Yeast2vs4	514	8	(cyt;me2)	(9.92, 90.08)	9.08
Yeast05679vs4	528	8	(me2;mit,me3,exc,vac,erl)	(9.66, 90.34)	9.35

Table 2 (*Continued*)

Vowel0	988	13	(hid;remainder)	(9.01, 90.99)	10.10
Glass016vs2	192	9	(ve-win-float-proc;build-win-float-proc,build-win-non_float-proc,headlamps)	(8.89, 91.11)	10.29
Glass2	214	9	(Ve-win-float-proc;remainder)	(8.78, 91.22)	10.39
Ecoli4	336	7	(om;remainder)	(6.74, 93.26)	13.84
Yeast1vs7	459	8	(vac;nuc)	(6.72, 93.28)	13.87
Shuttle0vs4	1829	9	(RadFlow;Bypass)	(6.72, 93.28)	13.87
Glass4	214	9	(containers;remainder)	(6.07, 93.93)	15.47
Page-blocks13vs2	472	10	(graphic;horiz.line,picture)	(5.93, 94.07)	15.85
Abalone9vs18	731	8	(18;9)	(5.65, 94.25)	16.68
Glass016vs5	184	9	(tableware;build-win-float-proc,build-win-non_float-proc,headlamps)	(4.89, 95.11)	19.44
Shuttle2vs4	129	9	(FpvOpen;Bypass)	(4.65, 95.35)	20.5
Yeast1458vs7	693	8	(vac;nuc,me2,me3,pox)	(4.33, 95.67)	22.10
Glass5	214	9	(tableware;remainder)	(4.20, 95.80)	22.81
Yeast2vs8	482	8	(pox;cyt)	(4.15, 95.85)	23.10
Yeast4	1484	8	(me2;remainder)	(3.43, 96.57)	28.41
Yeast1289vs7	947	8	(vac;nuc,cyt,pox,erl)	(3.17, 96.83)	30.56
Yeast5	1484	8	(me1;remainder)	(2.96, 97.04)	32.78
Ecoli0137vs26	281	7	(pp,imL;cp,im,imU,imS)	(2.49, 97.51)	39.15
Yeast6	1484	8	(exc;remainder)	(2.49, 97.51)	39.15
Abalone19	4174	8	(19;remainder)	(0.77, 99.23)	128.87

The Imbalance Ratio (IR) (Orriols-Puig et al, 2009), defined as the ratio of the number of instances of the majority class and the minority class, is used as a threshold to categorise the different imbalanced scenarios: data-sets with a *low imbalance* when the instances of the positive class are between 10 and 40% of the total instances (IR between 1.5 and 9) and data-sets with a *high imbalance* where there are no more than 10% of positive instances in the whole data-set compared to the negative ones (IR higher than 9).

Specifically, we have considered forty-four data sets from UCI repository (Asuncion and Newman, 2007) with different IR. Table 2 summarises the data employed in this study and shows, for each data set, the number of examples (#Ex.), number of attributes (#Atts.), class name of each class (minority and majority), class attribute distribution and IR. This table is ordered by the IR, from low to highly imbalanced data-sets.

In order to develop the study, we use a five fold cross validation approach, that is, five partitions for training and test sets, 80% for training and 20% for test, where the five test data-sets form the whole set. For each data-set we consider the average results of the five partitions.

5.2 Statistical Tests for Evaluation

In this paper, we use the hypothesis testing techniques to provide statistical support to the analysis of the results (García et al, 2009; Sheskin, 2006). Specifically, we will use non-parametric tests due to the fact that the initial conditions that guarantee the reliability of the parametric tests may not be satisfied, causing the statistical analysis to lose credibility with these parametric tests (Demšar, 2006).

We will use the Wilcoxon signed-rank test (Wilcoxon, 1945) as a non-parametric statistical procedure for performing pairwise comparisons between two algorithms. For multiple comparisons we use the Iman-Davenport test (Sheskin, 2006) to detect statistical differences among a group of results, and the Holm post-hoc test (Holm, 1979) in order to find which algorithms are distinctive among a $1 \times n$ comparison.

The post-hoc procedure allows us to know whether a hypothesis of comparison of means could be rejected at a specified level of significance α. However, it is very interesting to compute the p-value associated to each comparison, which represents the lowest level of significance of a hypothesis that results in a rejection. In this manner, we can know whether two algorithms are significantly different and how different they are.

Furthermore, we consider the average ranking of the algorithms in order to show graphically how good a method is with respect to its partners. This ranking is obtained by assigning a position to each algorithm depending on its performance for each data set. The algorithm which achieves the best accuracy on a specific data set will have the first ranking (value 1); then, the algorithm with the second best accuracy is assigned rank 2, and so forth. This task is carried out for all data sets and finally an average ranking is computed as the mean value of all rankings.

These tests are suggested in the studies presented in (Demšar, 2006; García and Herrera, 2008; García et al, 2009, 2010), where their use in the field of Machine Learning is highly recommended. For a wider description of the use of these tests, any interested reader can find additional information on the Website http://sci2s.ugr.es/sicidm/, together with the software for their application.

5.3 Experimental Study

In this part of the study we will focus on determining whether the HFRBCS is robust in the framework of imbalanced data-sets and if it improves the performance of other FRBCSs approaches and the well known C4.5 algorithm. Following this idea, Table 3 shows the results for the test partitions for each FRBCS method with its associated standard deviation. Specifically, by columns we include the Chi et al.'s method with 3 and 5 labels (Chi-3 and Chi-5), the Ishibuchi et al.'s method (Ishibuchi05), the E-Algorithm and the HFRBCS. Additionally, we include the results for the C4.5 decision tree. This table is divided by the IR, on the one hand data-sets with low imbalance and, on the other hand, data-sets with high imbalance. The best global result for test is stressed in **boldface** in each case.

Table 3 Detailed table of results for FRBCSs in imbalanced data-sets. Only test results are shown.

Data-set	Chi-3	Chi-5	Ishibuchi05	E-Algorithm	HFRBCS	C4.5
Data-Sets with Low Imbalance						
Glass1	64.90 ± 6.91	64.91 ± 6.87	59.29 ± 10.33	0.00 ± 0.00	73.66 ± 4.66	75.11 ± 3.74
Ecoli0vs1	92.27 ± 5.93	95.56 ± 5.15	96.70 ± 2.40	95.25 ± 4.75	93.63 ± 6.45	97.95 ± 2.20
Wisconsin	88.91 ± 2.13	43.58 ± 5.86	95.78 ± 1.38	96.01 ± 1.55	88.24 ± 1.63	95.44 ± 2.01
Pima	66.80 ± 5.93	66.78 ± 2.28	71.10 ± 4.45	55.01 ± 4.64	68.72 ± 5.26	71.26 ± 4.05
Iris0	100.0 ± 0.00	98.97 ± 2.29	100.0 ± 0.00	100.0 ± 0.00	100.0 ± 0.00	98.97 ± 2.29
Glass0	64.06 ± 3.51	63.69 ± 1.80	69.39 ± 7.70	0.00 ± 0.00	76.57 ± 8.05	78.14 ± 2.21
Yeast1	67.69 ± 1.91	69.66 ± 1.52	51.41 ± 12.18	0.00 ± 0.00	71.71 ± 2.39	70.86 ± 2.95
Vehicle1	70.92 ± 4.34	71.88 ± 1.25	64.89 ± 4.37	3.09 ± 6.90	71.76 ± 2.64	69.28 ± 3.41
Vehicle2	85.54 ± 3.36	87.19 ± 3.04	67.82 ± 4.95	43.83 ± 13.17	90.61 ± 2.17	94.85 ± 1.68
Vehicle3	69.22 ± 4.89	63.13 ± 1.95	63.12 ± 4.06	0.00 ± 0.00	66.80 ± 3.34	74.34 ± 1.08
Haberman	58.91 ± 6.03	60.40 ± 2.40	62.65 ± 2.84	4.94 ± 11.06	57.08 ± 4.09	61.32 ± 3.85
Glass0123vs456	85.83 ± 3.04	85.94 ± 1.66	88.56 ± 5.18	82.09 ± 6.96	88.37 ± 3.97	90.13 ± 3.17
Vehicle0	86.41 ± 3.06	84.93 ± 1.61	75.94 ± 1.42	39.07 ± 16.49	88.92 ± 1.96	91.10 ± 2.70
Ecoli1	85.28 ± 9.77	86.05 ± 8.57	85.71 ± 2.86	77.81 ± 7.90	84.18 ± 12.69	76.10 ± 9.58
New-Thyroid2	89.81 ± 10.77	96.34 ± 6.65	94.21 ± 4.23	88.57 ± 3.82	99.72 ± 0.63	96.51 ± 4.87
New-Thyroid1	87.44 ± 8.11	95.38 ± 8.80	89.02 ± 13.52	88.52 ± 8.79	98.58 ± 2.48	97.98 ± 3.79
Ecoli2	88.01 ± 5.45	87.64 ± 4.96	87.00 ± 4.43	70.35 ± 15.36	87.62 ± 8.24	91.60 ± 4.86
Segment0	94.99 ± 0.45	95.88 ± 1.21	42.47 ± 2.79	95.33 ± 1.14	97.51 ± 1.11	99.26 ± 0.61
Glass6	83.87 ± 9.82	78.13 ± 7.78	86.27 ± 8.19	90.23 ± 3.77	86.95 ± 10.84	83.00 ± 9.05
Yeast3	90.13 ± 4.09	89.33 ± 3.30	77.06 ± 17.73	81.99 ± 2.28	90.41 ± 2.34	88.50 ± 3.66
Ecoli3	87.58 ± 4.08	91.61 ± 4.95	85.39 ± 3.70	78.54 ± 8.68	90.81 ± 4.43	88.77 ± 7.65
Page-Blocks0	79.91 ± 4.29	87.25 ± 1.94	32.16 ± 9.61	64.51 ± 2.79	91.40 ± 0.67	94.84 ± 1.52
Mean	81.29 ± 4.90	80.19 ± 3.90	74.81 ± 5.83	57.05 ± 5.46	84.69 ± 4.09	**85.70 ± 3.68**
Data-Sets with High Imbalance						
Yeast2vs4	86.80 ± 5.53	86.39 ± 7.35	70.85 ± 23.45	80.92 ± 9.09	89.32 ± 4.18	85.09 ± 10.15
Yeast05679vs4	78.91 ± 5.99	75.99 ± 6.39	79.49 ± 9.54	59.99 ± 16.44	73.18 ± 7.47	74.88 ± 10.88
Vowel0	98.37 ± 0.61	97.87 ± 1.84	89.03 ± 6.63	89.63 ± 6.09	98.82 ± 1.62	94.74 ± 5.22
Glass016vs2	40.84 ± 7.62	56.17 ± 5.16	41.18 ± 15.40	0.00 ± 0.00	58.37 ± 20.04	48.91 ± 29.44
Glass2	47.67 ± 10.16	49.24 ± 8.19	43.55 ± 15.70	9.87 ± 22.07	54.84 ± 20.57	33.86 ± 32.29
Ecoli4	91.27 ± 7.43	92.11 ± 8.35	86.92 ± 8.65	92.43 ± 8.24	93.02 ± 8.17	81.28 ± 11.67

Table 3 (*Continued*)

Yeast1vs7	80.05 ± 6.43	63.02 ± 12.62	53.15 ± 10.35	27.55 ± 26.06	70.74 ± 12.40	67.73 ± 2.28
Shuttle0vs4	99.12 ± 1.15	98.71 ± 1.18	99.16 ± 1.15	98.40 ± 1.26	99.12 ± 1.15	99.97 ± 0.07
Glass4	84.96 ± 13.80	81.75 ± 11.24	78.27 ± 17.70	83.38 ± 19.89	70.39 ± 40.49	83.71 ± 10.78
Page-Blocks13vs4	91.92 ± 4.76	92.93 ± 9.48	94.53 ± 4.88	94.12 ± 10.33	98.64 ± 0.65	99.55 ± 0.47
Abalone9-18	63.93 ± 11.00	66.47 ± 10.67	65.78 ± 9.23	32.29 ± 20.61	67.56 ± 14.01	53.19 ± 8.25
Glass016vs5	71.48 ± 40.17	75.59 ± 42.27	88.77 ± 2.48	65.14 ± 39.41	77.96 ± 43.61	72.08 ± 42.33
Shuttle2vs4	89.99 ± 8.61	78.34 ± 43.87	99.17 ± 1.13	100.0 ± 0.00	97.49 ± 2.71	99.15 ± 1.90
Yeast1458vs7	62.40 ± 4.55	58.76 ± 8.57	40.80 ± 16.58	0.00 ± 0.00	62.49 ± 6.26	41.19 ± 6.06
Glass5	81.56 ± 12.65	64.33 ± 38.40	89.96 ± 2.43	50.61 ± 47.17	68.73 ± 39.56	86.70 ± 15.44
Yeast2vs8	72.75 ± 14.99	78.76 ± 8.60	72.83 ± 14.97	72.83 ± 14.97	72.47 ± 15.10	78.23 ± 13.05
Yeast4	82.99 ± 3.10	83.07 ± 2.58	71.36 ± 23.29	32.16 ± 20.59	82.64 ± 2.29	65.00 ± 8.94
Yeast1289vs7	76.12 ± 7.24	69.26 ± 4.57	48.55 ± 16.86	50.00 ± 13.62	69.37 ± 4.37	64.13 ± 9.00
Yeast5	93.41 ± 5.35	93.64 ± 2.70	94.94 ± 0.38	88.17 ± 7.04	94.20 ± 2.59	92.04 ± 4.99
Ecoli0137vs26	71.04 ± 41.38	49.57 ± 46.41	71.31 ± 41.65	73.65 ± 43.09	71.48 ± 41.80	71.21 ± 41.31
Yeast6	87.50 ± 10.55	87.73 ± 9.32	88.42 ± 6.06	51.72 ± 13.76	84.92 ± 12.88	80.38 ± 15.47
Abalone19	62.96 ± 8.27	66.71 ± 10.21	66.09 ± 9.40	0.00 ± 0.00	70.19 ± 8.56	15.58 ± 21.36
Mean	78.00 ± 10.51	75.75 ± 13.63	74.28 ± 11.72	56.95 ± 15.44	**78.45 ± 14.11**	72.21 ± 13.70
All Data-Sets						
Mean	79.65 ± 7.71	77.97 ± 8.77	74.55 ± 8.78	57.00 ± 10.45	**81.67 ± 9.10**	78.95 ± 8.69

This study is divided into two parts. First, we will analyze the results globally for all imbalanced data-sets and then, we will study the two imbalance scenarios defined in this paper. Furthermore, our aim is to test the HFRBCS against the FRBCSs approaches and C4.5 separately.

Global Analysis of the Hierarchical Fuzzy Rule Based Classification System

First of all, we will study the performance of the HFRBCS with the remaining FRBCSs approaches. In order to compare the results, we will use a multiple comparison test to find the best approach in this case, considering the results in the test partitions (GM_{Tst}). The results of Iman-Davenport tests informs us of the rejection of the null hypothesis of equality of means (p-value near to zero), telling us of the existence of significant differences among the observed results in all data-sets. Next, Table 4 shows the rankings of the 5 algorithms considered.

Now, we apply a Holm test to compare the best ranking method (HFRBCS) with the remaining fuzzy methods. The result of this test is shown in Table 5, in which the algorithms are ordered with respect to the z value obtained. Thus, by using the normal distribution, we can obtain the corresponding p-value associated with each comparison and this can be compared with the associated α/i in the same row of the table to show whether the associated hypothesis of equal behaviour is rejected in favour of the best ranking algorithm or not.

Table 4 Rankings obtained through a Friedman test for FRBCSs in all imbalance data-sets.

Algorithm	Ranking
HFRBCS	2.09091
Chi-5	2.77273
Chi-3	3.00000
Ishibuchi05	3.02273
E-Algorithm	4.11364

Table 5 Holm test Table for FRBCSs in all imbalanced data-sets. HFRBCS is the control method.

i	algorithm	z	p	α/i	Hypothesis
4	E-Algorithm	6.00038	1.96858E-9	0.01250	Rejected for HFRBCS
3	Ishibuchi05	2.76422	0.00576	0.01667	Rejected for HFRBCS
2	Chi-3	2.69680	0.00700	0.02500	Rejected for HFRBCS
1	Chi-5	2.02260	0.04311	0.05000	Rejected for HFRBCS

Therefore, analyzing the results presented in Table 3 and the statistical study shown in Tables 4 and 5 we conclude that our model is a solid FRBCS approach to deal with imbalanced data-sets, as it has shown to be the best performing algorithm when comparing with the remaining fuzzy rule learning methods applied in this study.

Finally, we use a Wilcoxon test for the comparison with the C4.5 algorithm, which is shown in Table 6. We can observe that our proposal achieves a higher ranking, but this is not enough to reject the null hypothesis. We may conclude that both approaches have a similar performance when treating all imbalanced data-sets as a whole, without taking into account the IR.

Table 6 Wilcoxon test to compare the HFRBCS against C4.5 in all imbalanced data-sets. R+ corresponds to HFRBCS and R- to C4.5.

Comparison	R^+	R^-	Hypothesis (α =0.05)	p-value
HFRBCS vs. C4.5	589	401	Not Rejected	0.273

Analysis of the Hierarchical Fuzzy Rule Based Classification System According to the Imbalance Ratio

In the final part of our study, we will analyze the behaviour of our hierarchical approach in each imbalanced scenario. Table 7 shows, by columns, the geometric mean in training and test of the different algorithms considered, for the two types of data-sets, that is, low and high imbalance (IR lower than 9 and higher than 9 respectively). The last column corresponds to the global results. Reader can refer to Table 3, presented in the previous part of this study, where we show the detailed results in each data-set.

Table 7 Results table for FRBCSs for the different degrees of imbalance

Algorithm	Low Imbalance		High Imbalance		All Data-Sets	
	GM_{Tr}	GM_{Tst}	GM_{Tr}	GM_{Tst}	GM_{Tr}	GM_{Tst}
Chi-3	85.50 ± 1.28	81.29 ± 4.90	83.64 ± 2.43	78.00 ± 10.51	84.57 ± 1.86	79.65 ± 7.71
Chi-5	91.31 ± 0.69	80.19 ± 3.90	89.04 ± 1.32	75.75 ± 13.63	90.17 ± 1.01	77.97 ± 8.77
Ishibuchi05	75.45 ± 3.04	74.81 ± 5.83	76.90 ± 6.35	74.28 ± 11.72	76.17 ± 4.70	74.55 ± 8.78
E-Algorithm	58.33 ± 4.09	57.05 ± 5.46	65.72 ± 5.06	56.95 ± 15.44	62.02 ± 4.57	57.00 ± 10.45
HFRBCS	94.30 ± 0.80	84.69 ± 4.09	93.35 ± 1.30	**78.45 ± 14.11**	93.82 ± 1.05	**81.57 ± 9.10**
C4.5	94.95 ± 0.87	**85.70 ± 3.68**	95.81 ± 1.77	72.21 ± 13.70	95.38 ± 1.32	78.95 ± 8.69

The main conclusion extracted from this table is that our HFRBCS is very robust in both imbalanced scenarios considered, as it obtains very competitive results independently of the IR. Next, we will analyze the results in each case, for data-sets with low and high imbalance. As we did in the previous section, we will compare the HFRBCS with the FRBCSs and with the C4.5 decision tree separately.

- **Data-sets with low imbalance:** This study is shown in Tables 9 and 10. First, we check for statistical differences an Iman-Davenport tests obtaining a p-value of 2.98974E-5. Table 9 shows the ranking for the algorithms and Table 8 contains a Holm test, which shows that the HFRBCS is better in performance than the remaining FRBCS unless the Chi et al.'s method with 5 labels.

Table 8 Rankings obtained through a Friedman test for FRBCSs in data-sets with low imbalance.

Algorithm	Ranking
HFRBCS	1.97727
Chi-5	2.63636
Chi-3	3.06818
Ishibuchi05	3.11364
E-Algorithm	4.20454

Table 9 Holm test table for FRBCSs in data-sets with low imbalance. HFRBCS is the control method.

i	algorithm	z	p	α/i	Hypothesis
4	E-Algorithm	4.67197	2.98329E-6	0.01250	Rejected for HFRBCS
3	Ishibuchi05	2.38366	0.01714	0.01667	Rejected for HFRBCS
2	Chi-3	2.28831	0.02212	0.02500	Rejected for HFRBCS
1	Chi-5	1.38252	0.16681	0.05000	Not Rejected

Now, we will compare the performance achieved by our proposal with C4.5 in low imbalanced data-sets by means of a Wilcoxon test, which is shown in Table 10. Furthermore, we compare the HFRBCS with the Chi et al.'s approach with 5 labels in order to check if we find differences between both algorithms.

Table 10 Wilcoxon test to compare the HFRBCS against Chi-5 and C4.5 in data-set with low imbalance. R+ corresponds to HFRBCS and R- to Chi-5 and C4.5 in each case.

Comparison	R^+	R^-	Hypothesis (α =0.05)	p-value
HFRBCS vs. Chi-5	219	34	Rejected for HFRBCS	0.003
HFRBCS vs. C4.5	84	169	Not Rejected	0.168

The main conclusion after this study is that the HFRBCS is better than the rest of the FRBCS methods. It outperforms the base Chi LRG-method, the Ishibuchi et al.'s approach and the E-Algorithm. When compared with C4.5, there are no statistical differences in this imbalance scenario.

- **Data-sets with high imbalance**: This part of the study is very important, since it includes the data-sets with a higher degree of imbalance. In this manner, we can analyze how the imbalance actually affects the different methods employed in this study. For this purpose, we use a Iman-Daverport test in order to find statistical differences obtaining a p-value of 0.00330. Next, Table 11 shows the ranking for the FRBCS algorithms, in which our HFRBCS proposal is the first one. Finally, we perform a Holm test, which is shown in Table 12, where we can only conclude that the HFRBCS is better than the E-Algorithm in data-sets with high imbalance.

Table 11 Rankings obtained through a Friedman test for FRBCSs in data-sets with high imbalance.

Algorithm	Ranking
HFRBCS	2.20454
Chi-5	2.90909
Chi-3	2.93182
Ishibuchi05	2.93182
E-Algorithm	4.02273

Table 12 Holm test table for FRBCSs in data-sets with high imbalance. HFRBCS is the control method.

i	algorithm	z	p	α/i	Hypothesis
4	E-Algorithm	3.81385	1.36818E-4	0.01250	Rejected for HFRBCS
3	Ishibuchi05	1.52554	0.12712	0.01667	Not Rejected
2	Chi-3	1.52554	0.12712	0.02500	Not Rejected
1	Chi-5	1.47787	0.13944	0.05000	Not Rejected

A Wilcoxon test (Table 13) will help us to make a pairwise comparison between our proposal and the remaining algorithms, including C4.5 in this case. Now, we detect differences between the HFRBCS and the Chi et al.'s method with 5 labels per variable, but it remains statistically similar to the Ishibuchi et al.'s algorithm and the Chi et al.'s method with 3 labels. Nevertheless, watching the results for the comparison with C4.5 we see that the null hypothesis is rejected in favour of our HFRBCS proposal.

Table 13 Wilcoxon test to compare the HFRBCS against the remaining FRBCS approaches and C4.5 in data-set with high imbalance. R+ corresponds to HFRBCS and R- to the remaining algorithms in each case.

Comparison	R^+	R^-	Hypothesis (α =0.05)	p-value
HFRBCS vs. C4.5	192	61	Rejected for HFRBCS	0.033

According to these results, we must emphasise the good behaviour achieved in highly imbalanced data-sets by the all fuzzy models studied here, particularly for our proposal. Furthermore, we can determine that it is very competitive, since it outperforms C4.5 algorithm in this type of data-sets, with a p-value of 0.033.

6 Concluding Remarks

In this work, we have presented a wide overview on hierarchical linguistic fuzzy systems, describing the main features of this type of systems, the learning methodology proposed to build such a model and the extensions developed in the literature.

The main aim of this hierarchical approach is to obtain a good balance among different granularity levels, applying a higher granularity in the areas where the RB has a bad performance in order to obtain a better coverage of that area of the space of solutions, and a lower granularity that provides a good generalization.

Regarding the interpretability-accuracy tradeoff of this methodology, we have stated that it is not always true that a set of rules with a higher granularity level performs a more accurate modeling of a problem than another with a lower one. The relationship between accuracy and interpretability does not only depend on granularity and specificity, but also on other factors, for example, rule weights, flexible rule consequents, and moreover, compasity of and cooperation policies between the rules.

As a case of study, we have shown the improvement obtained by applying this methodology in the framework of imbalanced data-sets. Specifically, the classification accuracy of the base FRBCS is enhanced in the overlapping areas between the minority and majority classes by combining both the fine and thick granularity. In the experimental study, we have shown statistically that our proposal performs better than well known FRBCSs approaches and that clearly outperforms the C4.5 decision tree, generally for all data-sets and particularly in data-sets with high imbalance.

Acknowledgment

This work had been supported by the Spanish Ministry of Science and Technology under Projects TIN2008-06681-C06-01/02, and the Andalusian Research Plan TIC-3928.

References

Alcalá, R., Cano, J.R., Cordón, O., Herrera, F., Villar, P., Zwir, I.: Linguistic modeling with hierarchical systems of weighted linguistic rules. Int. J. Approx. Reason. 32(2–3), 187–215 (2003)

Alcalá, R., Alcalá-Fdez, J., Herrera, F.: A proposal for the genetic lateral tuning of linguistic fuzzy systems and its interaction with rule selection. IEEE T. Fuzzy Syst. 15(4), 616–635 (2007)

Asuncion, A., Newman, D.: UCI machine learning repository (2007), `http://www.ics.uci.edu/~mlearn/MLRepository.html`

Barandela, R., Sánchez, J.S., García, V., Rangel, E.: Strategies for learning in class imbalance problems. Pattern Recogn. 36(3), 849–851 (2003)

Bastian, A.: How to handle the flexibility of linguistic variables with applications. Int. J. Uncertain Fuzz 2(4), 463–484 (1994)

Batista, G., Prati, R.C., Monard, M.C.: A study of the behaviour of several methods for balancing machine learning training data. SIGKDD Explor. 6(1), 20–29 (2004)

Bonissone, P.P., Decker, K.: Selecting uncertainty calculi and granularity: An experiment in trading-off precision and complexity. In: Kanal, L.N., Lemmer, J.F. (eds.) UAI, pp. 217–248. Elsevier, Amsterdam (1985)

Carse, B., Fogarty, T.C., Munro, A.: Evolving fuzzy rule based controllers using genetic algorithms. Fuzzy Set Syst. 80, 273–294 (1996)

Chawla, N.V., Bowyer, K.W., Hall, L.O., Kegelmeyer, W.P.: Smote: Synthetic minority over-sampling technique. J. Artif. Intell. Res. 16, 321–357 (2002)

Chawla, N.V., Japkowicz, N., Kolcz, A.: Editorial: special issue on learning from imbalanced data sets. SIGKDD Explor. 6(1), 1–6 (2004)

Chi, Z., Yan, H., Pham, T.: Fuzzy algorithms with applications to image processing and pattern recognition. World Scientific, Singapore (1996)

Cordón, O., Herrera, F.: A proposal for improving the accuracy of linguistic modeling. IEEE T. Fuzzy Syst. 8(4), 335–344 (2000)

Cordón, O., del Jesus, M.J., Herrera, F.: A proposal on reasoning methods in fuzzy rule-based classification systems. Int. J. Approx Reason 20(1), 21–45 (1999)

Cordón, O., Herrera, F., Hoffmann, F., Magdalena, L.: Genetic Fuzzy Systems. In: Evolutionary tuning and learning of fuzzy knowledge bases. World Scientific, Singapore (2001a)

Cordón, O., Herrera, F., Villar, P.: Generating the knowledge base of a fuzzy rule-based system by the genetic learning of data base. IEEE T. Fuzzy Syst. 9(4), 667–674 (2001b)

Cordón, O., Herrera, F., Zwir, I.: Linguistic modeling by hierarchical systems of linguistic rules. IEEE T. Fuzzy Syst. 10(1), 2–20 (2002)

Cordón, O., Herrera, F., Zwir, I.: A hierarchical knowledge-based environment for linguistic modeling: Models and iterative methodology. Fuzzy Set Syst. 138(2), 307–341 (2003)

Demšar, J.: Statistical comparisons of classifiers over multiple data sets. J. Mach. Learn. Res. 7, 1–30 (2006)
Domingos, P.: Metacost: a general method for making classifiers cost sensitive. In: Proceedings of the 5th International Conference on Knowledge Discovery and Data Mining, pp. 155–164 (1999)
Estabrooks, A., Jo, T., Japkowicz, N.: A multiple resampling method for learning from imbalanced data sets. Comput. Intell. 20(1), 18–36 (2004)
Fernández, A., García, S., del Jesus, M.J., Herrera, F.: A study of the behaviour of linguistic fuzzy rule based classification systems in the framework of imbalanced data-sets. Fuzzy Set Syst. 159(18), 2378–2398 (2008)
Fernández, A., del Jesus, M., Herrera, F.: Hierarchical fuzzy rule based classification system with genetic rule selection for imbalanced data-set. Int. J. Approx. Reason. 50, 561–577 (2009)
Gacto, M.J., Alcalá, R., Herrera, F.: Adaptation and application of multi-objective evolutionary algorithms for rule reduction and parameter tuning of fuzzy rule-based systems. Soft Comput. 13(5), 419–436 (2009)
García, S., Herrera, F.: An extension on "statistical comparisons of classifiers over multiple data sets" for all pairwise comparisons. J. Mach. Learn. Res. 9, 2677–2694 (2008)
García, S., Fernández, A., Luengo, J., Herrera, F.: A study of statistical techniques and performance measures for genetics–based machine learning: Accuracy and interpretability. Soft Comput. 13(10), 959–977 (2009)
García, S., Fernández, A., Luengo, J., Herrera, F.: Advanced nonparametric tests for multiple comparisons in the design of experiments in computational intelligence and data mining: Experimental analysis of power. Inform. Sciences 180, 2044–2064 (2010)
García, V., Mollineda, R., Sánchez, J.S.: On the k-NN performance in a challenging scenario of imbalance and overlapping. Pattern Anal. Appl. 11(3–4), 269–280 (2008)
He, H., Garcia, E.A.: Learning from imbalanced data. IEEE T. Knowl. Data En. 21(9), 1263–1284 (2009)
Herrera, F., Martínez, L.: A model based on linguistic 2-tuples for dealing with multigranular hierarchical linguistic contexts in multi-expert decision making. IEEE T. Syst. Man. CY B 31(2), 227–234 (2001)
Herrera, F., Herrera-Viedma, E., Martínez, L.: A fusion approach for managing multigranularity linguistic terms sets in decision making. Fuzzy Set Syst. 114(1), 43–58 (2000)
Holm, S.: A simple sequentially rejective multiple test procedure. Scand. J. Stat. 6, 65–70 (1979)
Ishibuchi, H., Nakashima, T.: Effect of rule weights in fuzzy rule-based classification systems. IEEE T. Fuzzy Syst. 9(4), 506–515 (2001)
Ishibuchi, H., Yamamoto, T.: Rule weight specification in fuzzy rule-based classification systems. IEEE T. Fuzzy Syst. 13, 428–435 (2005)
Ishibuchi, H., Nozaki, K., Tanaka, H.: Efficient fuzzy partition of pattern space for classification problems. Fuzzy Set Syst. 59, 295–304 (1993)
Ishibuchi, H., Nozaki, K., Yamamoto, N., Tanaka, H.: Selecting fuzzy if-then rules for classification problems using genetic algorithms. IEEE T. Fuzzy Syst. 9(3), 260–270 (1995)
Ishibuchi, H., Nakashima, T., Nii, M.: Classification and modeling with linguistic information granules: Advanced approaches to linguistic Data Mining. Springer, Heidelberg (2004)
Kuncheva, L.: Fuzzy classifier design. Springer, Berlin (2000)
Magdalena, L., Monasterio-Huelin, F.: A fuzzy controller with learning through the evolution of its knowledge base. Int. J. Approx. Reason. 16(3), 335–358 (1997)
Michalewicz, Z.: Genetic algorithms + data structures = evolution programs. Springer, Heidelberg (1996)

Orriols-Puig, A., Bernadó-Mansilla, E.: Evolutionary rule-based systems for imbalanced data-sets. Soft Comput. 13(3), 213–225 (2009)
Orriols-Puig, A., Bernadó-Mansilla, E., Goldberg, D.E., Sastry, K., Lanzi, P.L.: Facetwise analysis of XCS for problems with class imbalances. IEEE T. Evolut. Comput. 13, 260–283 (2009)
Palm, R., Driankov, D., Hellendoorn, H.: Model based fuzzy control. Springer, Berlin (1997)
Pedrycz, W.: Fuzzy modelling: paradigms and practice. Kluwer Academic Press, Dordrecht (1996)
Quinlan, J.R.: C4.5: Programs for Machine Learning. Morgan Kaufmann Publishers, San Mateo (1993)
Peña Reyes, C.A., Sipper, M.: Fuzzy CoCo: a cooperative coevolutionary approach to fuzzy modeling. IEEE T. Fuzzy Syst. 9(5), 727–737 (2001)
Sanz, J., Fernández, A., Bustince, H., Herrera, F.: Improving the performance of fuzzy rule-based classification systems with interval-valued fuzzy sets and genetic amplitude tuning. Inform. Sciences 180(19), 3674–3685 (2010)
Sheskin, D.: Handbook of parametric and nonparametric statistical procedures, 2nd edn. Chapman & Hall/CRC (2006)
Su, C.T., Hsiao, Y.H.: An evaluation of the robustness of MTS for imbalanced data. IEEE T. Knowl. Data En. 19(10), 1321–1332 (2007)
Su, C.T., Chen, L.S., Yih, Y.: Knowledge acquisition through information granulation for imbalanced data. Expert Syst. Appl. 31, 531–541 (2006)
Sun, Y., Kamel, M.S., Wong, A.K., Wang, Y.: Cost-sensitive boosting for classification of imbalanced data. Pattern Recogn. 40, 3358–3378 (2007)
Sun, Y., Wong, A.K.C., Kamel, M.S.: Classification of imbalanced data: A review. Int. J. Pattern Recogn. 23(4), 687–719 (2009)
Thrift, P.: Fuzzy logic synthesis with genetic algorithms. In: 4th International Conference on Genetic Algorithms (ICGA 1991), pp. 509–513. Morgan Kaufmann, San Mateo (1991)
Wang, L.X., Mendel, J.M.: Generating fuzzy rules by learning from examples. IEEE T. Syst. Man CYB 25(2), 353–361 (1992)
Weiss, G., Provost, F.: Learning when training data are costly: The effect of class distribution on tree induction. J.Artif. Intell. Res. 19, 315–354 (2003)
Wilcoxon, F.: Individual comparisons by ranking methods. Biometrics 1, 80–83 (1945)
Wu, G., Chang, E.: KBA: Kernel boundary alignment considering imbalanced data distribution. IEEE T. Knowl. Data En. 17(6), 786–795 (2005)
Xu, L., Chow, M.Y., Taylor, L.S.: Power distribution fault cause identification with imbalanced data using the data mining-based fuzzy classification e-algorithm. IEEE T. Power Syst. 22(1), 164–171 (2007)
Yager, R.R., Filev, D.: Essentials of Fuzzy Modeling and Control. John Wiley & Sons, Chichester (1994)
Yang, Q., Wu, X.: 10 challenging problems in data mining research. Int. J. Inf. Tech. Decis. 5(4), 597–604 (2006)
Yen, J., Wang, L., Gillespie, C.W.: Improving the interpretability of TSK fuzzy models by combining global learning and local learning. IEEE T. Fuzzy Syst. 6(4), 530–537 (1998)
Zadeh, L.A.: The concept of a linguistic variable and its application to approximate reasoning. Inform. Sciences 8(3), 199–249 (1975)
Zhou, Z., Liu, X.: Training cost-sensitive neural networks with methods addressing the class imbalance problem. IEEE T. Knowl Data En. 18(1), 63–77 (2006)

Fuzzy Information Granulation with Multiple Levels of Granularity

Giovanna Castellano, Anna Maria Fanelli, and Corrado Mencar

Abstract. Granular computing is a problem solving paradigm based on information granules, which are conceptual entities derived through a granulation process. Solving a complex problem, via a granular computing approach, means splitting the problem into information granules and handling each granule as a whole. This leads to a multi-level view of information granulation, which permeates human reasoning and has a significant impact in any field involving both human-oriented and machine-oriented problem solving. In this chapter we examine a view of granular computing as a paradigm of human-inspired problem solving and information processing with multiple levels of granularity, with special focus on fuzzy information granulation. To support the importance of granulation with multiple levels, we present a multi-level approach for extracting well-defined and semantically sound fuzzy information granules from numerical data.

Keywords: Fuzzy Information Granulation, Information Granules, Interpretability, Multi-level Granulation, Conditional Fuzzy C-Means, Double Clustering.

1 Introduction

In the last decade, information granulation has emerged as a powerful tool for data analysis and information processing, which is in line with the way humans adopt to process information. We perceive the world by structuring our knowledge, perceptions, and acquired evidence in terms of information granules that offer abstractions of the complex world and phenomena. Being abstract constructs, information granules and the processing of them, referred to as Granular Computing (GrC), provide problem solvers with a conceptual and algorithmic framework to deal with several real-world problems.

The term GrC spans a variety of disciplines, thus it is often loosely defined as an umbrella term covering any theories, methodologies, techniques, and tools that

---

Giovanna Castellano · Anna Maria Fanelli · Corrado Mencar
Department of Informatics, University of Bari "Aldo Moro"
Via Orabona, 4 – 70125 Bari, Italy

W. Pedrycz and S.-M. Chen (Eds.): Granular Computing and Intell. Sys., ISRL 13, pp. 185–202.
springerlink.com

make use of granules in complex problem solving (Yao 2005). Various frameworks for information granulation have been proposed so far, with a growing diversity of formalisms. One of the main formalism proposed for GrC is the theory of fuzzy sets (Zadeh 1997) introduced by Zadeh, who considers granular computing as basis for computing with words, i.e., computation with information described in natural language (Zadeh 1997; Zadeh 2006). Besides, many other formalisms can be considered fundamental for GrC, stemming from rough sets (Pawlak 1998; Yao 2001; Peters et al. 2003), interval analysis (Bargiela and Pedrycz 2003), shadowed sets (Pedrycz 2005), etc.

Beyond the theory underlying a GrC framework, two main features are desirable for any information granulation approach:

1. the ability to determine the granularity level that better represents the nature of data;
2. the ability to provide users with information granules that both represent the data accurately and carry a clear semantic meaning, i.e. granules that are *interpretable* for human users.

These features bring a shift in GrC, from a paradigm of machine-centric information processing to a paradigm of human-inspired problem solving. This shift is considered one of the recent trends in GrC research (Yao 2008). In this endeavor, it is well accepted that the theory of fuzzy sets has largely contributed to the emergence of granular computing as a paradigm of human-inspired problem solving and information processing (Bortolan and Perdrycz 2002). Actually, human-centered information processing was initiated with the introduction of fuzzy sets, which successively led to the development of the GrC paradigm (Bargiela and Pedrycz 2008; Zadeh 1997). The object of fuzzy information granulation is to build models by means of information granules that are quantified in terms of fuzzy sets, i.e. conceptual entities with well-defined semantics that are interpretable by humans.

To stress the shift from machine-centric computing toward human-inspired computing, a common theme is converging to a view of GrC as a paradigm of information processing with an underlying notion of multiple levels of granularity. The introduction of multiple levels of granularity corresponds to consider multiple levels of abstractions for a given problem, with each level capturing particular aspects of the problem. To some extent, this may avoid limitations of a single level of representation.

Along with these ideas, in this chapter we examine the current research directions on multi-level granular computing, with special focus on fuzzy information granulation. To give evidence of the advantage of multi-level granulation, we present an approach to perform fuzzy granulation with multiple levels of granularity in a hierarchical fashion. In particular, a framework of representation with two levels of granularity is described. At the first level, the whole dataset is granulated. At the second level, data embraced in each first-level granule are further granulated taking into account the context generated by that granule. The derived hierarchical collection of granules can be used to construct a committee of fuzzy models providing a good balance between interpretable representation and precise approximation.

2 Multi-level Granular Computing

The basic components of GrC are information granules, intended as chunks of knowledge made of different objects "drawn together by indistinguishability, similarity, proximity or functionality" (Zadeh 1997). Information granules yield an abstraction of the reality in form of concepts depending on the context. For such reason, a granulation process, i.e. the process of constructing granules, serves not only to design a model, but also to simplify our understanding of it. As computing units, granules can decompose a complex problem into some simpler or smaller problems, so that the computing costs are reduced and the problem can be better understood.

Based on complexity, abstraction grade and size, granules can be evaluated at different levels. The problem domain, i.e., the universe of discourse, represents the coarsest granule at the highest level. Granules at the lowest level are composed of elements or basic particles of the particular model that is used (Yao, 2007). According to the level of granularity taken into account, i.e. to the point of view, a granule "may be an element of another granule and is considered to be a part forming the other granule. It may also consist of a family of granules and is considered to be a whole" (Yao 2008). This leads to the notion of levels. While each granule provides a local view, a level provides a global view.

An important property of granules and levels is their granularity, i.e. the size of information granules and their distribution. By changing the granularity, we can control the amount of details so as to hide or reveal more or less details about the problem at hand (Bargiela and Pedrycz 2003). The lower level of granularity can yield the most detailed information, but some useful knowledge may be buried into unnecessary details. On the other hand, the higher level of granularity might reduce some information, but it can provide users with a better insight into the essence of data. From one level we can pass to a lower level of granularity by means of a granulation process that decomposes a whole into parts; this corresponds to "analytical thinking" whereas, going to an upper level merging parts into wholes, corresponds to "synthetic thinking" (Yao 2007).

Granularity enables us to properly arrange granules and levels, so as to derive a hierarchical view of the problem at hand. In building a hierarchical structure, we discover a vertical separation of levels and a horizontal separation of granules at the same hierarchical level (Yao 2009). Usually, the two separations must ignore information that is irrelevant to the current interest or does not greatly affect our solution. Furthermore, a single hierarchy only represents one view. As illustrated by Yao, granular structures enable both a multi-level view (given by a single hierarchy) and a multi-view understanding (given by many hierarchies) (Yao 2009). The latter stresses the consideration of diversity in modelling, for which we look at the same problem from many perspectives. This is useful when, in order to understand a problem, we need to explore multiple representations of it, in terms of multiple views and multiple levels of abstraction.

Summarizing, a fundamental key to GrC is representing and working with different levels of granularity in every stage of problem solving. We may view a problem through many different facets, and associate a representation with a particular view capturing specific aspects of the problem. For each view, we may consider multiple levels of abstractions, each representing the problem at a particular level of details. This particular issue of granulating information by using high-order properties, i.e. properties of collections of granules formed at higher levels, lies behind every - not necessarily computational - task involving problem solving: it describes a way of thinking, named "granular reasoning" (Zhong at al. 2008), that relies on the human ability to perceive the real world under various levels of granularity. In (Yao 2006), some fundamental issues related to the notion of multiple levels of granularity are addressed and analyzed from three different perspectives, namely philosophical, methodological and computational perspective leading to the so-called *triarchic theory* of GrC (Yao 2005; Yao 2007).

The philosophical facet of GrC offers a worldview in terms of structures as represented by multiple levels. This leads to a way of structured thinking, made of levels of abstraction, which is applicable to many branches of natural and social sciences. We may consider, for example, levels of understanding in education, levels of interpretation in history and language understanding, levels of organization in ecology and social sciences, levels of processing in modeling human memory, and many others. For example, Jeffries and Ransford proposed a multiple hierarchy model to integrate class, ethnicity, gender, and age for the study of social stratification (Jeffries and Ransford 1980). They show how the traditional single hierarchy approach based on social classes limits one's understanding of the complexities of modern societies, while a multiple hierarchy approach could increase one's understanding and be more comprehensive and valid for studying social inequality.

The methodological perspective of GrC raises quite natural. As a general method of structured problem solving, GrC provides practical strategies and effective principles that are used by humans for solving real-world problems. Those principles of granular computing have, in fact, been extensively used in different disciplines under different names and notations. For example, many principles of structured programming or software design can be readily adopted for granular computing (Han and Dong 2007). As another example, in (Belkhouche and Lemus-Olalde 2000) an abstract interpretation of multiple views in software design is formalized. In the process of modeling a system, the designer always generates a set of designs, such as functional, behavioral, structural and data designs. Each design focuses on a view that describes a subset of relevant features of a system and is expressed by one or more notations. The authors argued that a multiple view analysis framework can be used to systematically compare, identify and analyze the discrepancies among different views, enhance design quality and provide a multi-angled understanding of a problem or a project.

The computational perspective of GrC underlies the other two if we consider the term "computing" in its broad meaning to include information processing in the abstract, in the brain and in machines. As a paradigm of structured information processing, granular computing focuses on computing methods based on granular structures. In particular, when computation is intended as information processing by machines, i.e. data analysis, the use of hierarchical granular structures leads to multi-view intelligent data analysis, which explores data from different perspectives to reveal various types of structures and knowledge embedded in the data. Each view may capture a specific aspect of the data and hence satisfy the needs of a particular group of users. Collectively, multiple views provide a comprehensive description and understanding of the data. According to this idea, Chen and Yao proposed a multi-view approach that provides a unified framework for integrating multiple views of intelligent data analysis (Chen and Yao 2008). Managing multiple representations of the same data at different levels of granularity is widely recognized as a relevant one also in the community of spatial database. In (De Fent et al. 2005) spatial data are modeled through conceptual models with two distinct, but related, granularity dimensions: a spatial one and a semantic one. In this case, a multi-level granulation enables to capture scale as well as semantic changes and to constrain the relationships between them.

From a different viewpoint, we may say that a multi-level information granulation approach turns out to be beneficial as soon as we need to represent and solve the problem through a set of linguistic concepts. Actually, multi-level granulation has been widely investigated in the realm of linguistic approaches, i.e. approaches using linguistic terms to represent information in a qualitative way (Glöckner and Knöll 2001). Linguistic representations, such as those based on fuzzy sets, are especially suitable when information is unquantifiable due to its imprecise nature (e.g., when evaluating the "comfort" of a car, only terms like "good", "fair", "poor" can be used), but they can be used as well when a quantitative representation of information cannot be stated because either it is unavailable or the cost for its computation is too high, thus an approximate value can be tolerated (e.g., when evaluating the speed of a car, linguistic terms like "fast", "very fast", "slow" can be used instead of numeric values). When using a linguistic representation of information, an important parameter to determine is the semantic granularity, i.e. the cardinality of the linguistic term set used to express the information. According to the uncertainty degree of a domain expert, the linguistic term set used to represent her knowledge may have more or less terms. When different experts have different uncertainty degrees in their knowledge, then several linguistic term sets with different granularities are necessary. To cope with this multiple source of uncertain information, multi-granularity linguistic term sets based on fuzzy set theory have been proposed and applied in several fields, such as decision making (Herrera et al. 2000; Mata et al. 2009) and information retrieval (Herrera-Viedma et al. 2003).

Multi-level GrC has been also applied to represent taxonomies of concepts. In (Qiu et al. 2007) a hierarchy of granules corresponding to ontological concepts is built by an information table using rough-set techniques. A granular space model for ontology learning is explored, to describe domain ontologies at different granularities and hierarchies. In (Gu et al. 2006) an approach for constructing hierarchy of granules based on fuzzy concept lattices is proposed. The knowledge granularity is discussed, and an algorithm for constructing a hierarchical structure of coarser granules is also illustrated.

The potential applications and implications of multi-level granulation can be also recognized in the area of Web intelligence. As shown in (Yao 2007), GrC may provide the necessary theory for designing and implementing new types of web-based information processing systems based on a conceptual model of the human brain. Yao and Yao discuss how Web information retrieval can benefit from grouped and personalized views provided by a GrC process (Yao and Yao 2003). The use of both single-level and multi-level granulations of web documents is investigated. In (Li et al. 2001) a technique for automatically constructing multi-granular and topic-focused site maps using trained rules on Web page URLs, contents, and link structure is presented. This type of multi-granular site maps can support better interaction for users to scale up and scale down the details. The system provides more detail on the regions relevant to the focused topic while keeping the rest of the map compact so that the users can visualize their current navigation positions relative to other landmark nodes in the Web site.

To conclude, we emphasize the crucial role of interpretability in the realm of GrC. As a paradigm for human-centric information processing, GrC should provide a common interface for communicating information between humans and machines. This human-oriented communication is partially achieved by representing perceptual information in a computer manageable form, i.e. by means of information granules. To make more effective this communication, information granules should be interpretable, i.e. semantically co-intensive with human knowledge. Interpretability of information granules is a complex requirement that needs a comprehensive analysis of all facets of the problem for which granules are developed and used. Therefore interpretable information granulation opens several methodological issues, regarding the representation and manipulation of information granules, the interpretability constraints and the granulation processes (Mencar and Fanelli 2008; Mencar 2009). We believe that addressing all such issues at multiple levels of granularity leads to a sight of GrC as an effective tool to design information processing systems characterized by a strong human-centric imprint. To support this idea, in the next section we present a multi-level GrC strategy to derive interpretable information granules from data.

3 A Multi-level Approach for Information Granulation

As an example of multi-level granulation strategy, in this section we describe a multi-level approach for fuzzy information granulation. The approach is based on DC*f* (Double Clustering framework), a framework to create interpretable granules from data by taking into account a number of interpretability constraints (Castellano et al. 2005). DC*f* extracts fuzzy information granules that can be easily labeled with semantically sound linguistic labels. Moreover, the number of information granules to be extracted can be kept small, so as to provide a compact (and hence readable) description of data. Nevertheless, DC*f* provides for a flat representation of information granules, and the user is committed to define the granularity level. This is due to the fact that DC*f* can extract a fixed number of information granules. A high number of information granules leads to an accurate yet unreadable description of data, while a small number of information granules provides a highly interpretable description of data, but the employment of such granules in fuzzy predictive models may not result satisfactory because of a possibly coarse accuracy.

In order to obtain a more accurate description of data while keeping interpretability, we have developed an extension of DC*f*, called ML-DC (Multi-Level Double Clustering), which is intended to provide a multi-level granulation of available data, i.e. a granulation of data at different levels, in a hierarchical fashion (Castellano et al. 2007). ML-DC exploits the DC*f* structure to provide for multiple view of data: a coarse qualitative view where main relationships are described by large granules, and a more refined, quantitative granulation that could be used for defining the predictive model. At the first level, the whole dataset is granulated, while, at the second level, data embraced in each first-level granule are further granulated taking into account the context generated by that granule. Based on the extracted multi-level granules, a hierarchical committee of Fuzzy Inference Systems (FISs) is constructed that can approximate a mapping with a good balance between accuracy and interpretability.

Roughly speaking, ML-DC operates information granulation at two levels:

- at first level, granulation of data is carried out according to a specific granularity, as in DC*f*;
- at second level, for each discovered information granule, data are re-aggregated for a further granulation process.

The process can be reiterated for a number of levels. However a two-level granulation is adequate to obtain two views of the problem (a qualitative one and a quantitative one), so as to achieve a balanced trade-off between accuracy and interpretability of data. Hence, through the application of ML-DC, two levels of fuzzy information granules are built from data: granules of the first level are used to roughly describe data through qualitative linguistic labels; granules of the second level are used to describe each information granule of the first level. This is aimed at finding a more accurate description of the hidden relationships lying among data and – a the same time – preserve interpretability of the extracted knowledge since interpretability constraints are satisfied for both levels of granulation.

The first-level granulation in ML-DC is made according to the double-clustering framework DC*f*, which involves two main steps:

1. multi-dimensional clustering on the whole dataset, providing a collection of multidimensional prototypes;
2. one-dimensional clustering of the projection of the derived prototypes along each dimension, yielding to one-dimensional prototypes.

In ML-DC we perform step 1 by means of fuzzy clustering and step 2 by means of hierarchical clustering. Fuzzification of the information granules is achieved by first fuzzifying the one-dimensional granules defined by the one-dimensional prototypes and then by aggregating one-dimensional fuzzy sets to form multi-dimensional fuzzy information granules. For each dimension, the extracted clusters are transformed into as many interpretable fuzzy sets. Fuzzy sets with

Gaussian membership function are considered, whose centers and widths are defined so as to take into account the information provided by the clustering stages and, at the same time, to meet the following interpretability constraints:

- *Normality*, *convexity* and *continuity*: these constraints are verified as soon as Gaussian membership functions are chosen for fuzzy sets;
- *Proper ordering*: this is verified by defining the order relation of fuzzy sets reflecting the order of the prototypes.
- *Justifiable number of elements*: this constraint is verified by an appropriate choice of the number of prototypes;
- *Distinguishability* and *completeness*: these constraints are verified by the construction of the fuzzy sets, which is made so as to not exceed an overlap threshold ε and to guarantee the ε-coverage.

The second-level granulation is carried out according to the same double-clustering schema, but taking into account the context generated from each first-level information granule. Indeed, if this context is ignored, the second-level granulation would be identical to first-level granulation and no additional information would be derived from data. To keep into account contextual information in the second-level granulation, multi-dimensional clustering is performed by the Conditional Fuzzy C-Means (CFCM) proposed by Pedrycz, which is an extension of the well-known FCM clustering algorithm (Pedrycz 1996). The CFCM clustering algorithm minimizes the following objective function:

$$J(\mathbf{U},\mathbf{V}) = \sum_{j=1}^{c}\sum_{i=1}^{N} u_{ij}^{m} \left\| \mathbf{x}_i - \mathbf{v}_j \right\|^2 \tag{1}$$

where $\mathbf{x}_i \in \mathbb{R}^n$, $i = 1,2,\ldots,N$ are the observational n-dimensional data to be clustered, c is the number of clusters, $\mathbf{U} = \left[u_{ij}\right]_{j=1,2,\ldots c}^{i=1,2,\ldots,N} \in [0,1]^{N\times c}$ is the partition matrix, being u_{ij} the membership degree of the i-th observation to the j-th cluster, $\mathbf{V} = \left[\mathbf{v}_j\right]_{j=1,2,\ldots,c} \in \mathbb{R}^{n\times c}$, is the matrix of the prototypes corresponding to the fuzzy clusters and m is a fuzzification parameter, here fixed to 2.0.

The objective function (1) is constrained as follows to avoid degenerate solutions:

$$\forall j = 1,2,\ldots,c : \sum_{i=1}^{N} u_{ij} > 0 \tag{2}$$

and:

$$\forall i = 1,2,\ldots,N : \sum_{j=1}^{c} u_{ij} = f_i \tag{3}$$

Constraint (3) differentiates CFCM from the standard FCM clustering algorithm. Such constraint requires that the sum of memberships of a point to each cluster is equal to a constant $f_i \in [0,1]$ that defines the *context* of the clustering process. For the first-level granulation process, no context is defined for ML-DC. Hence, CFCM is reduced to standard FCM by setting:

$$\forall i = 1,2,\ldots,N : f_i = 1 \tag{4}$$

For the second-level granulation process, a context is defined by each first-level fuzzy information granule:

$$\forall i = 1,2,\ldots,N : f_i = \mu_k^{\mathrm{I}}(\mathbf{x}_i) \tag{5}$$

where $\mu_k^{\mathrm{I}}(\mathbf{x}_i)$ is the membership degree of the i-th observation to the k-th fuzzy information granule discovered in the first-level granulation process. The quantities f_i establish the connection between first-level and second-level information granules. Information granules at the second level are indeed forced to focus their location in the fuzzy sub-region of the domain where each first-level information granule is placed.

The CFCM clustering algorithm follows the Alternating Optimization strategy for the minimization of the objective function (1). This strategy is iterative. At each iteration the prototypes and the partition matrix are updated according to the following formulas:

$$\mathbf{v}_j[\tau+1] = \frac{\sum_{i=1}^{N} u_{ij}^m[\tau]\mathbf{x}_i}{\sum_{i=1}^{N} u_{ij}^m[\tau]} \tag{6}$$

and:

$$u_{ij}[\tau+1] = \frac{f_i}{\left(\sum_{k=1}^{c} \frac{\|\mathbf{x}_i - \mathbf{v}_i\|^2}{\|\mathbf{x}_i - \mathbf{v}_j\|^2}\right)^{\frac{1}{m-1}}} \tag{7}$$

Summarizing, ML-DC is a double-clustering approach where CFCM is used in the multi-dimensional clustering and hierarchical clustering is used in the one-dimensional clustering. For first-level granulation, ML-DC is applied to data with constant context. The result of first-level granulation is a set of K^{I} fuzzy information granules G_k^{I} with membership functions $\mu_k^{\mathrm{I}} : \mathbb{R}^n \to [0,1]$, $k = 1,2,\ldots,K^{\mathrm{I}}$.

ML-DC exploits the advantages of multi-level information granulation to reach a tradeoff between accuracy and interpretability. Here we describe a granulation methodology based on ML-DC, which is specifically designed for classification problems, even though the extension to function approximation problems is straightforward. Let $\mathbf{X} \subseteq \mathbb{R}^n$ be the Universe of Discourse of data (assumed as hyper-box) and $\mathbf{C} = \{1,2,\ldots,C\}$ a set of class labels. Let

$D=\{\langle \mathbf{x}_i, c_i \rangle \in \mathbf{X}\times\mathbf{C}, i=1,2,\ldots,N\}$ a set of pre-classified observational data. The first-level granulation of data provides for K^I fuzzy information granules with membership functions μ_k^{I}, $k=1,2,\ldots,K^I$. The value K^I can be fixed as small as desired (e.g. less than 7±2 according to (Valente de Oliveira 1999)). As shown above, each information granule G_k^I satisfies a number of interpretability constraints so that it can be labeled linguistically.

For each G_k^I and each class label $c\in\mathbf{C}$, we compute the relative frequency of observations of class c belonging to the fuzzy granule G_k^I, as follows:

$$\pi_{k,c}^{I}=\frac{\sum_{i=1,c_i=c}^{N}\mu_k^I(\mathbf{x}_i)}{\sum_{i=1}^{N}\mu_k^I(\mathbf{x}_i)} \tag{8}$$

These values can be used to build a set of K^I fuzzy rules with the following schema:

$$\begin{aligned}\text{IF}\,\mathbf{x}\,\text{is}\,G_k^I\ \text{THEN}\ & P(class=1)=\pi_{k,1}^I\\ & P(class=2)=\pi_{k,2}^I\\ & \ldots\\ & P(class=C)=\pi_{k,C}^I\end{aligned} \tag{9}$$

Given an input $\mathbf{x}$, the outputs of the classifier are computed according to the following formula:

$$\pi_c^I(\mathbf{x})=\frac{\sum_{k=1}^{K^I}\mu_k^I(\mathbf{x})\pi_{k,c}^I}{\sum_{k=1}^{K^I}\mu_k^I(\mathbf{x})} \tag{10}$$

for $c=1,2,\ldots,C$. If one class only must be assigned to $\mathbf{x}$, then the class with highest $\pi_c^I(\mathbf{x})$ is chosen (tiers are selected arbitrarily).

The FIS classifier designed through first-level information granulation is very compact (and hence highly interpretable) but expectedly not very accurate. Second-level information granules can be exploited to improve accuracy of the estimated mapping as follows. For each first-level information granule, ML-DC provides a set of second-level information granules that can be used to generate a corresponding FIS with the same rule schema of the first-level FIS. As a result, a set of K^{I} FISs are generated. All such FISs are connected to form a hierarchical committee of FISs from which the input/output mapping is inferred. The outputs of the hierarchical FIS are defined as the weighted sum of the outputs of each FIS belonging to the committee. Formally, given an input $\mathbf{x}\in\mathbb{R}^n$, the output of the FIS committee is:

$$\pi_c^{II}(\mathbf{x}) = \frac{\sum_{k=1}^{K^{\mathrm{I}}} \pi_{k,c}^{II}(\mathbf{x}) w_k}{\sum_{k=1}^{K^{\mathrm{I}}} w_k} \tag{11}$$

where $\pi_{k,c}^{II}(\mathbf{x})$ is the output of the k-th FIS belonging to the committee relatively to class c, while w_k is the weight assigned to the k-th FIS, corresponding to the degree of membership of the input $\mathbf{x}$ to the antecedent part of the first-level information granule:

$$w_k = \mu_{X,k}^{\mathrm{I}}(\mathbf{x}) \tag{12}$$

In this way, the weight of a FIS in the committee is high (and hence the corresponding output is very relevant in determining the final output) when the input falls within the associated first-level information granule. On the contrary, the weight becomes as small as far the input is from the first-level granule prototype.

In summary, two models are derived from the application of ML-DC (Fig. 2). A simple FIS generated from first-level granulation that can be used mostly for representation purposes, and a hierarchical committee of different FISs that can be used to model the input/output mapping. In this way, the trade-off between accuracy and interpretability can be well balanced.

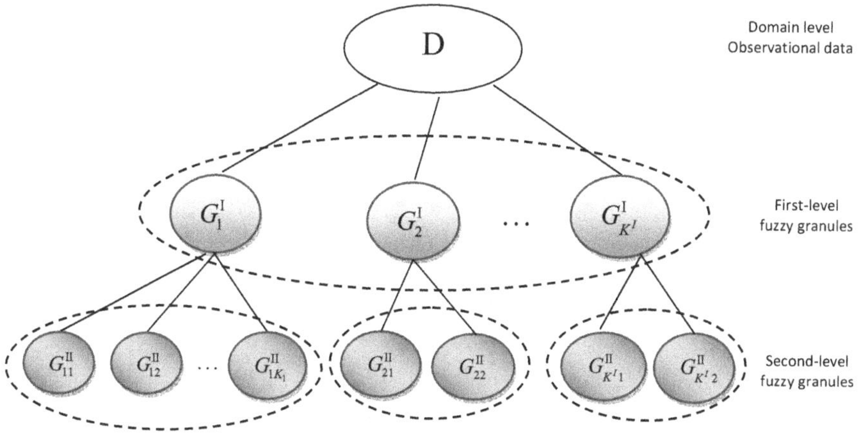

Fig. 1 The multi-level granulation obtained by ML-DC.

A Illustrative Example

In this section we report some simulation results from the application of ML-DC to solve the benchmark classification problem provided by the Cone-Torus (CT) dataset[1]. The CT dataset is available as a collection of a training set and a test set,

[1] http://www.bangor.ac.uk/~kuncheva/Z.txt

both consisting of 400 examples. Each example is two-dimensional and labeled with one out of three class labels. Data are generated synthetically so as to provide the distribution depicted in Fig. 2. This classification problem is inherently difficult, because of the nonlinear class boundaries and high overlapping of data belonging to different classes. According to the literature, simple classifiers are unable to provide an acceptable classification accuracy (classification error about 20-30%), while more complex models (such as FISs with more than 100 rules) provide a classification error not below 10% about (Kuncheva 2000). As a consequence, it is very hard to provide both a linguistically interpretable and accurate description of the Cone-Torus dataset.

ML-DC was applied to the data to perform a two-level granulation. At both levels of granulation, the number of information granules to be generated is not greater than five. Starting from the first-level granules, a FIS classifier was derived, while a committee of FIS classifiers was derived from the second-level granules. The rule base obtained for the first-level FIS is reported in Table 1, while the fuzzy sets generated by the granulation procedure are depicted in Fig. 3. At this level, information granules are quite rough, hence qualitative linguistic terms are more appropriate to represent knowledge. Fuzzy inference with this FIS provides a classification error of 28.25% on the training set and 26.75% on the test set. While the fuzzy model is highly interpretable, its classification ability is quite rough, especially in comparison with other black-box models known in literature.

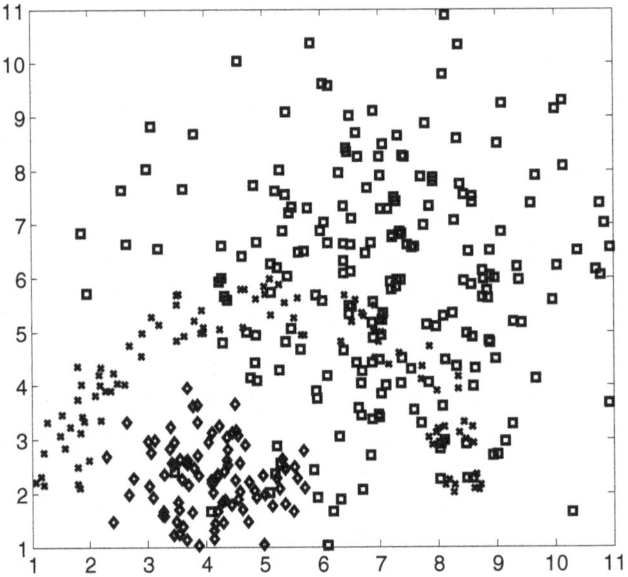

Fig. 2 The Cone-Torus dataset. Class labels are represented as diamonds, squares and crosses.

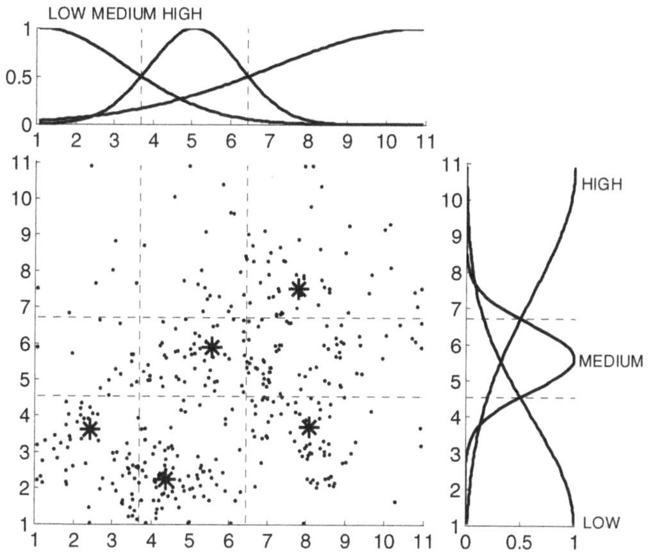

Fig. 3 First-level fuzzy information granulation. Stars represent multi-dimensional prototypes

Table 1 The fuzzy rule base of the first-level FIS

IF x is LOW AND y is LOW THEN P(class=1)=0.53, P(class=2)=0.38, P(class=3)=0.09
IF x is MEDIUM AND y is LOW THEN P(class=1)=0.6, P(class=2)=0.11, P(class=3)=0.29
IF x is MEDIUM AND y is MEDIUM THEN P(class=1)=0.01, P(class=2)=0.45, P(class=3)=0.54
IF x is HIGH AND y is LOW THEN P(class=1)=0.21, P(class=2)=0.27, P(class=3)=0.52
IF x is HIGH AND y is HIGH THEN P(class=1)=0.02, P(class=2)=0.11, P(class=3)=0.87

The committee of FISs derived from the second-level granulation process provides a classification error of 17.25% on the training set and 13.34% on the test set. These results are comparable with other models known in literature, as reported in Table 2. Each FIS in the committee describes a context, i.e. an information granule derived in the first-level granulation process. Because of their finer granularity, second-level information granules are labeled as fuzzy quantities. In Table 3, the rule-base of the second-level FIS derived for the context "x is LOW and y is LOW" is reported, while the fuzzy sets derived by the granulation procedure are depicted in Fig. 4. It should be noted that for each dimension, leftmost and rightmost fuzzy sets have constant membership value one for elements outside the context. This is in coherence with the semantics of the linguistic terms associated to these fuzzy sets.

Table 2 Classification results for the Cone-Torus dataset of different models known in literature

Model	Class. error on the training set	Class. error on the test set
Nearest mean	29.50%	26.25%
Linear Discriminant	25.50%	23.00%
Quadratic Discriminant	19.50%	16.75%
Parzen	8.75%	12.25%
Nearest Neighbor	17.25%	15.25%
Multi-Layer Perceptron (2-15-3)	13.50%	12.00%
LVQ1 (20 prototypes)	15.50%	14.50%
Wang-Mendel (10 fuzzy sets per input 100 rules)	13.50%	13.00%
ML-DC	17,25%	13,34%

Table 3 The rule base of a second-level FIS

```
Context: x is LOW and y is LOW
IF x is about 1.1 or less AND y is about 2.6 THEN P(class=1)=0.31, P(class=2)=0.38, P(class=3)=0.01
IF x is about 1.1 or less AND y is about 4.5 or more THEN P(class=1)=0.12, P(class=2)=0.88,
P(class=3)=0.00
IF x is about 3.7 or more AND y is about 1.0 or less THEN P(class=1)=0.96, P(class=2)=0.02,
P(class=3)=0.02
IF x is about 3.7 or more AND y is about 2.6 THEN P(class=1)=0.88, P(class=2)=0.06, P(class=3)=0.05
```

Based on the respective features of the two systems, the first-level FIS and the committee of second-level FISs can be used synergistically for tackling complex problems: the first-level FIS can be used to provide a rough description of data, with the primary focus of providing a first understanding of the hidden relationships laying among data. The committee of second-level FISs can be effectively used to accurately classify patterns. Each FIS in the committee locally describes a piece of the Universe of Discourse in an interpretable fashion, so as to provide a more detailed representation of the acquired knowledge. The locality of the description is determined by the contexts obtained by the first-level granulation process. As a result, the interpretability of the overall system is preserved.

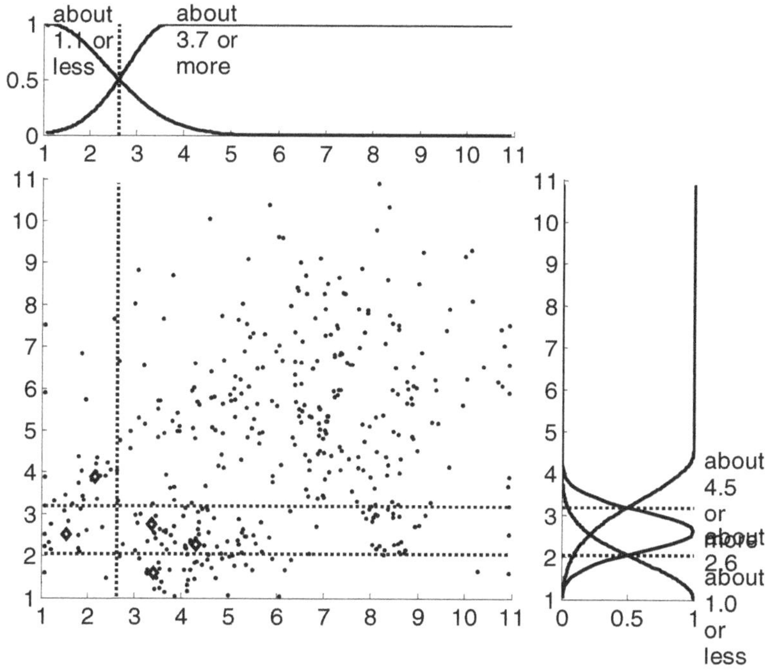

Fig. 4 Second-level information granules for the context "x is low and y is low". Diamonds represent multidimensional prototypes

4 Conclusions

The necessity and the benefits of using GrC in solving real-world problems have been widely pointed out in the recent literature, with an increasing proposal of concrete models and innovative methodologies. As a philosophy and a general methodology, GrC empowers problem solving in many fields; as a paradigm of structured information processing, it supports the development of human-inspired information processing systems. The chance of dealing with different levels of granularity in every stage of problem solving, makes GrC a powerful paradigm for representing and solving problems by means of multi-level strategies. A multi-level granulation process provides several granulated views of the same problem, enabling the focus on useful information structures without looking into too much details.

In this chapter we have emphasized the new perspective of multi-level GrC, as a way to better represent and understand knowledge. As an example of multi-level granulation strategy that improves understanding of data, we have presented ML-DC, a framework to perform a multi-level granulation of data with a balance between accuracy and interpretability. A complex problem is sliced into contexts, which can be used to provide a high-level –yet highly interpretable– description of available data. The contexts are used to locally acquire more accurate fuzzy models, which still preserve interpretability constraints for a readable representation of knowledge. The resulting structure of granules can provide a comprehensible as well as accurate knowledge base. In principle, MD-DC can derive more than two levels of granulation. However, the deeper the granulation level is, the more difficult the comprehensibility of the acquired knowledge. Two levels are deemed enough for a good balance between interpretability and accuracy, since the first level describes the data from a qualitative view, while the second level provides a more quantitative description of the data. Anyway three or more levels might be considered according to the specific application domain.

References

Bargiela, A., Pedrycz, W.: Recursive information granulation: aggregation and interpretation issues. IEEE T. Syts. Man Cyb. 33, 96–112 (2003)

Bargiela, A., Pedrycz, W.: Toward a theory of granular computing for human-centered information processing. IEEE T. Fuzzy Sys. 16, 320–330 (2008)

Belkhouche, B., Lemus-Olalde, C.: Multiple views analysis of software designs. Int. J. Softw. Eng. Know. 10, 557–579 (2000)

Bortolan, G., Pedrycz, W.: Fuzzy descriptive models: an interactive framework of information granulation. IEEE T. Fuzzy Sys. 10(6), 743–755 (2002)

Castellano, G., Fanelli, A.M., Mencar, C.: DCf: A Double Clustering framework for fuzzy information granulation. In: Proceedings of the IEEE International Conference on Granular Computing, Beijing, China, pp. 397–400 (2005)

Chen, Y.H., Yao, Y.Y.: A multiview approach for intelligent data analysis based on data operators. Inform. Sciences 178(1), 1–20 (2008)

De Fent, I., Gubiani, D., Montanari, A.: Granular GeoGraph: a Multi-Granular Conceptual Model for Spatial data. In: Proceedings of 13th Italian Symposium on Advanced Database Systems, Bressanone, Italy, pp. 368–379 (2005)

Glöckner, I., Knöll, A.: A formal theory of fuzzy natural language quantification and its role in granular computing. In: Pedrycz, W. (ed.) Granular Computing: An Emerging Paradigm. Physica-Verlag, Wurzburg (2001)

Gu, S.M., Zhu, S.X., Ye, Q.H.: An Approach for Constructing Hierarchy of Granules Based on Fuzzy Concept Lattices. In: Proceedings of the 5th International Conference on Fuzzy Systems and Knowledge Discovery, Jinan, China, pp. 679–684 (2006)

Han, J., Dong, J.: Perspectives of granular computing in Software Engineering. In: Proceedings of the IEEE International Conference on Granular Computing, pp. 66–71. Silicon Valley, California (2007)

Herrera, F., Herrera-Viedma, E., Martínez, L.: A fusion approach for managing multi-granularity linguistic term sets in decision making. Fuzzy Set Syst. 114, 43–58 (2000)

Herrera-Viedma, E., Cordòn, O., Luque, M., Lopez, A.G., Muñoz, A.M.: A model of fuzzy linguistic IRS based on multi-granular linguistic information. Int. J. Approx. Reason. 34(2-3), 221–239 (2003)

Jeffries, V., Ransford, H.E.: Social Stratification: A Multiple Hierarchy Approach. Allyn and Bacon Inc., Boston (1980)

Kuncheva, L.I.: Fuzzy Classifier Design. Physica Verlag, Berlin (2000)

Liang, J.Y., Li, D.Y.: Uncertainty and Knowledge Acquisition in Information Systems. Science Press, Beijing (2005)

Li, W.S., Ayan, N.F., Kolak, O., Vu, Q., Takano, H.: Constructing multi-granular and topic-focused web site maps. In: Proceedings of the 10th International Conference on the World Wide Web, Hong Kong, pp. 343–354 (2001)

Liang, J.Y., Shi, Z.Z.: The information entropy, rough entropy and knowledge granulation in rough set theory. Int. J. Uncertain Fuzz 12(1), 37–46 (2004)

Mata, F., Martınez, L., Herrera-Viedma, E.: An Adaptive Consensus Support Model for Group Decision-Making Problems in a Multigranular Fuzzy Linguistic Context. IEEE T. Fuzzy Syst. 17(2), 279–290 (2009)

Mencar, C.: Interpretability of Fuzzy Information Granules. In: Bargiela, A., Pedrycz, W. (eds.) Human-Centric Information Processing Through Granular Modelling. Springer, Berlin (2010)

Mencar, C., Castellano, G., Fanelli, A.M.: Balancing interpretability and accuracy by multi-level fuzzy information granulation. In: Proceedings of the IEEE International Conference on Fuzzy Systems, Vancouver, Canada, pp. 2157–2163 (2006)

Mencar, C., Fanelli, A.M.: Interpretability constraints for fuzzy information granulation. Inform. Sciences 178(24), 4585–4618 (2008)

Pawlak, Z.: Granularity of knowledge, indiscernibility and rough sets. In: Proceedings of the IEEE International Conference on Fuzzy Systems, pp. 106–110. Piscataway, New Jersey (1998)

Pedrycz, W.: Conditional Fuzzy C-Means. Pattern Recogn. Lett. 17, 625–632 (1996)

Pedrycz, W.: Granular Computing with Shadowed Sets. In: Ślęzak, D., Wang, G., Szczuka, M.S., Düntsch, I., Yao, Y. (eds.) RSFDGrC 2005. LNCS (LNAI), vol. 3641, pp. 23–32. Springer, Heidelberg (2005)

Pedrycz, W., Gomide, F.: Fuzzy systems engineering: toward human-centric computing. Wiley & Sons, Hoboken (2007)

Peters, J.F., Skowron, A., Synak, P., Ramanna, S.: Rough sets and information granulation. In: De Baets, B., Kaynak, O., Bilgiç, T. (eds.) IFSA 2003. LNCS, vol. 2715, pp. 370–377. Springer, Heidelberg (2003)

Qiu, T., Chen, X., Liu, Q., Hang, H.: A granular space model for ontology learning. In: Proceedings of the IEEE International Conference on Granular Computing, San José, California, pp. 61–65 (2007)

Valente de Oliveira, J.: Semantic Constraints for Membership Function Approximation. IEEE T. Syst. Man Cy. A 29(1), 128–138 (1999)

Zadeh, L.A.: Toward a theory of fuzzy information granulation and its centrality in human reasoning and fuzzy logic. Fuzzy Set Syst. 90(2), 111–127 (1997)

Zhong, N., Yao, Y.Y., Qin, Y., Lu, S., Hu, J., Zhou, H.: Towards granular reasoning on the Web. In: Proceedings of the International Workshop on New Forms of Reasoning on the Semantic Web, Bangkok, Thailand (2008)

Yao, J.T.: A ten-year review of granular computing. In: Proceedings of the IEEE International Conference on Granular Computing, San Jose, California, pp. 734–739 (2007)

Yao, J.T., Yao, Y.Y.: Information granulation for web based information retrieval support systems. In: Proceedings of SPIE, Orlando, Florida, pp. 138–146 (2003)
Yao, Y.Y.: Information granulation and rough set approximation. Int. J. Intell. Syst. 16(1), 87–104 (2001)
Yao, Y.Y.: Three perspectives of Granular Computing. J. Nanchang Institute of Technology 25(2), 16–21 (2006)
Yao, Y.Y.: Granular Computing for Web Intelligence and Brain Informatics. In: Proceedings of the IEEE/WIC/ACM International Conference on Web Intelligence, pp. 21–24. Silicon Valley, California (2007)
Yao, Y.Y.: Granular computing: past, present and future. In: Proceedings of the IEEE International Conference on Granular Computing, Hangzhou, China, pp. 80–85 (2008)

A Rough Set Approach to Building Association Rules and Its Applications

Junzo Watada*, Takayuki Kawaura, and Hao Li

Abstract. Data mining is a process or method of finding information, evidence, insight, knowledge and hypotheses in a huge database, such as marketing data.

Recently, the association rule presented by R. Agrawal in 1983 has been used to rapidly expand a data mining method. This method is general and flexible and can be applied to both general data analysis and very wide surveys. In addition, the rules for this method are complicated. On the other hand, when the support value is minimal and the confidence value is high, the obtained value is already known and trivial. A breakthrough method is needed.

The objective of this paper is to present a rough set model to overcome such issues. Employing the rough set model, we analyzed three different scales of databases and compared the results of simulation experiments using proposed and conventional models. The rough set model obtained an efficient number of association rules and usually took less computation time.

Keywords: Rough set, data mining, association rule.

1 Introduction

Data mining is a process or method of mining information, evidence, insight, knowledge, hypotheses, issues and so on by mining in a huge database,

Junzo Watada · Takayuki Kawaura
Graduate School of Information, Production and Systems, Waseda University,
2-7 Hibikino, Wakamatsu, Kitakyushu 808-0135, Fukuoka, Japan
e-mail: junzow@osb.att.ne.jp

Hao Li
Kansai Medical University

* Corresponding author.

W. Pedrycz and S.-M. Chen (Eds.): Granular Computing and Intell. Sys., ISRL 13, pp. 203–218.
springerlink.com

such as marketing data . Research in this field has begun to employ various techniques developed in machine learning and pattern recognition as well as methods based on conventional statistics. Data-mining systems were recognized as central to strategic information systems in marketing, management, and bioinformatics when R. Agrawal built the association rule method in the early 1990s. The present standard in data mining allows the acquisition of information from more complex data structures, such as texts, multimedia, time-series data, or geographic information.

Recently, the association rule presented by R. Agrawal in 1983 (Agrawal, 1994) has been rapidly expanding a data mining method. It is a general and flexible method that can be applied to general data analysis and very wide surveys. In addition, the rules for this method are complicated. On the other hand, when the support value is minimal and the confidence value is high, the obtained value is already known and trivial. The objective of this paper is to present a rough set method to overcome such issues.

On the other hand, Rough set is proposed by Pawlak (Pawlak 1982). This approach is fit to build the association rule. Rough set theory was developed by Pawlak (Pawlak 1982, 1984, 2004). It has been applied to many issues, including: medical diagnosis, engineering reliability, expert systems, empirical study of materials data (Leclair and Zairko 1996), machine diagnosis (Zhai et al. 2002), travel demand analysis (Goh and Law 2003), business failure prediction (Beynon and Peel 2001), solving linear programs (Azibi and Vanderpooten 2002), data mining (Li, Wang, 2004) and a-RST (Quafafou 2000). Imai et al. apply rough sets analysis to abstract rules in IT corporation labor management (Imai et al. 2008a, 2008b). Matsumoto and Watada analyze stock price movements based on rough sets analysis (Matsmoto, Watada, 2009, 2010). Tai and Watada analyze management problems of technology and engineering among corporate collaborations (Tai, Watada 2010). Watada and Li evaluate the rough sets efficiency comparing Apriori (Watada, Li 2006). Futhermore, Watada et al. employ fuzzy random variables in rough sets (Watada et al. 2010). Previous research has discussed the preference order of attribute criteria needed to extend the original rough set theory, such as sorting, choice and ranking (Greco et al. 2001), insurance markets (Shyng et al. 2007), and rough set theory combined with fuzzy theory (Polkowski 2003). Rough set theory is a useful method for analyzing data and reducing information simply. Rough set theory provides a new mathematical approach to analyze uncertainty and easily classify imperfect data or information with results presented in the form of decision rules. Therefore, in this research, we use rough set theory to analyze the human resource problem.

2 Association Rule

The association rule is a technique to discover rules to express the strength of connections among phenomena. For instance, if the rule obtained is, "when a customer has bought item A, he/she also will buy item B with high probability" from the accumulated data, then this rule clarifies how much item A contributes to sales of item B. In addition, it leads to forecasting the sales volume to a certain degree that a customer will buy item B when he/she has bought item A.

Moreover, in an association rule, a support value denotes the frequency with which all elements included in the association rule appear in the database, and a confidence value denotes the accuracy of the association rule in the database.

The objective of association rule analysis is to retrieve, as rules, the relationships among items that co-occur in many transactions. Association rule analysis is employed for basket analysis, marketing analysis, catalogue design and clustering.

2.1 Definition of Associate Rule

Let us denote a set of transactions by $D = \{t_1, t_2, \cdots, t_k, \cdots, t_n\}$ and the jth transaction t_k consists of items $t_k = \{i_1, i_2, \cdots, i_j, \cdots, j_p\}$, where each t_i indicates some items i_j and $k = 1, 2, 3, \cdots, n$.

The association rule is defined as follows:

$$B \Rightarrow H. \tag{1}$$

where B denotes the body of an association rule, that is, the condition part and H denotes the head of an association rule, that is, the conclusion part. "Body" and "Head" are a set of items. Equation 1 demonstrates that if a transaction has all the items included in the body, then the transaction should also have the items in the head. The association rule expresses the co-occurrence pattern of a group of items.

In searching for association rules, support values and confidence values are used to evaluate each association rule.

Out of all the transactions considered, the degree of support denotes the ratio of the number of transactions that fulfill the association rule. The degree of support indicates the co-occurrence ratio of the transactions that satisfy the association rule. Therefore, the higher the support becomes, the more frequently the pattern included in the body of the association rule is found in all the transactions.

The degree of support $sup(R)$ for the association rule R is defined by the following:

$$sup(R : B \Rightarrow H) = \frac{n(B \cup H)}{N}. \tag{2}$$

The confidence value of an association rule shows the occurrence rate of the conclusion (H) when the body (B) of the association rule is satisfied. In other words, the confidence value indicates the concurrency of the body and the head in all the transactions considered. Therefore, the higher the confidence value becomes, the more reliable the expression of the conclusion of the association rule.

$$conf(R : B \Rightarrow H) = \frac{n(B \cup H)}{sup(B)} \tag{3}$$

where $n(B \cup H)$ denotes the number of all the transactions that include all the items of B and H, and $sup(B)$ denotes the number of all the transactions that include all the items.

In evaluating an association rule, it is important not only for the rule to have a high confidence value but also for the support to be greater than some value . If the support is low, the rule will not occur often. In other words, such a combination of items will not often be bought.

If we search for association rules with low support values, we have to search a huge number of patterns; it reaches combinatorial explosion. Thus, it is better to use a cut-off support value to search for association rules.

2.2 Evaluation of Rules

The above-described support and confidence values are an index with a continuous value to evaluate the rule. The support and confidence values are formulated as follows:

Let us assume that the set of all items is, that the subset is called an item set and that D is a set of all transactions. Each transaction T is a subset of I. If $\Phi \neq X$, $Y \in I$ and $X \cap Y = \phi$, it calls association rule $X \Rightarrow Y$. In D the support for an association rule is defined by the ratio that contains both transactions. The confidence of association rule $X \Rightarrow Y$ is defined by the ratio to contain transaction Y.

Mathematically, it is denoted as follows:

Support= {number of transactions containing both itemsets X, Y}/ {number of all transactions}
Confidence= {number of transactions containing both itemsets X, Y}/ {number of transactions containing itemset X}

where x_i and x_j denote the value of i-th and j-th attribute. Then association rule is denoted $x_i \in I_{ik} \Rightarrow x_j \in I_{jk}$. Moreover, confidence c and support s are denoted as follows.

$$c(I_{ik}) \Rightarrow I_{jk} = \frac{|D(I_{ik}) \cap D(I_{jk})|}{|D(I_{jk})|} \tag{4}$$

$$s(I_{ik}) \Rightarrow I_{jk} = \frac{|D(I_{ik}) \cap D(I_{jk})|}{|D|} \tag{5}$$

2.3 Apriori Algorithm

Apriori algorithm is proposed by R. Agrawal (Agrawal, 1994) may generate an association rule at high speed. It is a high-speed algorithm for discovering association rules with support and confidence by setting thresholds for the minimal support value and minimal confidence value. The apriori algorithm is explained as follow.

Supposed that a database is denoted as D, a large item set is as T and the candidate item set is as C_i. Then, the database is scanned and the support in each candidate item set is examined. Assumed that is fulfilled minimum support in L_i. Finally, it generates L_1 from C_1. This process is continued until the candidate item set becomes empty.

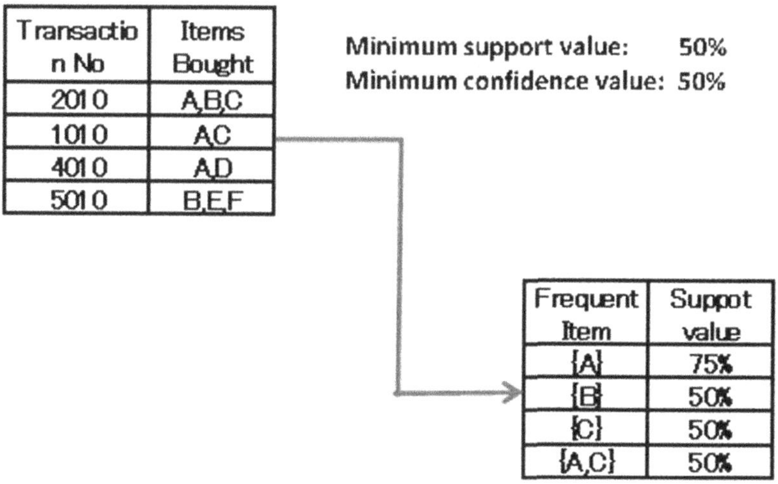

Fig. 1 Example of Associate Rules

In the example shown in Figure 1, let us consider rule $A \Rightarrow C$. The support value for the rule has 50% for $\{A, C\}$, and the confidence value of the rule is 66.6% for $\{A, C\}$.

The set of frequent items means the set of items with support values greater than the minimum support value. The extraction of association rules is the process of finding all the rules that satisfy the minimum support value and the minimum confidence value chosen.

The following are types of association rules:

boolean associateion rules
quantitative association rule
single-dimensional association rule

multi-dimensional association rule
multi-level association rule
single-level association rule.

Figure 2 illustrates the example of an apriori algorithm.

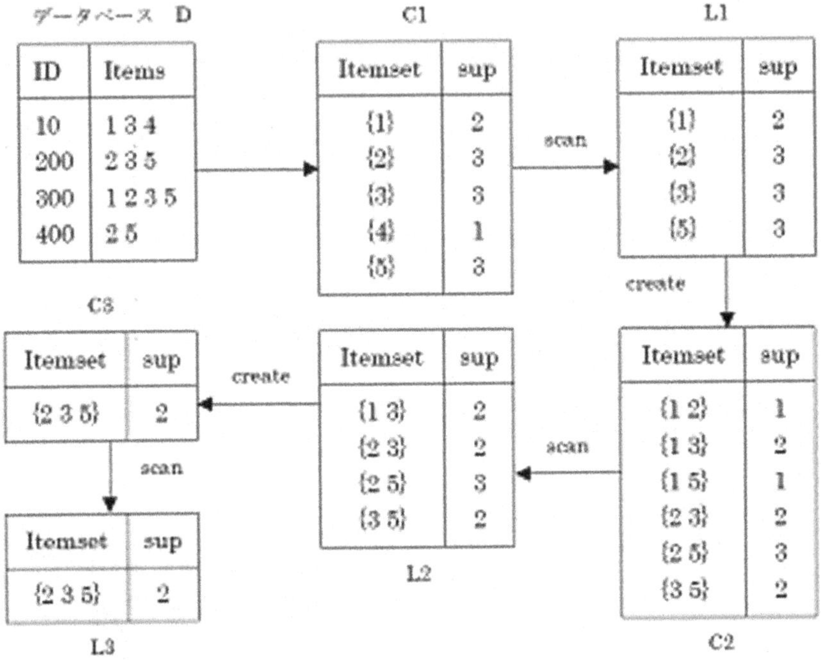

Fig. 2 Example of Apriori Algorithm

3 Rough Set Theory

A rough set expresses a set of subsets classified by equivalent relation using lower and upper approximations. It was proposed by Z. Pawlak in 1982 (Pawlak, 1982). In other words, the rough sets deal with vagueness included in the fact that classification cannot be done well. The rough set theory is closely related to data mining. For example, some targets expressed with a subset of known data can be approximated. That is, that indicates the features of the data that can identify customers who bought some goods. The rough set theory provides a method to approximately reduce the required feature (called **reduction**). Furthermore, a vague phenomenon is expressed using upper and lower approximations. This concept is called **rough**.

3.1 Information System

Let us consider a database defined using the set, U, of objects and the set, N, of features. This is called an information table.

Equivalent Relation. A relation is an equivalent relation if and only if the following relations hold:

$$\begin{array}{l} Reflectivity: {}_aR_a = 1 \\ Symetricity: {}_aR_b =_b R_a \\ Transitivity: {}_aR_b \ and \ {}_bR_c \Rightarrow_a R_c \end{array} \tag{6}$$

Lower and Upper Approximations. The lower approximation $apr_*(X)$ is defined as follows:

$$apr_*(X) = \{x \in U | R(x) \in X\} \tag{7}$$

The upper approximation $apr^*(X)$ is defined as follows:

$$apr^*(X) = \{x \in U | R(x) \cap X \neq \Phi\} \tag{8}$$

3.2 Reduction by a Rough Set

An important concept of a rough set is the rough approximation. From a data-mining viewpoint, the reduction of features is useful in mining. Reduction is a method to obtain the minimal number of features that can express the equivalent knowledge of experts. If few features can express the equivalent knowledge of experts, it is possible to recognize such a knowledge using few features. For example, in the case where the features that define preferable goods among various generations are color and shape, we can distinguish such knowledge using only the two features color and shape. We can design products that have such features.

In rough set theory, reduction is defined as a subset of the minimal features that can equivalently distinguish the objects that the set of all features can discriminate on a given information table. In other words, when an information table is given, a subset of features in the information table can obtain equivalent basic sets to the original basic sets. This subset of features is called the reduction.

Table 1 shows the names of goods as A, B, C, D and E on the top row. This is called a feature. Each entry shows whether a person buys a good with 1 and not with 0. The universal set is denoted as $U = \{L_1, L_2, L_3, L_4, L_5, L_6\}$, the set of all features is denoted as $AT = \{A, B, C, D, E\}$.

Let us simplify the explanation. We take the subset, N of the universe AT as $N = \{A, B\}$, as shown in Table 2. Then, we can understand that samples L_2 and L_6 take the same value for each feature. Samples L_3 and L_5 behave

Table 1 Example 1 of an information table

TID	A	B	C	D	E
L1	1	1	0	0	1
L2	1	0	0	1	0
L3	0	0	1	1	0
L4	0	1	1	0	1
L5	0	0	1	0	0
L6	1	0	0	1	1

Table 2 Information table 1 with two features

TID	A	B
L1	1	1
L2	1	0
L3	0	0
L4	0	1
L5	0	0
L6	1	0

Table 3 Information table 2 with three features

TID	A	B	C
L1	1	1	0
L2	1	0	0
L3	0	0	1
L4	0	1	1
L5	0	0	1
L6	1	0	0

the same as well. On the other hand, samples L_1 and L_4 behave differently. Therefore, the equivalent classes are as follows: $\{L_1\}$, $\{L_2, L_6\}$, $\{L_3, L_5\}$ and $\{L_4\}$.

This example shows the granulated samples in terms of features A and B. This equivalent class is called a basic set for the set of features $N = \{A, B\}$. Therefore, the subset of features $N = \{A, B\}$ cannot distinguish individual samples. When we employ three features, $N = \{A, B, C\}$, then we obtain the equivalent classes as $\{L_1\}$, $\{L_2\}$, $\{L_3\}$, $\{L_4\}$, $\{L_5\}$, $\{L_6\}$. All samples can be discriminated using three features A, B and C.

From this discussion, if we employ all features, we can distinguish individual samples, but we can also express all samples behaviors using only three features. Therefore, Tables 1 and 3 are equivalent.

In the rough set theory, the minimal subset of features is called the reduction of the set of the whole features AT when it can distinguish as equivalently all samples as AT.

3.3 Approximation by a Rough Set

In the rough set theory, sets or classes induced by equivalent relation or similarity relation are taken as knowledge, and two approximation methods are employed as lower and upper approximations. From the fuzzy-set viewpoint, the upper approximation gives an understanding of necessity and the lower approximation gives an understanding of possibility. The pair of these expressions is called a rough set. Uncertain and vague phenomena should be expressed using the lower and upper approximations. The pair of both expressions is called a rough approximation.

When we deal with incomplete information, an interval value is used instead of a rigid value. For example, we can define the age of a person [20,40] as shown in Table 4.

Table 4 Incomplete Information

Name	Age
a_1	$X_1 = [23, 26]$
a_2	$X_1 = [20, 22]$
a_3	$X_1 = [30, 36]$
a_4	$X_1 = [20, 23]$
a_5	$X_1 = [27, 31]$

Considering an example given in Table 4, let us retrieve Table 4. The age is given with incomplete information. Let us retrieve all people whose age is in $S = [20, 25]$. The upper approximation $R^*(S)$ is defined as follows:

$$R^*(S) = \{a_i | S \cap X_i \neq \Phi\} = \{a_1, a_2, a_4\} \tag{9}$$

where Φ is an empty set, $\cap$ denotes an intersection among sets. Set $R^*(S)$ is defined using $S \cap X_i \neq \Phi$. The result obtained is $\{a_1, a_2, a_4\}$.

The lower approximation $R_*(S)$ is defined as follows:

$$R_*(S) = \{a_i | S \subseteq X_i\} = \{a_2, a_4\} \tag{10}$$

where $\subseteq$ denotes an inclusion relation. In this definition, only persons whose ages are within the given interval $S = [20, 25]$ are retrieved, an example of a necessity relation

As understood in the example $\{a_2, a_4\} \subseteq \{a_1, a_2, a_4\}$, the upper and lower approximations hold the following relationship:

$$R_*(S) \subseteq R^*(S) \tag{11}$$

As mentioned above, incomplete information in a database can be treated with pairs of upper and lower approximations. These pairs are called a rough set $\{R_*(S), R^*(S)\}$, as mentioned above.

4 Analysis by a Rough Set Approach to Associate Rule Analysis

The objective of this paper is to mine a database using a rough set approach for association rule analysis.

4.1 Experimental Environment

We employed the following system:

- CPU: Pentium(R) 4 CPU 2.80GHz
- Main memory: 512 Mbyte
- Operation System: Microsoft Windows XP Professional Version 2002
- Program language: C

4.2 Experimental Data

We analyzed four datasets.

Simulation Data *A*. is the United States Congressional Voting records for 1984, a dataset that is widely employed in research on machine learning and knowledge acquisition. This data consists of 17 features, and the number of samples is 435.

Simulation Data *B*. is a randomly generated consumer behavior. The number of samples is 2000, and the number of features (goods) is 50, denoted by *A*,*B*,*C* and so on. The value 1 denotes buying, and 0 denotes not buying by 0.

Simulation Data *C*. is the Census-Incom Database from 1994 to 1995 from the UCI repository. This database includes 40 features, such as demography, employment and income. The number of samples is 199,523.

Table 5 Features defined in Data *A*

No	Feature	Values
1	Class Name:	2 (democrat, republican)
2	handicapped-infants:	2 (y,n)
3	water-project-cost-sharing:	2 (y,n)
4	adoption-of-the-budget-resolution:	2 (y,n)
5	physician-fee-freeze:	2 (y,n)
6	el-salvador-aid:	2 (y,n)
7	religious-groups-in-schools:	2 (y,n)
8	anti-satellite-test-ban:	2 (y,n)
9	aid-to-nicaraguan-contrast:	2 (y,n)
10	mx-missile:	2 (y,n)
11	immigration:	2 (y,n)
12	synfuels-corporation-cutback:	2 (y,n)
13	education-spending:	2 (y,n)
14	superfund-right-to-sue:	2 (y,n)
15	crime:	2 (y,n)
16	duty-free-exports:	2 (y,n)
17	Export-administration-act-south-africa:	2 (y,n)

	A	B	C	D	E	F	G	H	I	J	K	L	M	N	O	P	Q	R	S	T	U	V	W	X	Y	Z
1	0	1	0	0	1	0	1	0	1	1	1	0	0	1	0	1	1	1	0	1	1	0	0	1	0	1
2	1	1	1	1	1	0	0	1	1	0	1	1	0	1	1	0	0	0	0	0	0	1	1	0	1	1
3	0	0	0	0	1	0	0	1	0	1	0	0	0	0	0	0	0	0	1	0	1	0	0	0	0	1
4	1	1	1	0	0	0	1	0	0	0	1	1	0	0	0	0	0	0	0	1	0	0	0	0	1	0
5	1	0	1	1	0	1	0	1	0	0	0	0	0	1	0	1	1	1	0	1	0	0	1	0	1	0
6	1	1	1	0	1	1	0	1	0	0	0	0	0	0	1	0	0	1	0	0	0	0	0	1	0	1
7	1	1	0	1	0	0	0	0	0	0	0	0	0	0	0	1	0	0	0	0	0	0	0	0	0	0
8	1	0	0	1	0	0	0	0	0	1	0	1	0	0	0	0	0	0	0	0	1	0	0	1	1	0
9	1	0	1	0	1	1	1	0	0	0	1	0	0	0	1	0	1	0	1	1	0	1	1	0	1	1
10	0	0	0	0	0	1	0	1	1	0	0	0	0	0	0	0	0	1	1	1	1	0	0	0	0	1
11	0	1	0	0	1	1	0	1	0	0	0	0	0	0	0	1	0	0	0	0	0	0	0	0	0	1
12	1	1	0	1	0	0	0	0	0	0	0	0	0	1	0	1	1	1	1	1	0	1	1	0	1	0
13	0	0	0	0	0	0	0	0	0	1	0	0	1	1	1	1	1	0	0	0	1	0	0	0	0	0
14	1	1	1	0	1	1	0	0	1	0	1	1	0	1	0	1	0	1	1	0	1	0	1	0	0	1
15	0	1	0	1	0	1	1	1	0	0	0	1	1	0	1	1	1	0	0	1	0	1	0	1	0	1

Fig. 3 Randomly generated consumers behaviors

Table 6 Features defined in Data A

	Old Variable	New Variable		Old Variable	New Variable		Old Variable	New Variable
1	Age	dAge	24	Income8	dIncome8	47	Rlabor	iRlabor
2	Ancstry1	dAncstry1	25	Industry	dIndustry	48	Rownchld	iRownchld
3	Ancstry2	dAncstry2	26	Korean	iKorean	49	Rpincome	dRpincome
4	Avail	iAvail	27	Lang1	iLang1	50	RPOB	iRPOB
5	Citizen	iCitizen	28	Looking	iLooking	51	Rrelchld	iRrelchld
6	Class	iClass	29	Marital	iMarital	52	Rspouse	iRspouse
7	Depart	dDepart	30	May75880	iMay75880	53	Rvetserv	iRvetserv
8	Disabl1	iDisabl1	31	Means	iMeans	54	School	iSchool
9	Disabl2	iDisabl2	32	Military	iMilitary	55	Sept80	iSept80
10	English	iEnglish	33	Mobility	iMobility	56	Sex	iSex
11	Feb55	iFeb55	34	Mobillim	iMobillim	57	Subfam1	iSubfam1
12	Fertil	iFertil	35	Occup	dOccup	58	Subfam2	iSubfam2
13	Hispanic	dHispanic	36	Othrserv	iOthrserv	59	Tmpabsnt	iTmpabsnt
14	Hour89	dHour89	37	Perscare	iPerscare	60	Travtime	dTravtime
15	Hours	dHours	38	POB	dPOB	61	Vietnam	iVietnam
16	Immigr	iImmigr	39	Poverty	dPoverty	62	Week89	dWeek89
17	Income1	dIncome1	40	Pwgt1	dPwgt1	63	Work89	iWork89
18	Income2	dIncome2	41	Ragechld	iRagechld	64	Worklwk	iWorklwk
19	Income3	dIncome3	42	Rearning	dRearning	65	WWII	iWWII
20	Income4	dIncome4	43	Relat1	iRelat1	66	Yearsch	iYearsch
21	Income5	dIncome5	44	Relat2	iRelat2	67	Yearwrk	iYearwrk
22	Income6	dIncome6	45	Remplpar	iRemplpar	68	Yrsserv	dYrsserv
23	Income7	dIncome7	46	Riders	iRiders			

4.3 Simulation Result

In this experiment, we employed four databases A, B and C mentioned above. Minimal support values employed were 1%, 5% and 10%; the minimal confidence values employed were 30%, 50% and 70%.

Table 8 shows the computational time for the mining of Database A. The rough set approach required a little bit more time than the conventional one.

Regarding Database A, the number of association rules obtained is shown in Table 9. The number obtained by the rough set approach is smaller than by the conventional model.

The comparison of computation time relating to Database B is shown in Table 10, and the number of associate rules obtained is illustrated in Table 11. The computation time is not much different, but the number of rules obtained is drastically lower.

Table 7 Features defined in Data C

Old Variable	NewVariable	Old Variable	NewVariable
1 age	AAGE	24 state of previous residence	GRINST
2 class of worker	ACLSWKR	25 detailed household and family stat	HHDFMX
3 industry code	ADTIND	26 detailed household summary in household	HHDREL
4 occupation code	ADTOCC	27 instance weight	MARSUPWT
5 adjusted gross income	AGI	28 migration code-change in msa	MIGMTR1
6 education	AHGA	29 migration code-change in reg	MIGMTR3
7 wage per hour	AHRSPAY	30 migration code-move within reg	MIGMTR4
8 enrolled in edu inst last wk	AHSCOL	31 live in this house 1 year ago	MIGSAME
9 marital status	AMARITL	32 migration prev res in sunbelt	MIGSUN
10 major industry code	AMJIND	33 num persons worked for employer	NOEMP
11 major occupation code	AMJOCC	34 family members under 18	PARENT
12 mace	ARACE	35 total person earnings	PEARNVAL
13 hispanic Origin	AREORGN	36 country of birth father	PEFNTVTY
14 sex	ASEX	37 country of birth mother	PEMNTVTY
15 member of a labor union	AUNMEM	38 country of birth self	PENATVTY
16 reason for unemployment	AUNTYPE	39 citizenship	PRCITSHP
17 full or part time employment stat	AWKSTAT	40 total person income	PTOTVAL
18 capital gains	CAPGAIN	41 own business or self employed	SEOTR
19 capital losses	CAPLOSS	42 taxable income amount	TAXINC
20 divdends from stocks	DIVVAL	43 fill inc questionnaire for veteran's admin	VETQVA
21 federal income tax liability	FEDTAX	44 veterans benefits	VETYN
22 tax filer status	FILESTAT	45 weeks worked in year	WKSWORK
23 region of previous residence	GRINREG		

Table 8 Comparison of computation time for Database A

No	Suport Value	Condence Value	Conventional Model	Rough Set Model
1	1%	30%	1.12	1.26
2	1%	50%	1.02	1.34
3	1%	70%	1.06	1.58
4	5%	30%	0.98	1.32
5	5%	50%	1.14	1.03
6	5%	70%	0.94	1.13
7	10%	30%	0.97	1.25
8	10%	50%	0.91	0.96
9	10%	70%	0.92	0.96

The results for database C are shown in Tables 12 and 13, respectively. The number of rules obtained by the rough set approach was smaller than that obtained by the conventional model.

Table 9 Comparison of the number of obtained associate rules for Database A

No	Suport Value	Condence Value	Conventional Model	Rough Set Model
1	1%	30%	59,153	37,471
2	1%	50%	28,458	18,432
3	1%	70%	11,997	9,234
4	5%	30%	28,712	16,236
5	5%	50%	15,594	10,235
6	5%	70%	7,742	4,876
7	10%	30%	15,048	9,151
8	10%	50%	9,290	5,697
9	10%	70%	5,134	3,651

Table 10 Comparison of computation time for Database B

No	Suport Value	Condence Value	Conventional Model	Rough Set Model
1	1%	30%	1.97	1.98
2	1%	50%	1.78	2.01
3	1%	70%	1.65	1.88
4	5%	30%	1.85	1.93
5	5%	50%	1.66	1.71
6	5%	70%	1.54	1.62
7	10%	30%	1.68	1.66
8	10%	50%	1.25	1.46
9	10%	70%	1.11	1.35

Table 11 Comparison of the number of obtained associate rules for Database B

No	Suport Value	Condence Value	Conventional Model	Rough Set Model
1	1%	30%	135,879	81,523
2	1%	50%	85,469	52,995
3	1%	70%	44,825	29,584
4	5%	30%	95,642	51,648
5	5%	50%	48,120	27,583
6	5%	70%	24,056	16,117
7	10%	30%	44,235	23,549
8	10%	50%	21,457	12,874
9	10%	70%	13,547	9,347

Table 12 Comparison of computation time for Database C

No	Suport Value	Condence Value	Conventional Model	Rough Set Model
1	1%	30%	3,212.75	2,184.67
2	1%	50%	3,109.75	1,987.25
3	1%	70%	3,001.30	1,521.24
4	5%	30%	1,047.39	624.52
5	5%	50%	981.85	587.41
6	5%	70%	966.23	513.29
7	10%	30%	601.39	341.75
8	10%	50%	528.24	283.21
9	10%	70%	531.55	228.76

Table 13 Comparison of the number of obtained associate rules for Database C

No	Suport Value	Condence Value	Conventional Model	Rough Set Model
1	1%	30%	99,191,484	63,488,497
2	1%	50%	86,407,089	55,300,539
3	1%	70%	71,613,085	4,3696,891
4	5%	30%	15,959,402	9,728,612
5	5%	50%	13,928,786	7,956,458
6	5%	70%	11831416	6,713,462
7	10%	30%	8,305,791	4,423,415
8	10%	50%	7,405,234	3,874,562
9	10%	70%	6,443,111	2,945,813

5 Conclusions

In this paper, we proposed the rough set model for association rule analysis. Employing the model, we analyzed three different scales of databases and compared the results of simulation experiments with the proposed and conventional models. The rough set model obtained an efficient number of association rules and usually took less computation time.

For the smaller scale database, the proposed model took a little bit more computation time than the conventional model. Regardless, the proposed model obtained an effective and efficient number of association rules.

References

Agrawal 1994. Agrawal, R.: Fast Algorithms for Data Mining Applications. In: Proceedings of the 20th International Conference on Very Large Databases, Santiago, Chile, pp. 487–489 (1994)

Apriori 2006. Apriori-Association Rule Induction / Frequent Item Set Mining (2006), `http://fuzzy.cs.uni-magdeburg.de/borgelt/apriori.html`. (accessed May 12, 2006)

Borgel 2002. Borgel, C.: `http://fuzzy.cs.uni-magdeburg.de/borgelt/apriori/` (accessed May 15, 2002)

Imai et al 2008. Imai, S., Lin, C.W., Watada, J., Tzeng, G.H.: Knowledge acquisition in human resource management based on rough sets. In: Proceedings, PICMET: Portland International Center for Management of Engineering and Technology, July 27-31, pp. 969–974 (2008)

Imai et al 2008. Imai, S., Lin, C.W., Watada, J.: Knowledge Acquisition in Human Resource Management Based on Rough Sets Analysis. International Journal of Simulation (2008)

Lin, Watada, 2010. Lin, L.C., Watada, J.: Restructuring of rough sets for fuzzy Random data of creative city evaluation. In: Huynh, V.-N., Nakamori, Y., Lawry, J., Inuiguchi, M. (eds.) Proceedings of Integrated Uncertainty Management and Applications, IUM 2010. AISC, vol. 68, pp. 523–534. Springer, Heidelberg (2010), ISSN 1867-5662

Matsumoto, Watada 2009. Matsumoto, Y., Watada, J.: Knowledge Acquisition from Time Series Data through Rough Sets Analysis. International Journal of Innovative Computing, Information and Control (IJICIC) 5(12(B)), 4885–4897 (2009)

Matsumoto, Watada 2010. Matsumoto, Y., Watada, J.: Prediction of Tick-wise Price Fluctuations for Rough Sets. In: Proceedings, ISME 2010, Kokura, Japan, August 28-30 (2010)

Mori et al. 1994. Mori, N., Tanaka, H., Inoue, K.: Rough Set and Kansei. Kaibundo Publishing (2004)

Pawlak 1982. Pawlak, Z.: Rough sets. International Journal of Information Computer Science 11(5), 341–356 (1982)

Salton and McgGill 1983. Salton, G., McgGill, M.: Introduction to Modern Information Retrieval. McGraw-Hill, New York (1983)

Tai, Watada 2010. Tai, Y.L., Watada, J.: Creating SMESf Innovation Capabilities Through Formation of Collaborative Innovation Network in Taiwan. In: Proceedings of IFMIP 2010, WAC 2010, Kobe, Japan, September 19-23 (2010)

UCI 2005. The UCI Knowledge Discovery in Databases Archive (2005), `http://kdd.ics.uci.edu/summary.data.date.html` (accessed 2005); [19]

Usage 2001. The Usage of Apriori Program (2001), `http://bruch.sfc.keio.ac.jp/course/DM01/man-a.html` (accessed June 20, 2001); [12]

Watada, Li 2006. Watada, J., Li, H.: A Rough Set Approach to Building Association Rules and Its Applications. In: Proceedings, ICAiET 2006, 3rd Int. Conf. on Artificial Intelligence in Engineering and Technology, November 22-24 (2006)

Watada et al 2010. Watada, J., Lin, L.C., Ding, L., Shapiai, M.I., Chew, L.C., Ibrahim, Z., Jau, L.W., Khalid, M.: A Rough-Set-Based Two-class Classifier for Large Imbalanced Dataset. In: Proceedings, KES-IDT 2010, Baltimore, USA, July 28-30 (2010)

Watada et al 2010. Watada, J., Bahar Yaakob, A., Takahashi, T., Okamoto, T.: The Diagnosis of Power Transformer Failures by Fuzzy Random Based Rough Sets Analysis. In: Proceedings, International Conference on Condition Monitoring and Diagnosis (CMD 2010), Shibaura, Japan, September 6-11 (2010)

Watada et al. 2010. Watada, J., Lin, L.C., Matsumoto, Y.: Fuzzy Random Based Rough Sets Analysis and Its Application. In: Proceedings, I FMIP 2010, WAC 2010, Kobe, Japan, September 19–23 (2010)

Watada, Yaakob 2010. Watada, J., Bahar Yaakob, S.: Diagnosis System Based on Rough Sets Analysis. In: Proceedings, ICGEC-2010-IS24-04, Fourth International Conference on Genetic and Evolutionary Computing (ICGEC-2010), Shenzhen, China, December 13–15(2010)

UCI 2005. The UCI Knowledge Discovery in Databases Archive (2005), `http://kdd.ics.uci.edu/summary.data.date.html` (accessed June 3, 2005)

Usage 2001. Usage of Apriori Program (2001), `http://bruch.sfc.keio.ac.jp/course/DM01/mana.html` (accessed June 20,2001)

Yokomori, Kobayashi 1994. Yokomori, T., Kobayashi, S.: Rough Sets and Decision Making, Relay Series: Rough Sets: Its Theory and Apllications. Journal of Mathematical Science 5, 77–83 (1994) (in Japanese); [11]

Fuzzy Modeling with Grey Prediction for Designing Power System Stabilizers

Y.T. Hsiao, T.L. Huang, and S.Y. Chang

Abstract. This work presents a novel method for designing power system stabilizers (PSS) by using a fuzzy PID controller tuned with grey prediction. The PSS design can be formulated as an optimal linear regulator control problem; however, implementing the PSS requires designing estimators, increasing implementation complexity and reducing control system reliability. Therefore, this work seeks a control scheme that adopts only the desired state variable, for example speed. The grey fuzzy PID control is integrated with grey prediction to determine the control signal of each generator thus simplifying the design and increasing system performance. The grey prediction uses forecast information regarding the output state variables of the generators to the fuzzy tuning PID controller to control power system behavior and thus achieve good performance, enabling the presented method to reduce power system oscillation and increase dynamic stability. Finally, the advantages of the proposed method are highlighted by simulating the detailed behavior of a multimachine power system.

Keywords: Power systems stability, fuzzy, PID control, grey prediction.

1 Introduction

Power systems are complex nonlinear systems and frequently exhibit low-frequency power oscillations because of insufficient damping. The use of supplementary excitation control signals to improve the dynamic stability of power systems has attracted much interest over the last two decades. Power system stabilizers (PSSs) provide this supplementary stabilizing signal and are used extensively to effectively suppress the electromechanical oscillations of the generator

Y.T. Hsiao
Department of Digital Technology Design, National Taipei University of Education, Taipei, Taiwan R.O.C

T.L. Huang
Department of Computer Science, National Taipei University of Education, Taipei, Taiwan R.O.C

S.Y. Chang
Department of Electrical Engineering, Lee-Ming Institute of Technology, Taipei, Taiwan R.O.C

W. Pedrycz and S.-M. Chen (Eds.): Granular Computing and Intell. Sys., ISRL 13, pp. 219–235.
springerlink.com

and improve power system stability (Moucod and Yu 1972, Hung and Yang 1991, Chung *et al.* 2002, Chow *et al.* 2004).

The conventional design of PSSs is based on a linearlized model of a power system near a nominal operating point, and aims to provide optimal performance at that point. Restated, the problem of designing the PSSs is formulated as an optimal linear regulator control problem with a complete state control scheme (Moucod 1972). Therefore, the implementation of "conventional" design methods depends including state estimators in control scheme. However, this inclusion complicates implementation and reduces control system's reliability. A simple control scheme must use only the desired state variable, such as speed, to control the generators and simplify the design of the PSSs. The design of conventional PSS considers previous system states to determine the control signal, and thus cannot easily be used to control the power system before the system state changes. Many attempts have been made to eliminate the need to estimate the system states using adaptive or observer controllers (Hosseinzadeh and Kalam 1999, You *et al.* 2003). However, such systems involve complex mathematical models or/and expensive computation. This work employs a mathematically simple and computationally efficient grey predictor to evaluate the output status of the generators and thus tune the controllers. It also presents fuzzy PID control to replace complete state control, thus reducing the computational complexity and improving the performance of the PSS.

The grey theory presented by Deng has been successfully applied to control systems and to control motors (Deng 1982, Deng 1989, Wong and Liang 1997). The grey prediction method requires relatively few system output data to construct a grey model, and forecasts a future value without complex calculations. The fuzzy method was developed to design adaptive PSS, and exhibits excellent potential to increase the damping of generator oscillations (Hosseinzadeh and Kalam 1999, You *et al.* 2003). This work integrates the grey predictor with the fuzzy PID controller to develop a predictive PSS. Simulations reveal that the proposed method reduces oscillation and increases dynamic stability in power systems. The following sections detail the proposed method. This work presents the results of the simulation of a two-machine system to demonstrate the effectiveness of the proposed method.

2 Grey Prediction

The black system illustrates that system internal structure, parameters and characteristics are unknown and can only be obtained from external behaviors. Restated, a black system is one about which information is lacking, while a white system is one whose internal properties are fully known. Meanwhile, a system whose properties are only partially known is termed a grey system. Grey theory attempts to use white system performance to predict grey system performance. Grey prediction establishes a grey model that makes predictions based on information about the past. The grey model can be used to predict trends in system output.

Grey theory is based on grey exponential law. Elucidating the trend in a non-negative sequence with an arbitrary distribution is difficult. However, such a sequence becomes monotonically nondecreasing when accumulated. Grey exponential law is then used to derive an optimal exponential curve to fit this sequence, and predictions are made using inverse transform. Prediction accuracy depends on the system characteristics. Grey systems generally perform well if they follow an exponential law (Su 2000). In the prediction, the accumulated generating operation (AGO) and inverse accumulated generating operation (IAGO) are the basic tools for building a grey model. If { $y^{(0)}(k)$ }, $y^{(0)}(k) \geq 0$, k = 0, 1, …, n are time-sequence data, then the ACO is

$$y^{(1)}(k) = AGO \bullet y^{(0)} = \sum_{m=1}^{k} y^{(0)}(m), \qquad k = 1,2,\ldots,n \tag{1}$$

where $y^{(1)}(k)$ denote the accumulated generating sequence data, and are monotonically increasing. Since $y^{(0)}(k)$ is not always positive, exponential or linear mapping should be used to alter the sequence behavior before performing the next generation operation.

The data sequence $z^{(1)}$ is defined by applying the following MEAN generating operation to $y^{(1)}$

$$z^{(1)}(k) = MEAN \bullet y^{(1)} = \frac{1}{2}[y^{(1)}(k) + y^{(1)}(k-1)], k = 2,3,\ldots,n \tag{2}$$

From $z^{(1)}(k)$, a grey model GM(1,1) was obtained using a first-order single-variable whitening differential equation:

$$\frac{dy^{(1)}}{dt} + az^{(1)}(k) = u \tag{3}$$

The parameters a and u can be determined using the least-square method:

$$\hat{\theta} = \begin{bmatrix} a \\ u \end{bmatrix} = (B^T B)^{-1} B^T y_N \tag{4}$$

where

$$B = \begin{bmatrix} -z^{(1)}(2) & 1 \\ -z^{(1)}(3) & 1 \\ \vdots & \vdots \\ -z^{(1)}(n) & 1 \end{bmatrix} \tag{5}$$

and

$$y_N = (y^{(0)}(2), y^{(0)}(3), \ldots, y^{(0)}(n)), \tag{6}$$

Based on the whitening equation (3), the predicted value of the GM(1,1) model is:

$$\hat{y}^{(1)}(t+1) = (y^{(0)}(1) - \frac{u}{a})e^{-at} + \frac{u}{a} \tag{7}$$

where ^ denotes the predicted values. The forecasted value of $\hat{y}^{(1)}(t+1)$ can be expressed in general form as,

$$\hat{y}^{(1)}(n+p) = (y^{(0)}(1) - \frac{u}{a})e^{-a(n+p-1)} + \frac{u}{a} \tag{8}$$

where p denotes the predicted step size. The operation of IAGO for $y^{(1)}(k)$ is

$$y^{(0)}(0) = y^{(1)}(0) \tag{9}$$

$$y^{(0)}(k) = y^{(1)}(k) = y^{(1)}(k-1) \tag{10}$$

Therefore, the predicted output at step (n + p) is obtained by differentiating the following equation:

$$\hat{y}^{(0)}(n+p) = (y^{(0)}(1) - \frac{u}{a})(1-e^{a})e^{-a(n+p-1)}, \quad n \geq 4 \tag{11}$$

The grey model prediction is a local curve-fitting extrapolation schema, so at least four data sets are required for the first-order single-variable grey prediction model to obtain an approximate grey prediction. This work uses the four most recent output data to predict the subsequent output $\hat{y}(k+1)$ by applying the grey model. The data sets are substituted into Eq. (4) to recursively determine the grey parameters a and u,. Subsequent system output can then be predicted by substituting the grey parameters into (11).

3 Fuzzy PID Controller

Proportional-integral-derivative (PID) control is a well-known technique for controlling industrial processes. PID control is simple and performs robustly under a broad range of operating conditions. The design of PID control depends on specifying three parameters - proportional gain (K_p), integral time constant (K_i) and derivative time constant (K_d). A problem has traditionally been solved by trial and error development. Design engineers must tune PID controllers manually, which is a time-consuming task. The transfer function of a PID controller is expressed as

$$G_C(s) = K_p + \frac{K_i}{S} + K_d S \tag{12}$$

where K_p, K_i and K_d represent the proportional, integral and derivative gains, respectively. The output $u(t)$ of the PID controller reads as

$$u(t) = K_p e(t) + K_i \int_0^t e(\tau)d\tau + K_d \dot{e}(t) \tag{13}$$

where $e(t)$ is the error between the reference input and the output at time t.

From a practical point of view, the K_p and K_d can be set in the ranges $[K_{p\min}, K_{p\max}] \subset R$ and $[K_{d\min}, K_{d\max}] \subset R$ such that $K_p \in [K_{p\min}, K_{p\max}]$ and $K_d \in [K_{d\min}, K_{d\max}]$. For convenience, K_p and K_d are normalized as follows.

$$K'_p = \frac{K_p - K_{p\min}}{K_{p\max} - K_{p\min}} \tag{14}$$

$$K'_d = \frac{K_d - K_{d\min}}{K_{d\max} - K_{d\min}} \tag{15}$$

The integral time constant T_i of the PID controller is assumed to be obtained from the derivative time constant T_d of the PID controller as

$$T_i = \alpha T_d \tag{16}$$

Therefore,

$$K_i = \frac{K_p}{\alpha T_d} = \frac{K_p^2}{\alpha K_d} \tag{17}$$

In other words, the parameters of the PID controller to be tuned using the fuzzy inference rules are K'_p, K'_d and α. Figure 1 illustrates the proposed structure of the fuzzy PID with a two-input-three-output schema. The inputs of the fuzzy system are the error $e(t)$ and the derivative of the error $\dot{e}(t)$, while the outputs are K'_p, K'_d and α. The fuzzy inference rules are constructed as a set of fuzzy IF-THEN rules.

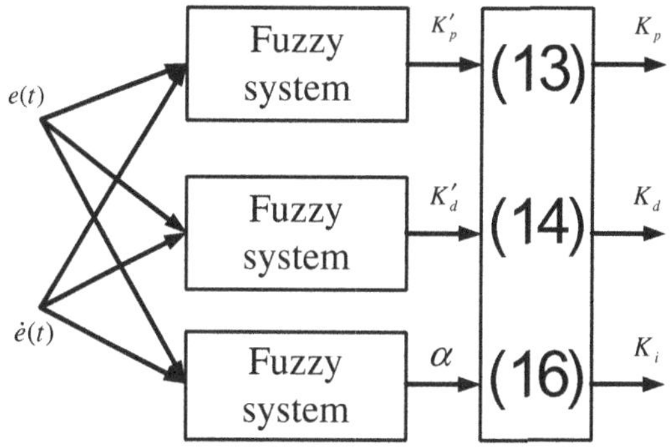

Fig. 1 The structure of the proposed fuzzy PID controller.

The overall block diagram of the fuzzy PID controller reveals that the error (e) and derivative of the error $\dot{e}(t)$ are inputted to the fuzzy systems to derive the control parameters of the PID controller, K'_p, K'_d and α. The control rule base comprises a group of if-then control rules; that is,

IF $e(t)$ is A^l and $\dot{e}(t)$ is B^l, THEN K'_p is C^l; K'_d is D^l, and α is E^l.

Where $e(t)$ and $\dot{e}(t)$ are the input fuzzy variables, and K'_p, K'_d and α are the output fuzzy variables. The fuzzy subsets of the input linguistic variables and the output linguistic variable are $e(t)$: {NB, NM, NS, ZO, PS, PM, PB}, $\dot{e}(t)$: {NB, NM, NS, ZO, PS, PM, PB}, K'_p: {Small, Big}, K'_d: {Small, Big}, α: {S, MS, M, B}. Figures 2, 3 and 4 plot the membership functions of the input and output fuzzy variables. In these figures, N denotes negative, P positive, ZO - approximately zero, S-small, M medium and B - big. Furthermore, NM denotes negative-medium, and so on. A^l, B^l, C^l, D^l and E^l are the fuzzy subsets of the corresponding supporting sets. Tables 1 to 3 list the rules for tuning K'_p, K'_d and α. For example, rule R(4, 2) is

R(4, 2): IF $e(t)$ is ZO and $\dot{e}(t)$ is NM, THEN K'_p is S; K'_d is B, and α is M.

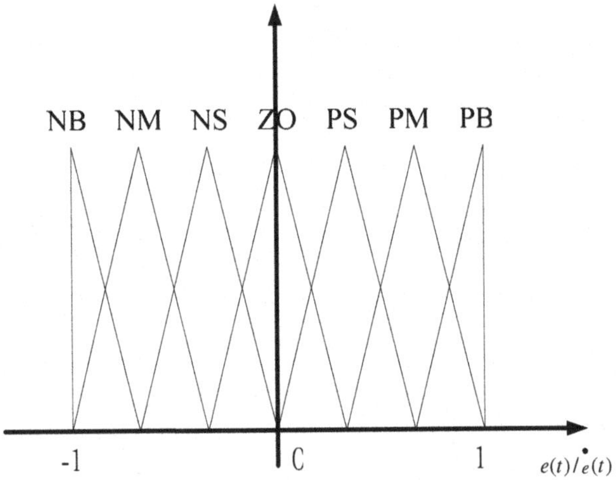

Fig. 2 Membership functions for $e(t)$ and $\dot{e}(t)$.

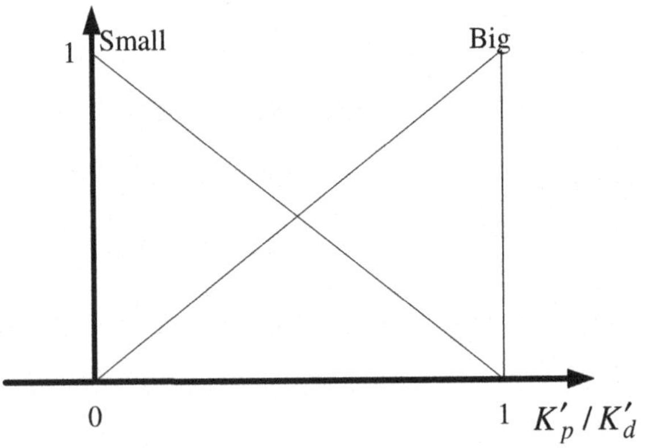

Fig. 3 Membership functions for K'_p and K'_d.

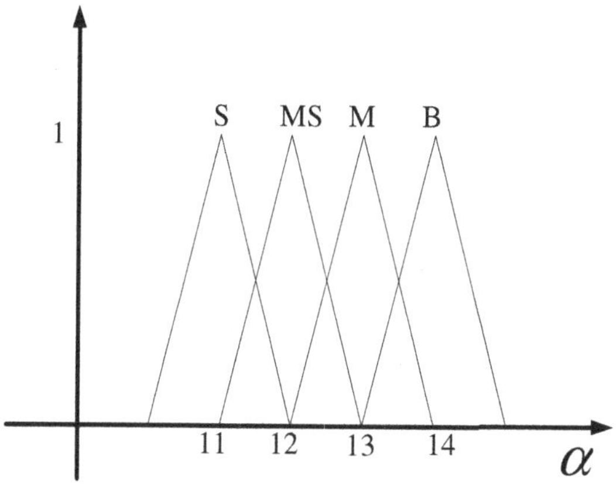

Fig. 4 Membership functions for α.

Table 1 fuzzy turning rules for K_p'

		$\dot{e}(t)$						
		NB	NM	NS	ZO	PS	PM	PB
$e(t)$	NB	B	B	B	B	B	B	B
	NM	S	B	B	B	B	B	S
	NS	S	S	B	B	B	S	S
	ZO	S	S	S	B	S	S	S
	PS	S	S	B	B	B	S	S
	PM	S	B	B	B	B	B	S
	PB	B	B	B	B	B	B	B

Table 2 fuzzy turning rules for K'_d

		$\dot{e}(t)$						
		NB	NM	NS	ZO	PS	PM	PB
$e(t)$	NB	S	S	S	S	S	S	S
	NM	B	B	S	S	S	B	B
	NS	B	B	B	S	B	B	B
	ZO	B	B	B	B	B	B	B
	PS	B	B	B	S	B	B	B
	PM	B	B	S	S	S	B	B
	PB	S	S	S	S	S	S	S

Table 3 fuzzy turning rules for α

		$\dot{e}(t)$						
		NB	NM	NS	ZO	PS	PM	PB
$e(t)$	NB	S	S	S	S	S	S	S
	NM	MS	MS	S	S	S	MS	MS
	NS	M	MS	MS	S	MS	MS	M
	ZO	B	M	MS	MS	MS	M	B
	PS	M	MS	MS	S	MS	MS	M
	PM	MS	MS	S	S	S	MS	MS
	PB	S	S	S	S	S	S	S

4 The Proposed System Structure

The proposed power system stabilizer structure shown in Fig. 5 comprises two subsystems - the grey predictor and the fuzzy tuning PID controller. The grey predictor adapts the last four data sets of the generator deviation speed $\Delta\omega$, to predict the subsequent value of the derivative of the angular speed of the generator $\Delta\hat{\omega}$. The fuzzy tuning PID controller generates the control signal of the control system based on the forecast values of the generator deviation speed.

From the experiments, the small forecasting step always causes a short rise time and a large overshoot of the system response. On the other way, the large forecasting step always produces the response of the system with a small overshoot and long rise time. This work utilizes the switch mechanism to get a suitable forecasting step for the grey predictor according to the response of the system. The control strategy of the switch mechanism is based on the forecasting error $\hat{e}$ of the generator deviation speed to decide a control signal. If the forecasting error is large, the forecasting step is set as small for outputting a larger control signal to speed up the response of the system. By the same way, while the forecasting error is small, the forecasting step is set a large value to prevent the overshoot of the system. The switching mechanism is defined by

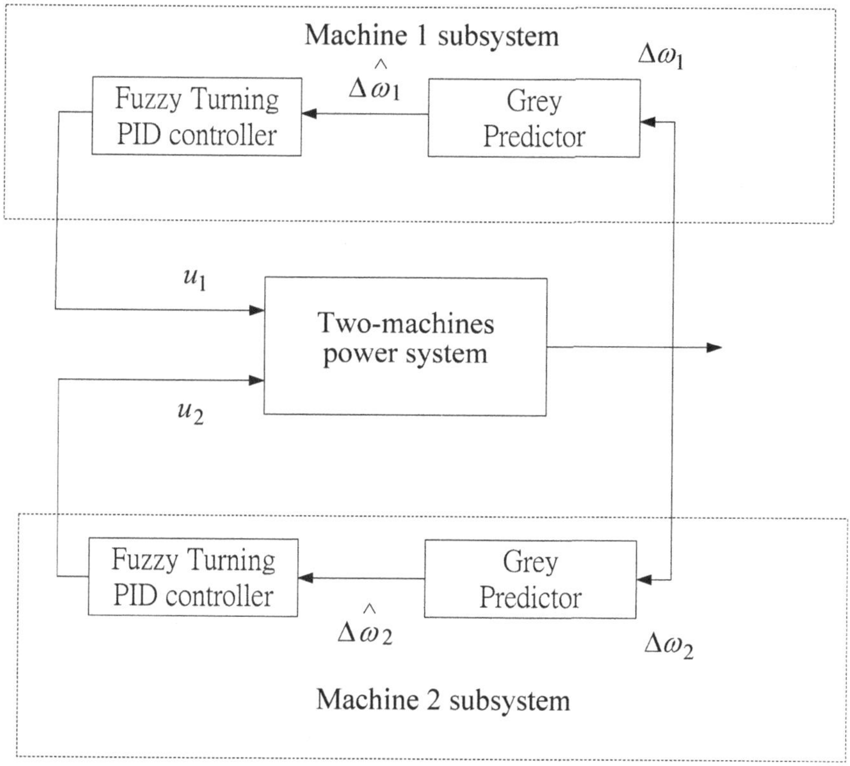

Fig. 5 The structure of the grey fuzzy tuning PID control power system stabilizer.

$$p = \begin{cases} p1 & ,If \ \hat{e} > E_s \\ \frac{1}{2}(ps + pl) & ,If \ \frac{1}{2}E_s < \hat{e} \le E_s \\ p2 & ,If \ \hat{e} \le \frac{1}{2}E_s \end{cases} \quad (18)$$

where *p1* denotes the forecasting step size for large forecasting error, *p2* is the forecasting step size for small forecasting error, and E_s is the switching time of the switch mechanism.

In order to avoid time-consuming, we apply the genetic algorithms to decide the appropriate forecasting step size and the switching time E_s. The switching grey prediction controller not only can reduce the overshoot of the system, but also can produce a shorter rise time than that of the traditional controller design methods.

5 Simulation Results

This study considered the two-machine-infinite-bus power system shown in Fig. 6 to demonstrate the capability of the proposed method to enhance the system stability. In the grey process, AGO is first performed to establish a grey model GM(1,1) (Feliachi *et al.* 1988, Hung and Yang 1991). Figure 7 shows the original sequence of data for the derivative of the angular speed of generator 1. Figure 8 illustrates the first and second-order AGO of the derivative of the angular speed of generator 1. Forecasting step size influences the performance of the grey predictor. Figures 9 and 10 illustrate the simulation results obtained using different sized forecasting steps in the grey predictor, for K_p=5, K_i=0.1 and K_d=2. This work selects forecasting step size p = 2 for the grey predictor.

For comparison, simulations of the two-machine power system were run, based on full-order optimal control, reduced-order optimal control (Feliachi *et al.* 1988) and the proposed method. Figures 11 and 12 plot the transient responses of the angular frequencies with a 5% change in the mechanical torque of both machines. Meanwhile, Figs. 13 and 14 plot the torque angle responses. The simulation results indicate that the proposed PSS offers better damping characteristics and transient responses than the conventional PSS.

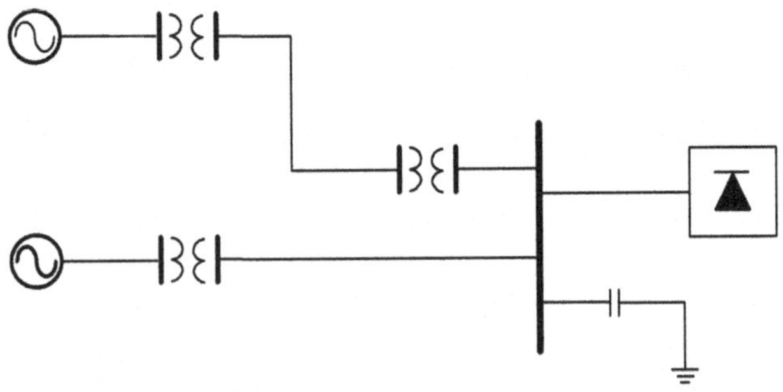

Fig. 6 The two machine-infinite-bus power system

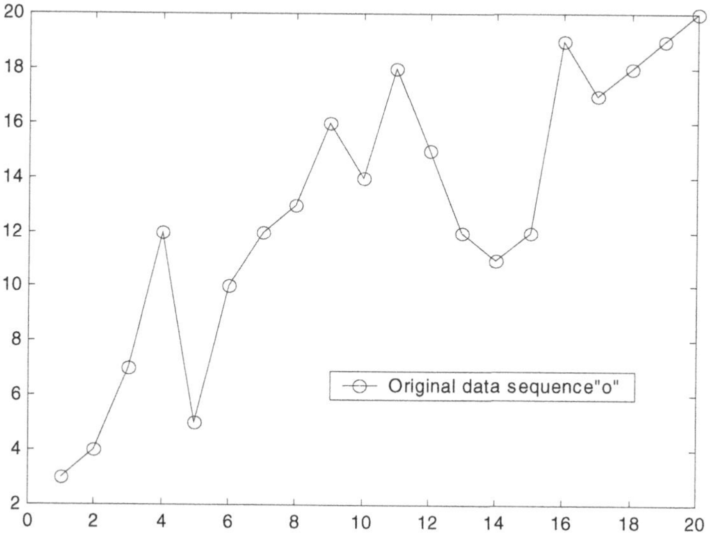

Fig. 7 The original data sequence of the derivative of the angular speed of the generator 1.

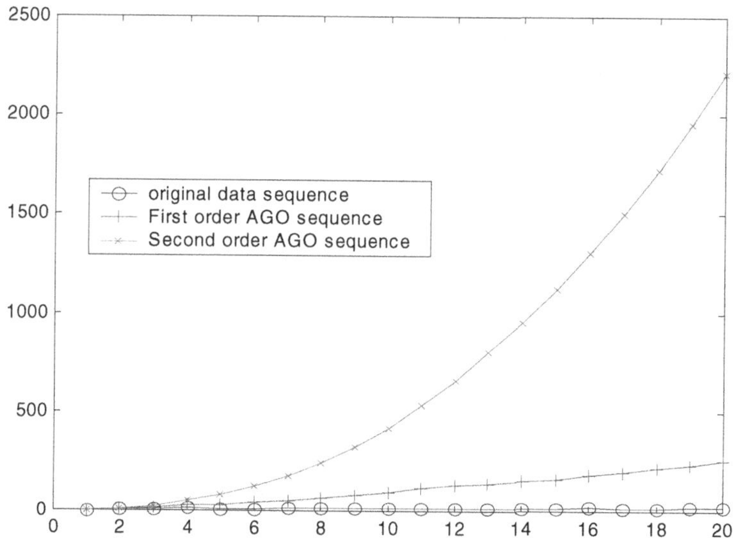

Fig. 8 The first and second order AGO of the derivative of the angular speed of the generator 1

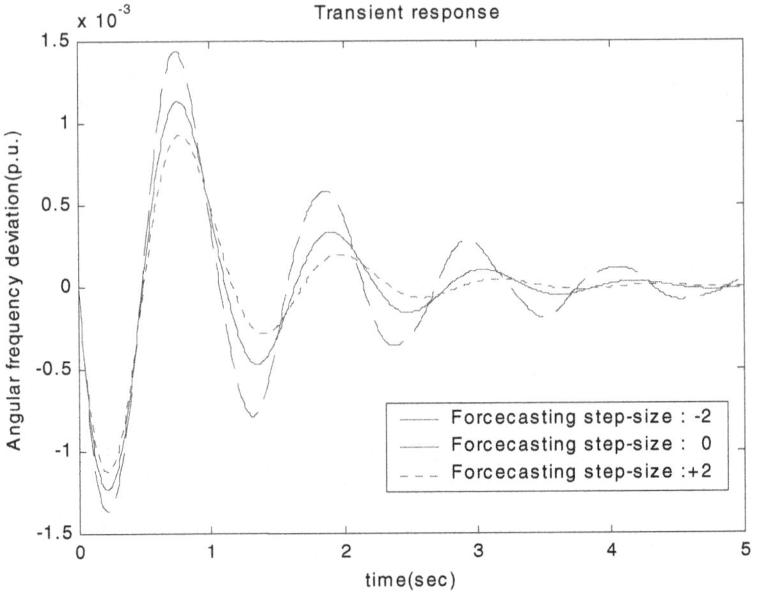

Fig. 9 The angular frequency responses of the machine 1 under various forecasting step size

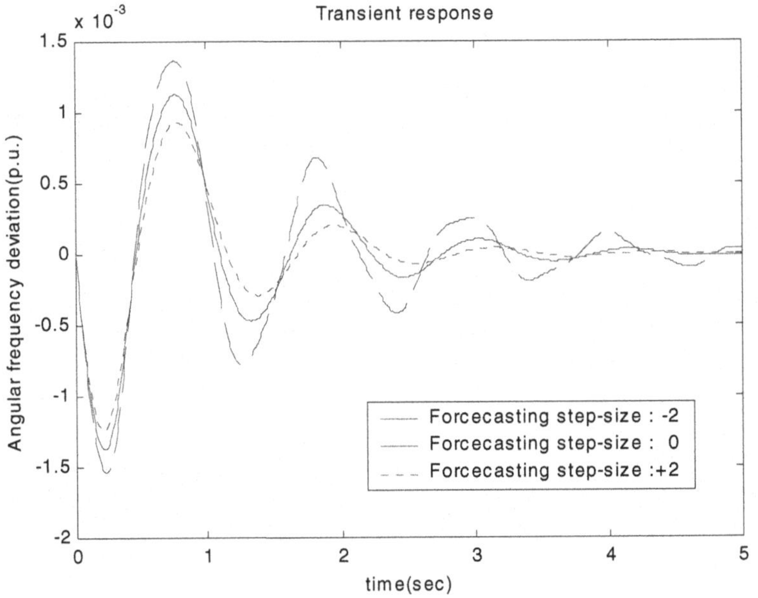

Fig. 10 The angular frequency responses of the machine 2 under various forecasting step size.

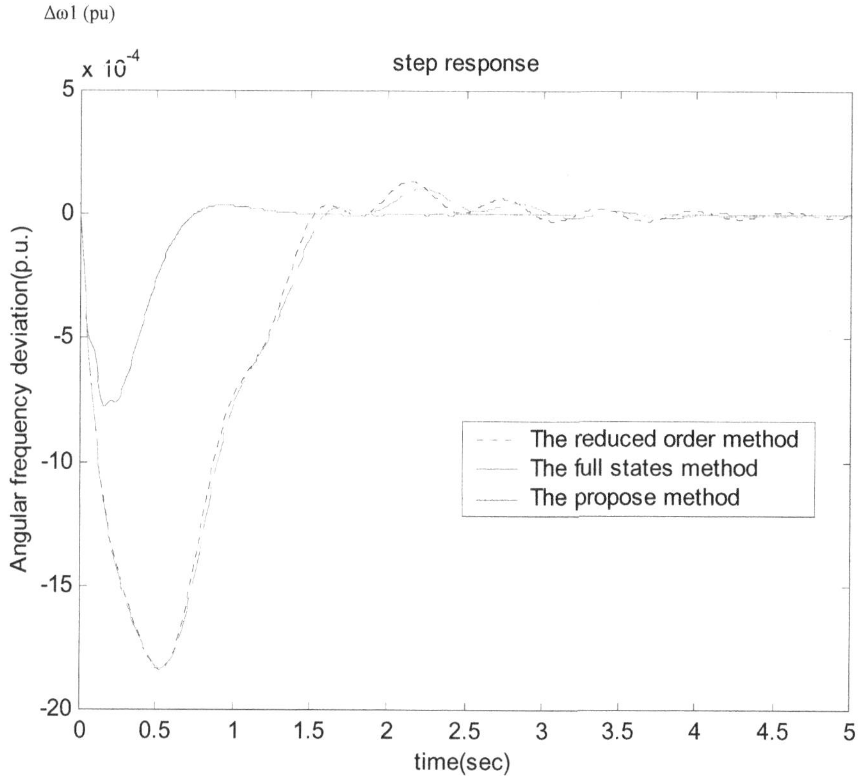

Fig. 11 The angular frequency response of machine 1 for the test system.

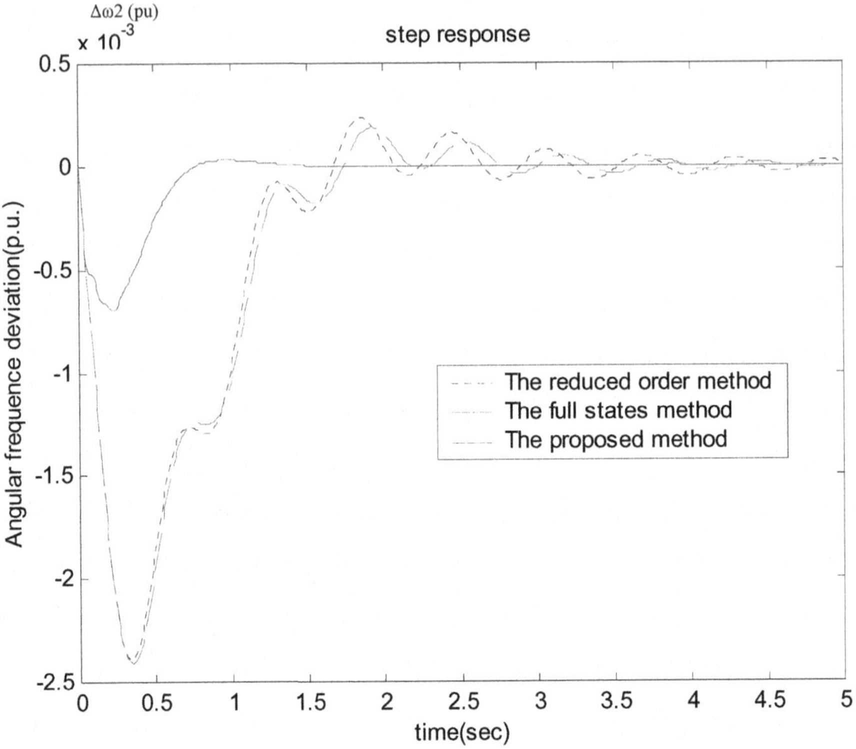

Fig. 12 The angular frequency response of machine 2 for the test system.

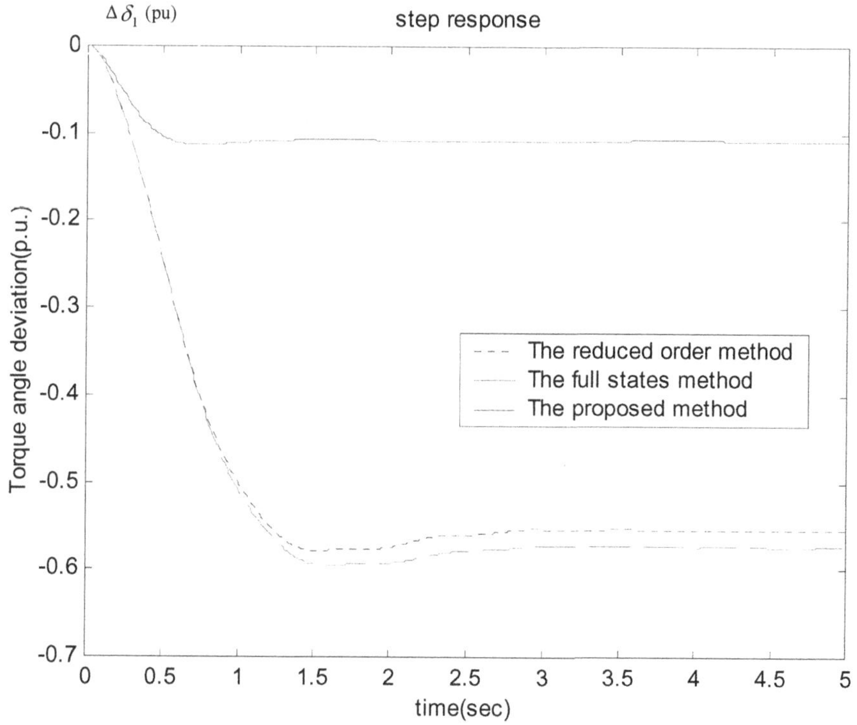

Fig. 13 The torque angle response of machine 1 for the test system

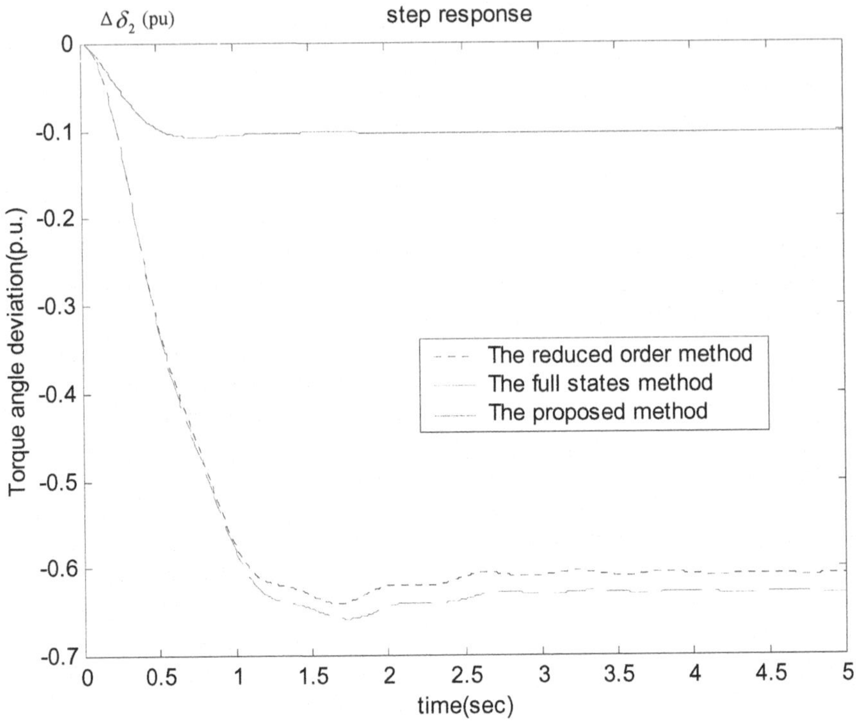

Fig. 14 The torque angle response of machine 2 for the test system

6 Conclusions

This work presented a novel method of designing power system stabilizer. The proposed approach integrates the grey prediction technique, fuzzy inference and PID controller to replace the traditional full-order optimal control method. This study considered and tested a two-machine-infinite-bus power system. The proposed method was then compared with traditional optimal control and optimal reduced methods, demonstrating the effectiveness of the grey fuzzy turning PID control power system stabilizer in improving dynamic performance stability, as confirmed by simulation. The results of this study demonstrate that in addition to its ability to be easily implemented, the proposed solution algorithm can reduce the oscillation and improve the dynamic stability of the considered power system.

References

Chow, J.H., Boukarim, G.E., Murdoch, A.: Power-system stabilizer as undergraduate control design projects. IEEE Trans. on Power Systems 19(1), 144–151 (2004)

Chung, C.Y., Wang, K.W., Tes, C.T.: Power-system stabilizer (PSS) design by probabilistic sensitivity indexes. IEEE Trans. on Power Systems 17(3), 688–693 (2002)

Deng, J.L.: Control problems of grey system. System & Control Letters 1(5), 288–294 (1982)
Deng, J.L.: Introduction to grey system theory. The Journal of Grey System 1(1), 1–24 (1989)
Feliachi, A., Zhang, X., Sims, C.S.: Power system stabilizer design using optimal reduced order models, Part I: Model Reduction. IEEE Trans. on Power system 3(4), 1670–1675 (1988)
Feliachi, A., Zhang, X., Sims, C.S.: Power system stabilizer design using optimal reduced order models, Part II: Design. IEEE Trans. on Power Systems 3(4), 1676–1684 (1988)
Hung, T.L., Yang, W.T.: Two-Level optimal output feedback stabilizer design. IEEE Trans. on Power Systems 6(4), 1042–1048 (1991)
Hosseinzadeh, N., Kalam, A.: A direct adaptive fuzzy power system stabilizer. IEEE Trans. on Energy Conversion 14(4), 1564–1571 (1999)
Moucod, H.A., Yu, Y.: Optimal power system stabilization through excitation and/or governor control. IEEE Trans. on Power App. Sits 91, 1166–1174 (1972)
Su, S.L., Su, Y.C., Huang, J.F.: Grey-based power control for DS-CDMA cellular mobile systems. IEEE Trans. on Vehicular Technology 49(6), 2081–2088 (2000)
Wong, C.C., Liang, W.C.: Design of switching grey prediction controller. The Journal of Grey System 9(1), 47–60 (1997)
Hung, T.L., Yang, W.T.: Two-Level optimal output feedback stabilizer design. IEEE Trans. on Power Systems 6(4), 1042–1048 (1991)
You, R., Eghbali, H.J., Nehrir, M.H.: An online adaptive neuro-fuzzy power system stabilizer for multimachine system. IEEE Trans. on Power Systems 18(1), 128–135 (2003)

A Weighted Fuzzy Time Series Based Neural Network Approach to Option Price Forecasting

Yungho Leu, Chien-Pang Lee, and Chen-Chia Hung

Abstract. Option price forecasting has become an important financial issue in recent years. However, it remains a challenging problem due to the fact that the option price is determined by many factors. This paper proposes a new method to forecast the option price. The proposed approach, a weighted fuzzy time series based neural network (WFTSNN) model, is a hybrid method composed of a fuzzy time series model and a neural network model. In the WFTSNN, a fuzzy time series model is used to select the training data set from the historical data set to train a neural network for option price forecasting. The experimental results show that the WFTSNN outperforms several existing hybrid methods in terms of the mean absolute error and the root mean squared error.

Keywords: Option price forecasting, Hybrid method, Fuzzy sets, Fuzzy time series, Neural networks.

1 Introduction

Recently, the option has become an important tool for risk management in financial markets (Ko 2009). For example, a producer may buy a *put* option to prevent a loss due the possible price decline on his products in the future. Similarly, a customer may buy a *call* option to buy his desired products for an expected price in the future. However, an option is like an insurance policy in that one has to pay premium for an option. The premium, also called the price, of an option is determined by many factors such as the current price of the underlying asset, the strike price, the time to expiration, the volatility of the price of the underlying asset and the risk-free interest rate (Black and Scholes 1973). Being affected by so many factors, option price forecasting remains a challenging problem. The well-known Black-Scholes model (B-S model) (Black and Scholes 1973) offers a deterministic equation for option pricing and is a standard for activity of financial practitioners (Morelli et al. 2004). Though it gives an exact solution for the European option, it

Yungho Leu · Chien-Pang Lee · Chen-Chia Hung
Department of Information Management, National Taiwan University of Science and Technology, 43, Keelung Road, Section 4, Taipei, Taiwan 10607, ROC

W. Pedrycz and S.-M. Chen (Eds.): Granular Computing and Intell. Sys., ISRL 13, pp. 237–248.
springerlink.com

is not well-suited for pricing options in other markets such as the American and the Asian option markets (Morelli et al. 2004; Labjcygier 2004). As a result, many new methods have been proposed for option price forecasting in the recent years.

Artificial neural networks (ANNs) are popular for financial forecasting. For example, Grudnitski and Osburn (Grudnitski and Osburn 1993) used the ANN to forecast futures prices. Shin and Han (Shin and Han 2000) and Panda and Narasimhan (Panda and Narasimhan 2007) used the ANN to forecast the exchange rates. The ANN is also widely used in option price forecasting. For example, Morelli et al. (Morelli et al. 2004) used both the multi-layer perceptron and the radial basis function neural networks to predict the prices of options. Recently, the fuzzy time series models have gained its popularity for financial forecasting such as the stock price indices forecasting (Yu 2005; Cheng et al. 2008), the foreign exchange rates forecasting (Leu et al. 2009) and the forecasting of the indices of the Taiwan Futures Exchange market (Lee et al. 2006).

Recently, Tseng et al. (Tseng et al. 2008) and Wang (Wang 2009) proposed several hybrid models for option price forecasting. According to the B-S option-pricing model (Black and Scholes 1973), the underlying asset price (the spot price) S, the strike price X, the time-to-expiration t, the risk-free interest rate r, and the volatility of the underlying asset σ are parameters to determine the price of the option of an underlying asset. Among these parameters, the asset volatility σ is difficult to estimate. Traditionally, the AutoRegressive Conditional Heteroskedasticity (ARCH) model and its variants, the generalized ARCH (GARCH), the Glosten-Jagannathan-Runkle GARCH (GJR-GARCH) and the Exponential General Autoregressive Conditional Heteroskedastic (EGARCH) (Robert 2001) are used for calculating the volatility of a financial time series. In the work of Tseng et al., they applied the grey model GM(1,1) to calculate the error terms in the GJR-GARCH model to come up with the Grey-GJR-GARCH model. Similarly, they applied the GM(1,1) grey model to calculate the error term in the EGARCH to come up with the Grey-EGARCH model. Having estimated the volatiltiy of the underlying asset, they used a neural network trained with backpropagation to calculate the option price of the underlying asset using S, X, t, r and σ as the inputs to the neural network.

In this paper, we proposed a hybrid model to predict the option price of "Taiwan Stock Exchange Stock Price Index Options (TXO)". The weighted fuzzy time series based neural network (WFTSNN) model is composed of a fuzzy time series model and a neural network model. In the WFTSNN, a fuzzy time series model is used to select the training data set from the historical option prices for the neural network model to build an option price forecasting model.

The remainder of this paper is organized as follows. Section 2 reviews the definitions of the fuzzy time series. Section 3 introduces the procedure of the WFTSNN. Section 4 describes the application of the WFTSNN on option price forecasting. Section 5 compares the performance of the WFTSNN with many existing hybrid methods. Section 6 concludes this paper.

2 Fuzzy Time Series

Since the proposed method is based on the fuzzy time series model, we will briefly review the definitions pertaining to the fuzzy time series model.
(Chen 2002; Song and Chissom 1993; Song and Chissom 1994),

Definition 1
Let $Y(t)$ (t =…,0,1,2,…), a subset of R^1, be the universe of discourse in which fuzzy sets $f_i(t)$ (i = 1,2...) are defined. If $F(t)$ is a collection of $f_i(t)$, $F(t)$ is called a fuzzy time series defined on $Y(t)$.

Definition 2
If for any $f_j(t) \in F(t)$, there exists an $f_i(t\text{-}1) \in F(t\text{-}1)$, such that there exists a fuzzy relation $R_{ij}(t, t\text{-}1)$ and $f_j(t)=f_i(t\text{-}1) \circ R_{ij}(t, t\text{-}1)$ where '$\circ$' is the max-min composition, $F(t)$ is said to be caused by $F(t\text{-}1)$ only, and is represented by $F(t\text{-}1) \rightarrow F(t)$.

Definition 3
If $F(t)$ is caused by $F(t\text{-}1)$, $F(t\text{-}2)$,…,and $F(t\text{-}n)$, $F(t)$ is called a 1-factor n-order fuzzy time series, and it can be represented by

$$F(t\text{-}n),\ldots, F(t\text{-}2), F(t\text{-}1) \rightarrow F(t).$$

Definition 4
If $F_1(t)$ is caused by $(F_1(t\text{-}1), F_2(t\text{-}1))$, $(F_1(t\text{-}2), F_2(t\text{-}2))$,…, $(F_1(t\text{-}n), F_2(t\text{-}n))$, $F_1(t)$ is called a 2-factor n-order fuzzy time series, which is represented by

$$(F_1(t\text{-}n), F_2(t\text{-}n)),\ldots, (F_1(t\text{-}2), F_2(t\text{-}2)), (F_1(t\text{-}1), F_2(t\text{-}1)) \rightarrow F_1(t).$$

Let $F_1(t)=X_t$ and $F_2(t)= Y_t$, where X_t and Y_t are fuzzy variables whose values are possible fuzzy sets of the first factor and the second factor of day t, respectively. Then, a 2-factor n-order fuzzy logic relationship (FLR) (Chen 2002) can be represented as follows,

$$(X_{t\text{-}n}, Y_{t\text{-}n}), \ldots, (X_{t\text{-}2}, Y_{t\text{-}2}), (X_{t\text{-}1}, Y_{t\text{-}1}) \rightarrow X_t,$$

where $(X_{t\text{-}n}, Y_{t\text{-}n})$, …, $(X_{t\text{-}2}, Y_{t\text{-}2})$ and $(X_{t\text{-}1}, Y_{t\text{-}1})$, are referred to as the left-hand side (LHS) of the fuzzy logic relationship, and X_t is referred to as the right-hand side (RHS) of the fuzzy logic relationship.

3 The WFTSNN Method

In this paper, we adopt a 2-factor fuzzy time series model as the time series model of the hybrid method. Each of the two factors of the fuzzy time series model plays a different role in the prediction. Since the first factor is the main factor for the prediction and the second factor assists the first factor in the prediction, we give the first factor a higher weight than that of the second one. Fig. 1 shows the flowchart of the WFTSNN and the details of the procedure are described in the following.

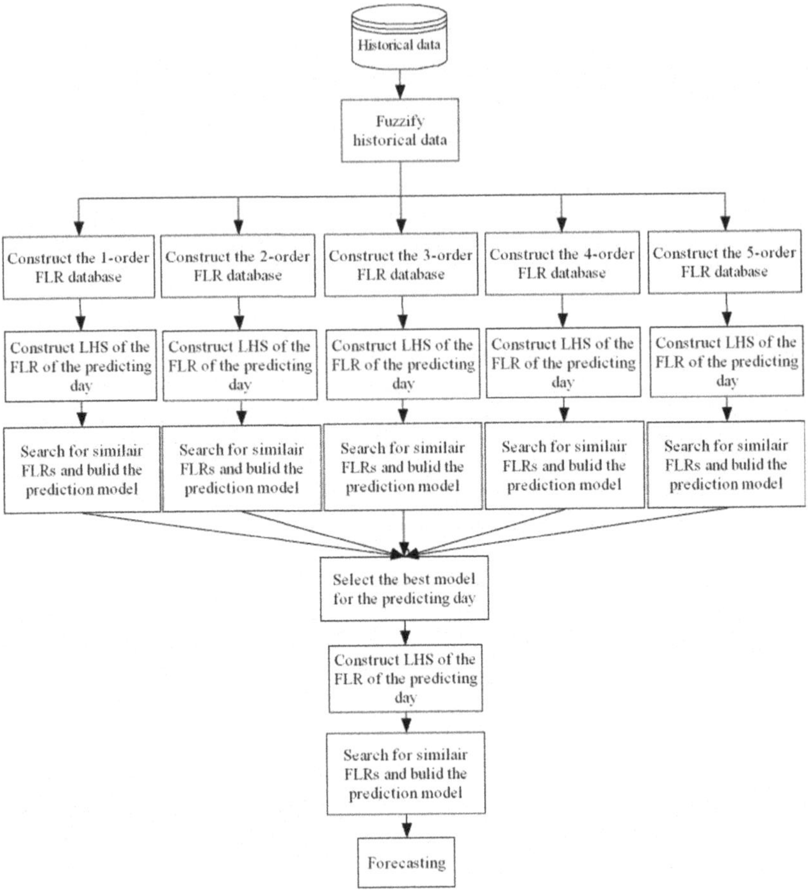

Fig. 1 The flowchart of the WFTSNN

Step 1: Divide the universe of discourse

The universe of discourse of the first factor is defined as U= $[D_{min}-D_1, D_{max}+D_2]$, where D_{min} and D_{max} are the minimum and maximum values, respectively, of the first factor; D_1 and D_2 are two positive numbers which help to divide the universe of discourse into n equal length intervals. Similarly, the universe of discourse of the second factor is defined as V= $[V_{min}-V_1, V_{max}+V_2]$, where V_{min} and V_{max} are the minimum and maximum values, respectively, of the second factor; V_1 and V_2 are two positive numbers used to divide the universe of discourse of the second factor into m equal length intervals. Note that the length of the intervals of each factor is determined by its maximum value of the factor in the historical data set.

Step 2: Define fuzzy sets

Linguistic terms A_i, $1\leq i \leq n$, are defined as fuzzy sets on the intervals of the first factor. They are defined as follows:

$$A_1 = 1/u_1 + 0.5/u_2 + 0/u_3 + \cdots + 0/u_{n-2} + 0/u_{n-1} + 0/u_n,$$
$$A_2 = 0.5/u_1 + 1/u_2 + 0.5/u_3 + \cdots + 0/u_{n-2} + 0/u_{n-1} + 0/u_n,$$
$$\vdots$$
$$A_{n-1} = 0/u_1 + 0/u_2 + 0/u_3 + \cdots + 0.5/u_{n-2} + 1/u_{n-1} + 0.5/u_n,$$
$$A_n = 0/u_1 + 0/u_2 + 0/u_3 + \cdots + 0/u_{n-2} + 0.5/u_{n-1} + 1/u_n,$$

where u_i denotes the ith interval of the first factor.

Similarly, linguistic terms B_j, $1 \le j \le m$, are defined as fuzzy sets on the intervals of the second factor. They are defined as follows:

$$B_1 = 1/v_1 + 0.5/v_2 + 0/v_3 + \cdots + 0/v_{m-2} + 0/v_{m-1} + 0/v_m,$$
$$B_2 = 0.5/v_1 + 1/v_2 + 0.5/v_3 + \cdots + 0/v_{m-2} + 0/v_{m-1} + 0/v_m,$$
$$\vdots$$
$$B_{m-1} = 0/v_1 + 0/v_2 + 0/v_3 + \cdots + 0.5/v_{m-2} + 1/v_{m-1} + 0.5/v_m,$$
$$B_m = 0/v_1 + 0/v_2 + 0/v_3 + \cdots + 0/v_{m-2} + 0.5/v_{m-1} + 1/v_m,$$

where v_i is the ith interval of the second factor.

Step 3: Determine the order of WFTSNN

The WFTSNN method uses a 2-factor n-order fuzzy time series model to select similar FLRs from the historical data set for training a neural network. Similar FLRs imply similar trends in the historical data set. The order of the WFTSNN can be regarded as the length of the trends. Different application favors different length of trends. To determine the length of trends, we choose to vary the order of the WFTSNN from 1 to 5 to build five different prediction models. Then, we choose the one with the best prediction accuracy as the final prediction model. Note that since the order n ranges from 1 to 5, Step 3(a) to Step 3(d) in the following will be performed for five times to build five different models.

(a) Construct the FLRs database

For the historical data of day i, let $X_{i\text{-n}}$, $Y_{i\text{-n}}$ denote the fuzzy set of $F_1(i\text{-n})$ and $F_2(i\text{-n})$ of the fuzzy time series, respectively. Let X_i denotes the fuzzy set of $F_1(i)$. The FLR of day i can be represented as follows:

$$(X_{i\text{-n}}, Y_{i\text{-n}}), \ldots, (X_{i\text{-}2}, Y_{i\text{-}2}), (X_{i\text{-}1}, Y_{i\text{-}1}) \rightarrow X_i.$$

(b) Construct the LHS of the FLR of the predicting day (assume that day t is the predicting day)

The LHS of the FLR of day t can be represented as follows:

$$(X_{t\text{-n}}, Y_{t\text{-n}}), \ldots, (X_{t\text{-}2}, Y_{t\text{-}2}), (X_{t\text{-}1}, Y_{t\text{-}1}).$$

(c) Select similar FLRs from the historical FLR database

Calculate the Euclidean distance (ED) of the LHS of the FLR of day t against the LHS of each candidate FLR in the historical FLR database. Then, we select the top five FLRs with the smallest Euclidean distance from the historical FLR database as the training data set. As mentioned above, a certain weight is given to different factors in prediction. For option price forecasting, we choose to give the first factor twice the weight as that of the second factor. The Euclidean distance between the FLR of day t and the FLR of day i can be calculated according to Formula 1 2, and 3.

$$ED_{Ai} = \sqrt{(IX_{t-n} - RX_{i-n})^2 + \cdots + (IX_{t-2} - RX_{i-2})^2 + (IX_{t-1} - RX_{i-1})^2}, \quad (1)$$

$$ED_{Bi} = \sqrt{(IY_{t-n} - RY_{i-n})^2 + \cdots + (IY_{t-2} - RY_{i-2})^2 + (IY_{t-1} - RY_{i-1})^2}, \quad (2)$$

$$ED_i = \frac{2 \times ED_{Ai} + ED_{Bi}}{3}. \quad (3)$$

In the above formulae, $IX_{t\text{-n}}$ and $IY_{t\text{-n}}$ are the subscripts of the fuzzy sets of the first factor and the second factor, respectively, of the LHS of day t's FLR. Similarly, $RX_{i\text{-n}}$ and $RY_{i\text{-n}}$ are the subscripts of the first factor and the second factor, respectively, of the LHS of day i's FLR.

(d) Construct the neural network model

With the top five similar FLRs, we can train a radial basis function neural network (RBFNN) model for forecasting. The training framework of the neural network model is shown in Fig. 2.

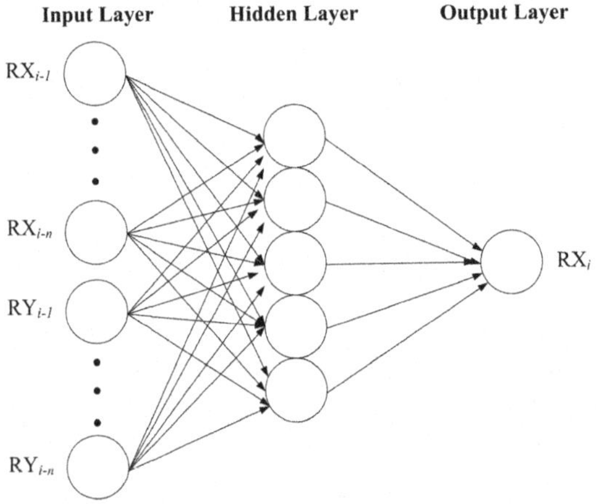

Fig. 2 The training framework of the RBF neural network model

(e) Model selection

Having constructed the candidate models, we can choose the best model for prediction. To do that, we use a test sample to test the prediction accuracy of the candidate models. The test sample, denoted by FLR_s, is the FLR with its LHS most similar to the LHS of the FLR of day t. To select the forecasting model, we use the LHS of FLR_s as the input to a trained model and calculate the error according to the following formula:

$$\text{Error function} = |\ \text{Forecasted RHS} - \text{Testing RHS}\ | . \qquad (4)$$

We then choose the model with the smallest error as the forecasting model. Note that in Formula 4, the forecasted RHS denotes the subscript of the forecasted fuzzy set of FLR_s, and the testing RHS denotes the actual subscript of the fuzzy set of the RHS of FLR_s.

Step 4: Forecasting
Having determined the forecasting model, we use the LHS of the FLR of the predicting day as the input to the neural network to get the forecasted subscript of the RHS of the prediction day. Then, we defuzzify the subscript into its corresponding forecasted value. We use the weighted average method as the defuzzification method described by the following expression:

$$forecast\_value = \begin{cases} \dfrac{M[1]+0.5\times M[2]}{1+0.5} & k=1, \\ \dfrac{0.5\times M[k-1]+M[k]+0.5\times M[k+1]}{0.5+1+0.5} & 1<k\le n-1, \\ \dfrac{0.5\times M[n-1]+M[n]}{0.5+1} & k=n, \end{cases} \qquad (5)$$

where $M[k]$ denotes the midpoint value of fuzzy set k. Note that an iteration of the above procedure (Step 1 through Step 4) predicts only one forecasted value.

4 Option Price Forecasting

To forecast the option price of TXO, we choose the closing price of TXO as the first factor and the "Taiwan Stock Exchange Capitalization Weighted Stock Index (TAIEX)" as the second factor. Parts of the historical data set are shown in Table 1. With this historical data set, U is set to [0, 1000] and is divided into 100 intervals. Accordingly, u_1=[0, 10], u_2=[10, 20],..., and u_{100}=[990, 1000]. For the second factor, V is set to [3000, 13000] and is divided into 240 intervals. Similarly, v_1=[3000, 3050], v_2=[3050, 3100], ...,and v_{240}=[12950, 13000]. Having defined the intervals, we fuzzify the historical data set into fuzzy sets. Table 2 shows parts of the fuzzified historical data set. Then, we construct the historical FLR database from the fuzzified historical data set. A 2-order FLR database for this historical data set is shown in Table 3.

Having constructed the FLR database, we can forecast the option price by the WFTSNN. For example, if we want to forecast the option price of day *11*, we first construct the LHS of the FLR of day *11* as follows:

$$(A_{37}B_{118}), \qquad (A_{36}B_{118}).$$

Then, we calculate the Euclidean distance between LHS of the FLR of day *11* and the LHS of each candidate FLR in the FLR database. Table 4 shows the calculated Euclidean distances. Then, we select the top five similar FLRs from the FLR database. In this example, FLR_4, FLR_5, FLR_8, FLR_3, FLR_6 are selected. Finally, we use these selected FLRs as the training data set to build a neural network model.

Table 1 Segments of the historical data set

Date	Day 1	Day 2	Day 3	Day 4	Day 5
Option Prices	615	500	455	408	413
Stock indices	6143.12	6060.46	5988.37	5935.99	5975.66
Date	Day 6	Day 7	Day 8	Day 9	Day 10
Option Price	310	267	309	376	365
Stock indices	5879.08	5853.94	5889.52	5945.27	5933.57

Table 2 Segments of the fuzzified data set

Day 1	Day 2	Day 3	Day 4	Day 5
(A_{61}, B_{122})	(A_{50}, B_{121})	(A_{45}, B_{119})	(A_{40}, B_{118})	(A_{41}, B_{119})
Day 6	Day 7	Day 8	Day 9	Day 10
(A_{31}, B_{117})	(A_{26}, B_{117})	(A_{30}, B_{117})	(A_{37}, B_{118})	(A_{36}, B_{118})

Table 3 The FLRs database

FLR_1	(A_{61}, B_{122})	(A_{50}, B_{121})	$\rightarrow$	A_{45}
FLR_2	(A_{50}, B_{121})	(A_{45}, B_{119})	$\rightarrow$	A_{40}
…	…	…	…	…
FLR_8	(A_{30}, B_{117})	(A_{37}, B_{118})	$\rightarrow$	A_{36}

Table 4 Obtained Euclidean distances

	ED_A	ED_B	ED
FLR_1	$\sqrt{(37-61)^2+(36-50)^2}$	$\sqrt{(118-122)^2+(118-121)^2}$	20.19
FLR_2	$\sqrt{(37-50)^2+(36-45)^2}$	$\sqrt{(118-121)^2+(118-119)^2}$	11.60
…			…
FLR_8	$\sqrt{(37-30)^2+(36-37)^2}$	$\sqrt{(118-117)^2+(118-118)^2}$	5.05

To select a model for forecasting, we use FLR_4, which is the most similar FLR to the FLR of day *11*, as the testing sample. To perform the testing, we use the LHS of FLR_4 as the input to the neural network to get the predicted subscript. Then, we calculate the error between the predicted subscript and the actual subscript. For this example, the predicted subscript is 31 and the actual subscript is also 31. Hence, the error is equal to 0. Note that the procedure above will be performed once for each of the 1-order, 2-order, 3-order, 4-order and 5-order fuzzy time series models. The one with the smallest error is selected for forecasting.

For the above example, assume that a 2-order neural network model is selected, and the forecasted subscript is 32 for day *11*. Substituting 315, 325 and 335 for

$M[31]$, $M[32]$, and $M[33]$, respectively in Formula 5, we get 325 as the forecasted option price for day *11*. Note that the actual option price of day *11* is 326 in this example.

5 Results and Performance

5.1 The Data Set

The data set used in this paper comprises the daily transaction data of TXO and TAIEX from January 3, 2005 to December 29, 2006. This paper investigates a sample of 23,819 *call* option prices. Call options can be divided into three categories according to their *S/K* ratios. The distribution of the data set is shown in Table 5. We refer to (Tseng et al. 2008) and (Wang 2009) for the definition of the categories. The data set comprises 30 different strike prices from 5,200 to 8,200 and 12 different expiration dates from January 2005 to December 2006.

Table 5 Data distribution according to moneyness

Categories	Moneyness	Number
In-the-money	S/K > 1.02	8938
At-the-money	0.95 < S/K ≦ 1.02	7508
Out-of-the-money	S/K ≦ 0.95	7373

Note: S denotes the spot price; K denotes the strike price.

Note that the option prices of the beginning 10 transaction dates of each option are not predicted due to lack of the required training data. In predicting the option price of a specific date, the option prices of the previous transaction dates become the training data set.

5.2 Performance Measures

Two different performance measures, mean absolute error (MAE) and root mean squared error (RMSE), are used to measure the forecasting accuracy of the WFTSNN. The formulae are shown in the following:

$$MAE = \frac{\sum_{t=1}^{n} |A_t - P_t|}{n}, \tag{6}$$

$$RMSE = \sqrt{\frac{\sum_{t=1}^{n} (A_t - P_t)^2}{n}}, \tag{7}$$

where A_t and P_t denote the actual option price and the forecasted option price of day t, respectively.

5.3 Performance

The performance of the WFTSNN is compared with other proposed methods published in the literatures (Leu et al. 2010; Tseng et al. 2008; Wang 2009). Table 6 shows parts of the results of the option prices forecasted by the WFTSNN. It is noted that the forecasted prices are very close to the actual option prices in Table 6, except those on the dates when the option prices change abruptly. The performance of the WFTSNN and the other methods are shown in Table 7. In Table 7, the difference between WFTSNN and FTSNN (Leu et al. 2010) is in that the WFTSNN gives different weight to the two factors, while the FTSNN give equal weight to both factors. Table 7 shows that the performance of the WFTSNN is better than that of the FTSNN in all cases except the case of in-the-money with the MAE measure, in which the difference is insignificant. Table 7 shows that, in terms of both RMSE and MAE, the forecasting accuracy of the WFTSNN is better than those of the other methods in all the three option categories. Fig. 3 shows the forecasted prices of an option with strike price equal to 7,400 and expiration date in December 2006.

Table 6 Parts of forecasted prices of an option with strike price equal to 6,300 and expiration date of November 2005

Transaction dates	Actual option prices	Forecasted option prices
2005/09/02	80	77.5
2005/09/05	77	72.5
2005/09/06	93	72.5
2005/09/07	96	82.5
2005/09/08	93	87.5
2005/09/09	84	82.5
2005/09/12	90	87.5
2005/09/13	91	92.5

Table 7 The performance in RMSE and MAE

	RMSE			MAE		
Moneyness	In	At	Out	In	At	Out
WFTSNN	67.94*	29.72*	8.96*	52.13	19.99*	6.05*
FTSNN	72.79	36.99	16.19	52.11*	23.98	9.78
GARCH	85.49	44.02	25.73	69.54	34.73	18.78
GJR	76.19	41.06	25.53	59.28	31.67	17.41
Gery-GJR	73.76	40.11	25.89	56.13	30.21	17.26
EGARCH	73.90	41.35	26.34	57.02	32.17	18.30
Gery-EGARCH	72.11	40.26	26.13	57.26	32.51	18.26

Notes: 1. * denotes the smallest value. 2. GJR denotes the GJR–GARCH model. 3. Grey-GJR denotes the Grey-GJR–GARCH model.

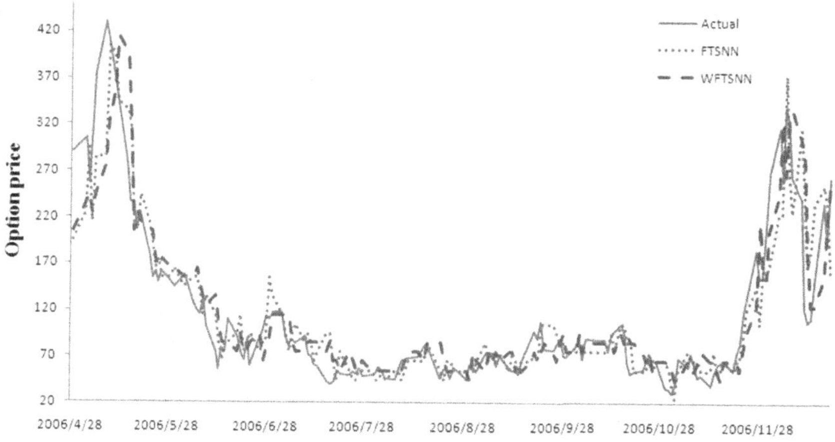

Fig. 3 Time series of the actual and the forecasted option prices of an option with strike price equal to 7,400 and expiration date in December 2006

6 Conclusions

Option price forecasting is difficult due to the fact that it is influenced by many factors in a complex manner. Hybrid methods have been shown successful in predicting the option price. In this paper, we propose the WFTSNN method for option price forecasting. In the WFTSNN, we use a weighted fuzzy time series model to select the training examples from the historical data set for a neural network to build a prediction model for option prices. According to the experimental results, the WFTSNN outperforms other existing hybrid models in terms of RMSE and MAE. Finally, since a real-life financial time series always exhibits non-deterministic fluctuation, the higher order fuzzy sets which can handle uncertainty may be useful for financial time series forecasting.

References

Black, F., Scholes, M.: The Pricing of Options and Corporate Liabilities. J. Polit. Econ. 81, 637–654 (1973)

Chen, S.M.: Forecasting Enrollments Based on High-Order Fuzzy Time Series. Cybern. Syst. 33, 1–16 (2002)

Cheng, C.H., Chen, T.L., Teoh, H.J., Chiang, C.H.: Fuzzy time-series based on adaptive expectation model for TAIEX forecasting. Expert Syst. Appl. 34, 1126–1132 (2008)

Grudnitski, G., Osburn, L.: Forecasting S&P and gold futures prices: An application of neural networks. J. Futures Markets 13, 631–643 (1993)

Ko, P.C.: Option valuation based on the neural regression model. Expert Syst. Appl. 36, 464–471 (2009)

Lajbcygier, P.: Improving option pricing with the product constrained hybrid neural network. IEEE Trans. Neural Networks 15, 465–476 (2004)

Lee, L.W., Wang, L.H., Chen, S.M., Leu, Y.: Handling forecasting problems based on two-factors high-order fuzzy time series. IEEE Trans. Fuzzy Systems 14, 468–477 (2006)

Leu, Y., Lee, C.P., Jou, Y.Z.: A distance-based fuzzy time series model for exchange rates forecasting. Expert Syst. Appl. 36, 8107–8114 (2009)

Leu, Y., Lee, C.P., Hung, C.C.: A fuzzy time series-based neural network approach to option price forecasting. In: Nguyen, N.T., Le, M.T., Świątek, J. (eds.) ACIIDS 2010. LNCS (LNAI), vol. 5990, pp. 360–369. Springer, Heidelberg (2010)

Morelli, M.J., Montagna, G., Nicrosini, O., Treccani, M., Farina, M., Amato, P.: Pricing financial derivatives with neural networks. Physica. A 338, 160–165 (2004)

Panda, C., Narasimhan, V.: Forecasting exchange rate better with artificial neural network. J. Policy Modeling 29, 227–236 (2007)

Robert, E.: GARCH 101: The Use of ARCH/GARCH Models in Applied Econometrics. Journal of Economic Perspectives 15(4), 157–168 (2001)

Song, Q., Chissom, B.S.: Forecasting enrollments with fuzzy time series - part I. Fuzzy Sets Syst. 54, 1–9 (1993)

Song, Q., Chissom, B.S.: Forecasting enrollments with fuzzy time series - part II. Fuzzy Sets Syst. 62, 1–8 (1994)

Shin, T., Han, I.: Optimal signal multi-resolution by genetic algorithms to support artificial neural networks for exchange-rate forecasting. Expert Syst. Appl. 18, 257–269 (2000)

Tseng, C.H., Cheng, S.T., Wang, Y.H., Peng, J.T.: Artificial neural network model of the hybrid EGARCH volatility of the Taiwan stock index option prices. Physica A 387, 3192–3200 (2008)

Yu, H.K.: Weighted fuzzy time series models for TAIEX forecasting. Physica. A 349, 609–624 (2005)

Wang, Y.H.: Nonlinear neural network forecasting model for stock index option price: Hybrid GJR-GARCH approach. Expert Syst. Appl. 36, 564–570 (2009)

A Rough Sets Approach to Human Resource Development in IT Corporations

Shinya Imai and Junzo Watada

Abstract. In IT corporations, it is essential to increase competitive advantages and organizational performance. Employees are critical to a company's success. A new research method is needed to quantify employees' influence on building relationships with customers and to facilitate human resource and customer relationship management. Rough sets theory is a mathematical approach to dealing with vagueness and uncertainty. It can change a qualitative problem into a quantitative one and produce a possible solution by providing useful and valuable information and guidelines for decision making. The objective of this study is to determine through the use of analysis analyzed with rough employee characteristics and behaviors that yield positive or negative relationships with customers. The rough set approach distinguishes between these two groups and leads us to suggest policies to improve human resource and customer relationship management and development. The proper management of employees and customers will ensure project success and good corporate performance.

Quality is an attribute that is important for products as well as for management and the company itself. The development and promotion of personnel resources is indispensable for improving the quality of a company's management. Management quality is closely related to corporate culture and a sense of social responsibility. Therefore, personnel resource development and personnel training for employees should be emphasized. In the main discussion of this paper, information was gathered from engineers at a regional IT company through questionnaires and their observable talents were analyzed. The research addressed questions such as what kinds of values should be promoted. An attempt was made to clarify the relation between QWL (Quality of Working Life) and personnel training. This paper suggests that the management quality and CSR (Corporate Social Responsibility) of regional companies is closely related with the quality and improvement of their growth.

Keywords: human resource development, rough set theory, customer relationship management.

Shinya Imai · Junzo Watada
Graduate School of Information, Production and Systems
Waseda University 2-7 hibikino, Wakamatsu-ku, Fukuoka 808-0135 Japan
e-mail: shinyaimai@ruri.waseda.jp,
watada@waseda.jp

W. Pedrycz and S.-M. Chen (Eds.): Granular Computing and Intell. Sys., ISRL 13, pp. 249–273.
springerlink.com

1 Introduction

Numerous studies have provided important methods for human resource management (HRM). In the 1980s, many companies were product-oriented. It was believed that employees just needed certain skills to be productive, and human resource management was approached from a job-based perspective. In 1973, Harvard University psychology professor McClelland challenged the idea of a "work as the center" evaluation. He pointed out that using competence instead of intelligence as an assessment criterion emphasized the concept of "work as the center" (Mirabile 1997). The concept of competency-based evaluation has led to a revolutionary change in the modern evaluation system that views an organization as a paragon of competence. This focus on competence and growth is especially suited to the dynamic and people-oriented features of the knowledge-economy era (Yan et al. 1997). Therefore, employees are now required not only to perform their jobs but also to maintain good relationships with customers.

When a company retains good relationship with its customers, it can increase its revenue and market share, respond quickly to market opportunities, increase customer loyalty and easily collect information to ensure that corporate resources are used in a suitable way. To achieve this goal, a manager will ensure that employees are in the right place at the right time to satisfy the company's customers. However, this can become a problem if the company cannot satisfy both its employees and its customers. When there is a conflict between employees and customers, managers often sacrifice employees' rights and satisfaction. This is no longer appropriate. The objective of this research is to identify a compromise between customers and employees and to determine what employee characteristics and behaviors of employees produce a good relationship with customers. The results of this research may guide organizations to adapt good strategies and policies for human resource and customer relationship management.

1.1 Human Resource Management

Storey (1995) defines human resource management as a distinctive approach to employment management that seeks to achieve a competitive advantage through the strategic deployment of a highly committed and capable workforce using an array of cultural, structural and personnel techniques.

In recent years, studies have connected human resource management (HRM) with strategic management and corporate performance to emphasize the positive link between HR practices and organizational performance (Truss 2001, Pfeffer and Veiga 1999). If a company has an effective HRM team, it will also have better organizational performance. Therefore, increasing employees' motivation and passion is the key to businesses' survival in the IT industry. Research by Pfeffer (Pfeffer 1994) showed that high motivation and strong commitment among employees leads to a long-term increase in business performance. HRM must be implemented for an extended period of time to have an effect on the success of its organization. It can be a part of the strategy of a company to facilitate important decision making among the staff, thereby improving organizational achievement and attaining the company's objectives.

Cascio (Cascio 1992)] suggests that organizations must gain a competitive advantage through the effective utilization of their human resources. The IT industry workforce is complex because employees strive to balance their personal development and their loyalty to a company. People change jobs frequently for higher salaries or more benefits. This is often very costly to the company in terms of training costs and the loss of time, knowledge, customer relationships and contact data, all of which are potentially transferred to a competitor when experienced workers are hired by another company. Therefore, human resource management should identify the employee characteristics that increase or reduce this risk.

Human resources, like a well-informed, capable citizenry, can improve an organization, a society, a government agency, a country or a nation (Khan 2003). For example, planning and needs-based recruitment keeps staff costs down, while merit-based selection procedures can improve the quality of staff (World Bank 1997). Companies can simultaneously have qualified, effective and low-cost employees though effective HRM.

In IT corporations, human resources are an important issue, but the best way to manage employees and customer relationships is often uncertain. Fuzzy set theory and rough set theory have shown to be particularly useful for the analysis of inexact, uncertain or vague data (Walczak and Massart 1999). Both theories deal with this indescribable knowledge.

1.2 IT Industry

Remarkable innovations based on communications and information technology (IT) are changing the way we live, the nature of economic activity and the policies of every country in the world. In recognition of this phenomenon, the G8 (the G7 countries plus Russia) recently adopted the Okinawa Charter on the Global Information Society. The idea that "IT is one of the most important forces shaping the twenty-first century" has thrust IT into the very center of this development.

The last thirty years have been characterized by rapid developments in the IT industry. Mainframe computers have been used since the early 1950s, and several manufacturers produced mainframe computers from the late 1950s through the 1970s. These manufacturers were known as "IBM and the Seven Dwarfs": Burroughs, Control Data, General Electric, Honeywell, NCR, RCA, and UNIVAC.

A greater use of computer applications may be traced to the introduction of the personal computer (PC) in the 1970s. A personal computer (PC) is a computer whose price, size, and capabilities make it useful for individuals. During the 1990s, the power of personal computers increased radically, blurring the formerly sharp distinction between personal computers and multiuser computers, such as mainframes. Today, higher-end computers are often distinguished from personal computers by greater reliability or a greater ability to multitask, rather than by CPU ability alone.

The 1980s were characterized by significant structural changes in the IT industry. These included the emergence of software as an independent and dynamic component segment of the IT industry and the growth of global production networks. Software generally falls into three different categories:

Table 1 Internet history

year	event
1958	US Government forms the Advanced Research Projects Agency (ARPA) to establish the United States' lead in science and technology for military purposes.
1970	The Internet is born—ARPANET commissioned by the US Department of Defense.
1971	Electronic mail (e-mail) is invented.
1973	First international connection to the ARPANET, with UK and Norway, occurs. File transfer protocol is specified. The ethernet is outlined. Internet ideas start.
1976	Networking becomes popular. Unix to Unix Copy is developed. E-mail and Internet become operational.
1981	Minitel deployed across France.
1986	The power of Internet is recognized: 5,000 hosts, 241 newsgroups.
1987	The commercialization of the Internet begins: 28,000 hosts.
1990	ARPANET ceases to exist.
1991	The world wide web (WWW) is established at the European Centre for Nuclear Research (CERN)—the most important development to date.
1992	Multimedia changes the use of Internet. The number of hosts exceeds 1 million mark.
1993	The WWW revolution truly begins. The White House and United Nations go online.
1994	Shopping malls and banks arrive on the Internet.
1995	Traditional dial-up systems begin to provide Internet access in the US. Internet-related companies go public, and commercial use begins. Domain registration is no longer free. Search engines emerge. Mobile code JAVA emerges.
1996	WWW browser war begins.
2000	Number of internet hosts exceeds 70 million. G8 adopts the Charter on the Global Information Society at Okinawa.

(1) Platform software includes firmware, device drivers, operating systems and (typically) graphical user interfaces.

(2) Application software is what most people think of when they think of software. Typical examples include office suites and video games. Application software is often purchased separately from the computer hardware.

(3) User-written software tailors systems to meet the user's specific needs. User software includes spreadsheet templates, word processor macros, scientific simulations, and scripts for graphics and animations.

During the 1990s, the Asia-Pacific region became one of the major players in world IT production in all of the core segments of the industry. Developing the labor force in Asia also became an important topic in every company and every country.

The principal components of IT include the following components:

(1) hardware, which is designed around the "microchip" (semiconductor integrated circuit) and its associated peripherals
(2) software, which includes programming languages and their applications
(3) communications devices, which comprises both terrestrial and wireless units and related equipment
(4) the Internet, which is based on a generation of new computer languages and protocols that link individual computers into a vast network through which information can flow unimpeded

The history of the Internet in terms of the growth of the Internet is presented in Table 1.

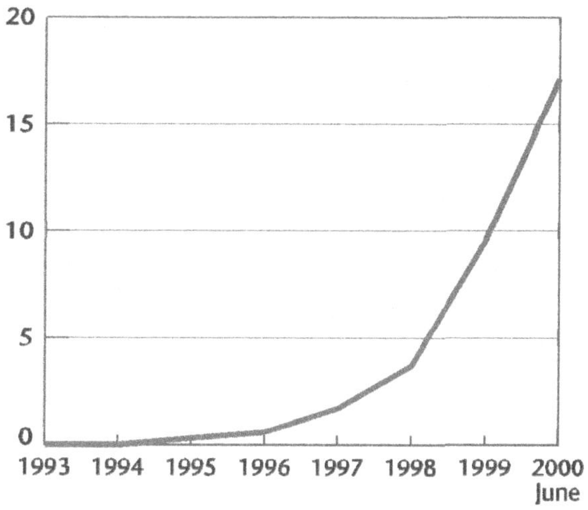

Fig. 1 Growth of World Wide Web Sites

In Figure 1, we can withness the rapid growth of World Wide Web sites from 1998 to 2000, demonstrating the importance of rapid developments for the IT industry.

The G8 Charter was a call to both public and private sectors to bridge the information and knowledge divide—the "digital divide." It underscores the fact that an effective partnership among stakeholders, through policy cooperation, is the key to the sound development of a truly global information society. The Charter emphasizes the importance of building on the following foundations:

(1) Economic and structural reforms to foster an environment of openness, efficiency, competition, and innovation
(2) Sound macroeconomic management to help economic agents plan confidently and exploit the advantages of new technology
(3) Information networks offering fast, reliable, secure, and affordable access through competitive market conditions
(4) Human resources capable of responding to the demands of the information age
(5) Active utilization of IT by the public sector

Human resources are present in all these facets outlined above, and human resources become necessary in the future to develop the IT industry.

1.3 Soft Computing

Because computer hardware and software allows for the management of large amount of data, that have been collected and stored in databases. Traditional ad hoc mixtures of statistical techniques and data management tools are no longer adequate to analyze this vast collection of data (Mitra 2002). Many businesses need to collect huge databases for financial investment, human resource management, customer relationship management, production and inventory management, and other purposes. However, a major problem is how to analyze such large amounts of data. Many studies use data mining and soft computing to identify meaningful information from a large-scale database (Mitra 2002). Current ways that data mining is used in practice include the following:

(1) Classification—classifies a data item into one of several predefined categorical classes
(2) Regression—maps a data item to a real valued prediction variable
(3) Clustering—maps a data item into one of several clusters, where clusters are natural groupings of data items based on similarity metrics or probability density models
(4) Rule generation—extracts classification rules from the data
(5) Discovering association rules—describes an association relationship among different attributes
(6) Summarization—provides a compact description for a subset of data
(7) Dependency modeling—describes significant dependencies among variables
(8) Sequence analysis—models sequential patterns, like time-series analysis

The goal is to model the states of the process generating the sequence or to extract and report deviation and trends over time.

Data mining has been very popular in many types of research. Recently, various soft computing methodologies, such as fuzzy logic, neural networks, genetic algorithms, and rough sets, have been used to handle the different challenges posed by data mining (Mitra 2002). Each of these methods can be used to analyze a problem. These methodologies are not homogeneous techniques, but they can be used together to solve complex questions. Studies increasingly combine these research methods to produce results that are more adapted to the current world than traditional techniques.

Current studies that use data mining have some limitations. These methods often focus on algorithms and visualization techniques. Data mining can identify a huge number of patterns in a database, but most of these patterns are useless or uninteresting to the user (Mitra 2002). Rough set theory is suitable for analyzing different types of uncertainty and can reduce superfluous information in large amounts of data.

In the IT industry, important business strategies include strengthening the skills of employees and promoting business collaboration. The objective of this collaboration is to balance the weaknesses and strengths of companies and to create a mutually beneficial situation. These companies enhance their value through "win-win business collaborations" or "chains of business collaborations."

IT companies are changing their focus from computer hardware to service. Important issues in the IT industry include producing IT technical experts, providing high quality, improving business relations, and strengthening maintenance services. The most important issue is improving business ability to strengthen competitiveness and increase customer satisfaction.

A company's executives enhance organizational ability and cultivate human resources to increase sales volume and profits and expand their achievements in service.

Marketing staff (sales and marketing, systems engineering) play a central role in business planning. Marketing staff are required to understand sales products and to apply market strategies. Therefore, when considering personnel resources, it is important to understand how marketing staff evaluate their own IT skills, career achievement and advancement and expectations.

In this paper, questionnaires were used to quantify employee satisfaction. We asked 167 randomly selected employees involved in front-end marketing in five regional partner companies of foreign IT Company A to complete questionnaires.

This information was used to clarify issues faced by IT companies, to analyze how employees can improve and meet expectations, and to illustrate a latent structure to improve both corporate and individual quality.

2 Rough Set Theory

Rough set theory comes with many advantages. It provides efficient algorithms for finding hidden patterns in data, determines minimal sets of data (data reduction), evaluates the significance of data, and generates minimal sets of decision rules from the data. It is easy to understand and as such it offers a straightforward interpretation of the produced results (Pawlak 1996). These advantages can facilitate analysis. For these reasons, a rough set approach is a frequently used research method. It is important in Artificial Intelligence (AI) and cognitive science, especially in the areas of machine learning, knowledge acquisition, decision analysis, knowledge discovery from databases, expert systems, decision support systems, inductive reasoning, and pattern recognition.

Rough set theory was developed by Pawlak (Pawlak 1982, 1984, 2004). It has been applied to many areas, including: medical diagnosis, engineering reliability, expert systems, empirical study of materials data (Leclair and Zairko 1996), machine diagnosis (Zhai et al. 2002), travel demand analysis (Goh and Law 2003), business failure prediction (Beynon and Peel 2001), solving linear programs (Azibi and Vanderpooten 2002), data mining (Li & Wang, 2004) and a-RST (Quafafou 2000). Previous research has discussed the preference order of attribute criteria needed to extend the original rough set theory, such as sorting, choice and ranking (Greco et al. 2001), insurance markets (Shyng et al. 2007), and rough set theory combined with fuzzy theory (Polkowski 2003). Rough set theory is a useful vehicle for analyzing data and reducing information.

The theory provides a new mathematical setting to analyze uncertainty and classify imperfect data or information with results presented in the form of decision rules.

2.1 Information Systems

Generally, an information system, *IS* for brief, is defined as $IS = (U, A)$, where U consists of finite objects and is a universe and A is a finite set of attributes $\{a_1, a_2, ..., a_n\}$. Each attribute a belongs to set A, that is, $a \in A$. $f_a: U \rightarrow V_a$ f_a means all V_a are in the U, where V_a is a set of values of attributes. It is a domain of attribute a.

2.2 Lower and Upper Approximations

The essence of rough sets relies on the two basic concepts of the lower and upper approximations of a set, as shown in Figure 2.

In Figure 2, some squares are included in the circle, but the others are not. The set of the squares included completely in the circle is called a lower approximation. The set of the squares partly and completely included in the circle is called an upper approximation.

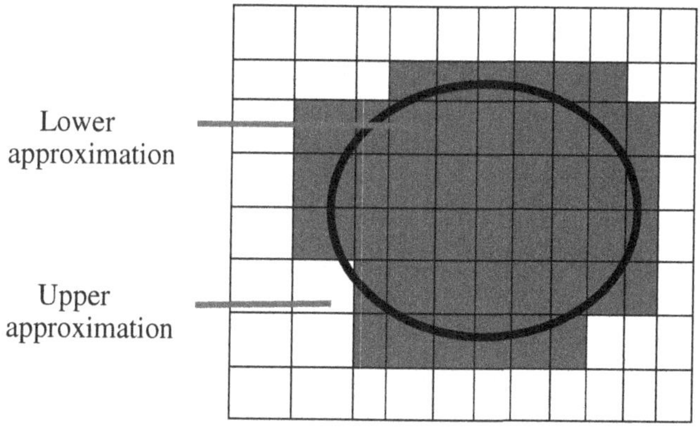

Fig. 2 Upper and Lower Approximations of Set *X*.

Let X be a subset of elements in universe U, that is, $X \subset U$. Let us consider a subset in $V_a, P \subseteq V_a$. The low approximation of P, denoted as $\underline{PX}$, can be defined by the union of all elementary sets X_i contained in X as follows:

$$\underline{\mathrm{PX}} = \{ x_i \in U \mid [x_i]_{\mathrm{ind}(p)} \subset X \} \tag{1}$$

where X_i is an elementary set contained in X, $i = 1, 2, \cdots, n$.

The upper approximation of P, denoted as PX, can be denoted by a non-empty intersection of all elementary sets X_i contained in X as follows:

$$\mathrm{PX} = \{ x_i \in U \mid [x_i]_{\mathrm{ind}(p)} \cap X \neq 0 \} \tag{2}$$

The boundary of X in U is defined in the following: PNX=PX-$\underline{\mathrm{PX}}$

Figure 2 illustrates a lower approximation and an upper approximation.

2.3 *Core and Reduct of Attributes*

Core and reduct attribute sets are two fundamental concepts of rough sets. The reduct is a minimal subset that makes the object classification the full set of attributes. The core is common to all reducts (Shyng et al. 2007). Reduct attributes can remove the superfluous attributes and give the decision maker simple information. There may be more than one reduct attribute. If the set of attributes is dependent, we are interested in finding all possible minimal subsets of attributes that have the same number of elementary sets, called the reducts (Walczak Massart 1999).

The reduct attribute set affects the process of decision making, and the core attribute is the most important attribute in decision making. If the set of attributes is indispensable, the set is called the core (Walczak Massart 1999).

$$\mathrm{RED(P)} \subseteq \mathrm{A} \tag{3}$$

$$\mathrm{COR(B)} = \cap\, \mathrm{RED(P)} \tag{4}$$

2.4 Decision Rules

Decision rules can be regarded as a set of decision (classification) rules of the form $a_{k_i} = d_j$,where a_{k_i} means that attribute a_k has value I, d_j means the decision attributes and the symbol '$\Rightarrow$' denotes propositional implication. In the decision rule $\theta \Rightarrow \phi$, formulas θ and ϕ are called the condition and decision, respectively (Walczak Massart 1999).

Through the decision rules, we can minimize the set of attributes, reduct the superfluous attributes and group elements. In this way, we can have many decision rules, each with meaningful features. Stronger rules will cover more objects, and the strength of each decision rule can be calculated to determine the appropriate rules.

3 Business Quality in the IT Industry

3.1 State of the Art of Business Activities

IT companies activate technologies for new businesses to expand the hardware business. They try to establish new business collaborations using their own technology to develop software packages. Computer makers restructure and organize special divisions and reallocate responsible personnel to promote this change in emphasis from hardware to software. Furthermore, computer makers provide seminars for each IT company and for each product brand. Nevertheless, there seems to be a difference in the intentions of IT companies and computer makers, even if the objective of both is to promote and sustain the business of IT companies.

It should be emphasized that the current IT industry needs to provide total solutions and businesses are changing by applying various skills from systems development. These skills are classified into three groups: (1) technical skills, (2) human skills and (3) management skills. Computer makers have a responsibility to promote this business direction toward a proposing type business and cultivate these three skills.

3.2 Issues at the Level of Companies

If manufacturers do not consider consumers' preferences and needs, it is not possible to use their products to increase customer satisfaction. Therefore, decision making about products has changed from a model in which manufacturers determine the functions, design, performance, and quality of products based on their own judgment or standards. It is important to understand customers' needs, to commercialize products and services faster than competitors, and to identify target markets and customers appropriately. For example, when a manufacturer announces a new product, it should arrange its advertisements and promotions simultaneously. A corporation increases overall customer satisfaction by improving its products, providing clear instructions, creating effective customer channels for managing inquiries and complaints, improving the response rate and providing better customer service after purchase.

An integrated power evaluation of a company requires "corporate quality" to realize customer satisfaction. One of the important issues for companies is how to cultivate and educate employees to become leaders. Companies have to increase the number of training days for employees in a year, and all employees should be encouraged to be leaders. Therefore, every company has a responsibility to promote and develop these talents from each employee. Companies must emphasize not only job sites and factory floors; they must also provide a sufficient education system for employees and a strategy for personnel training and education. This research on the behavioral features of managers illustrates important issues about corporate quality, as identified in the survey of employees.

3.3 Issues of Individual Employees

"A corporation cannot develop without the development of individuals." These are the words of T.J. Watson, Sr., the founder of IBM. Watson believed that the aggregation of individual abilities could support a company. When individual employees can improve their abilities, the company can develop at the same time. Likewise, the company's development leads to the growth of individuals. That is, the goal of individual employees should be realized as well as that of the company.

The qualities necessary to each individual employee's working life can be summarized as follows:

(1) Job satisfaction
(2) The successful application of the employee's ability and development
(3) Taking responsibility and authority for himself or herself
(4) Appraisal of the employee's job by other people
(5) Appraisal equivalent to achievement (Labart et al. 1984)

Therefore, each employee expects to be properly compensated for the achievement of his or her own targets, job functions and responsibilities. It is said that job experience and responsibility help people mature. On the other hand, an employee's own spirit and enthusiasm toward the job may be the most important factors. It is necessary to develop employees' abilities within the framework of personnel management strategies and to improve employees' skills according to their own career plans. There have been many cases where employees have developed their own skills through on-the-job training (OJT) either inside or outside of their company. In the competitive technology environment of the IT industry, OJT is one of the easiest methods to help an individual employee understand flexible approaches and problem-solving methods. After obtaining on-the-job training experiences, education hours must be provided for employees to learn theories. It is the responsibility of individual employees themselves to ensure steady knowledge acquisition and to direct their passion and enthusiasm in the education process.

4 Corporate Quality and Individual Employees in the IT Industry

4.1 Corporate Quality

Quality control starts with the detection of problems and deficits. Effective quality control requires employees to be eager and enthusiastic in trouble-shooting. The most important way to find defects in a system is to start from the facts. Therefore, quality control is a method that depends on facts. All quality control activities begin by recording and interpreting facts.

"The remedy to cure the illness of Nissan should be found within the company." These are the words of Mr. Carlos Ghosn, who led Nissan to recovery. He committed himself to talking with his employees thoughtfully and thoroughly after being appointed as the CEO (Terano et al. 1992). We can relate this to our own organization with the following questions: Did your company change from the past? Did your company successfully respond to changes in markets, competitive structures and customers' needs?

Companies survive by changing and innovating in accordance with the changing environment. The V-curve of recovery for Nissan came primarily from Ghosn's leadership, but we should not forget the role of experts from the middle lines and the general employees who embodied the rebuilding of Nissan (Terano et al. 1992).

Employees are classified into line specialists and staff specialists. The most important goal of an organization is to obtain maximum interests. It is not possible for the sales staff to increase profits for the entire company, and it is not possible to build the competitive power of a company unless employees in each division master their job skills. The responsibility of an organizational sales division is to realize that employees and staff specialists eagerly and sincerely consider the needs of their customers. Staff specialists support line employees. Sales divisions have to collaborate with other related divisions in sales activities and approach problems and opportunities flexibly to achieve the goals that will satisfy all employees in the company, including those in sales divisions. Corporate quality should be enhanced. A company distinguishes itself from its competition if it can produce at least one employee who can imagine the satisfaction of the customer when doing their job. To realize organizational sales, the required behavioral elements for all employees include skill and knowledge. A well-organized sales department should respond to customers' requests as a team and take final responsibility for the customer's needs. It is of primary importance to increase employees' loyalty toward the company.

Figure 3 shows marketing employees' satisfaction with their work stations. Affirmative answers including "yes" and "partly satisfied" represented 57.5% of all the answers obtained from the five companies for Question 41, "Are you satisfied with the work environment of your company?" An individual employee expects the company's support to increase the quality of his or her work life and to enrich his or her individual ability. It is important to increase employee satisfaction. Employees cannot sufficiently demonstrate their own abilities without a satisfying work environment, even if educational menus are provided for employees to flexibly and rapidly address customer requests. This is because employees want a safe and stable work environment. A stable work environment is essential to nurture employees' abilities sufficiently in an environment in which they spend the entire day.

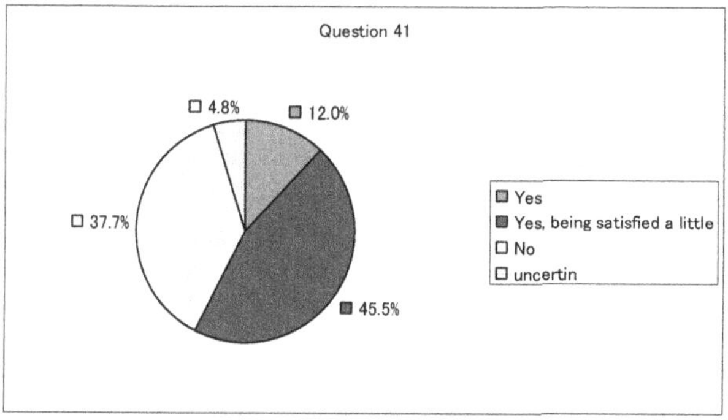

Fig. 3 Question 41: "Are you satisfied with the work environment of your company?"

How do employees view their compensation and their promotion opportunities? Are they satisfied with morale and motivation as well as education and career development potential? The results of Question 37 and 38 are similar, as shown in the following figures:

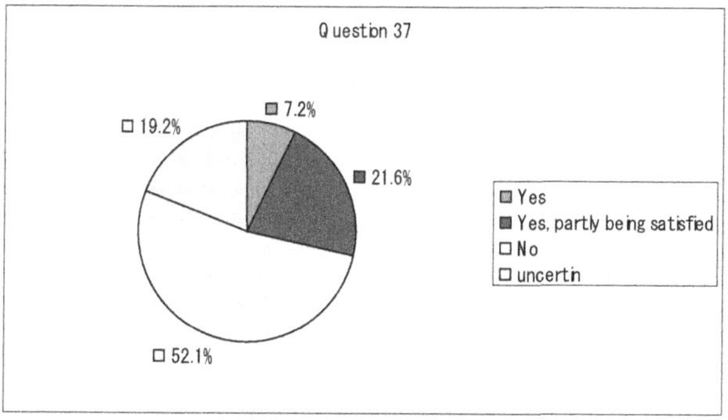

Fig. 4 Question 37: Satisfaction rating of in-house programs to improve morale/motivation

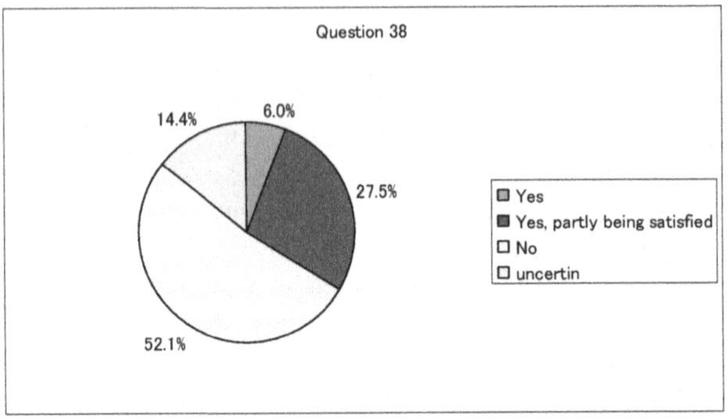

Fig. 5 Question 38: Satisfaction rating for systems and programs of in-house promotion and recognition

Figures 4 and 5 show that employee satisfaction is strongly correlated with customer satisfaction. If employees are dissatisfied with a work environment or a company, they will not take sufficient responsibility, leading to an increase in errors and a decrease in efficiency. Accordingly, the quality of products and services will worsen, leading to customer dissatisfaction. Therefore, improving employee satisfaction enhances quality and productivity (JPC=SED 2003)

4.2 Individual Quality

Figure 6 shows the responses to the question about the understanding of sales territories and customers in Question 6. The total number of affirmative responses in Figure 6 is 62.8%. This figure is not sufficient for marketing staff on the front line. The response "out of duty" was disappointingly high, at 34.7%. The questionnaires were primarily geared toward marketing staff, but application system engineers and maintenance staff chose this response as well. Of the respondents, 27.1% were system engineers and maintenance staff, and 10.2% were operational staff.

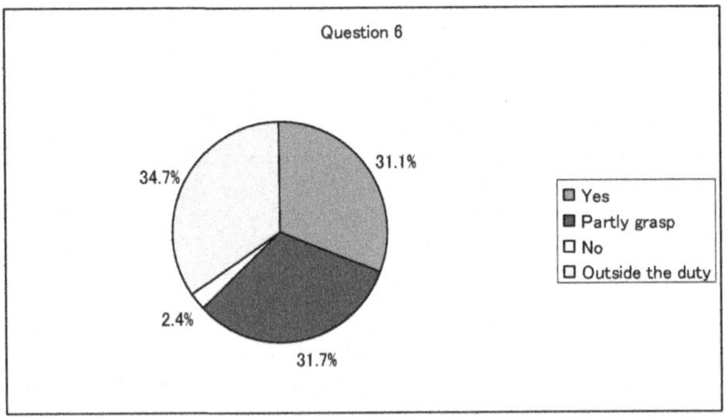

Fig. 6 Question 6: "Do you understand your own targeted sales territory and customers?"

Question 7 asks, "Do you understand the annual target figure/annual action target?" Figure 7 shows that 23.4% of employees responded "out of duty." This number is too high in comparison with the total of "yes" or "partly grasped" responses, at 73.7%. About one-fourth of all employees selected "out of duty" or "no." This fact conflicts with employee satisfaction. Marketing staff have to work hard to realize the target figure. Otherwise, the company cannot realize the target set.

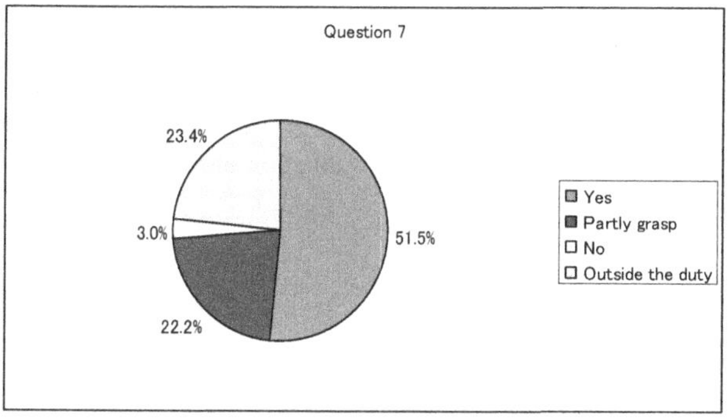

Fig. 7 Question 7: "Do you understand the setting of the action target and the target sales amount for a year?"

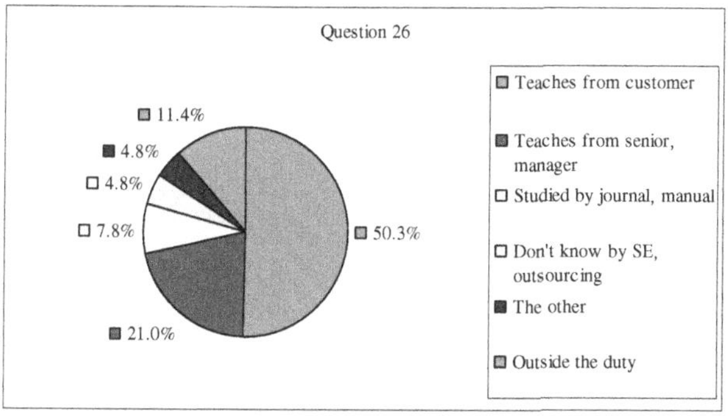

Fig. 8 Question 26: "How can employees obtain the application knowledge of operations?"

Next, we discuss how employees can obtain the applied knowledge of operations, as shown in Figure 8. This is the most important answer in this research and includes the skills and qualities required for a quality marketing staff. What customers buy from manufacturers and agents depends on marketing ability and business applications, and 50.3% of marketing staff learned this knowledge from their customers. This is important for individual growth and career development. Marketing staff can understand and improve customer loyalty by acquiring new knowledge through lectures and seminars and implementing this knowledge to improve overall quality.

"Think" is a motto frequently used at IBM. It is a word that Watson, the founder of IBM, used. Watson used to tell his employees to read books, listen to other people's opinions, discuss with everyone, and then think. IBM's corporate engine is based on education and training, and they invest 3% of their entire sales volume in employee education and training.

5 Analysis of Quality by Pattern Classification

5.1 Application of Pattern Classification

The objective of this paper is to clarify the issues in improving individual quality in personnel development and employees' career development by understanding the present state of employees in the IT industry. The policies involving these issues should be linked to the company's management quality. The questionnaires were analyzed using a pattern classification model.

5.2 Scatter Graphs of Attributes

(1) Axis 1 in Figure 9
In a scatter graph, positive values of the fifth item are plotted in the positive region of Axis 1 (+). In particular, all of Evaluation A is included in the first quadrant.

Regarding the skill evaluation of four products, both evaluations A and B are in the positive region of Axis 1 (+). In particular, the skill of Product X has high value.

(2) Axis 2 in Figure 9
In the scatter graph, five items have a strong correlated movement with "the evaluation of increasing skills." On Axis 2 of the scatter graph, evaluations A and B of five item attributes are positive. The positive values are E-mo and E-pro.

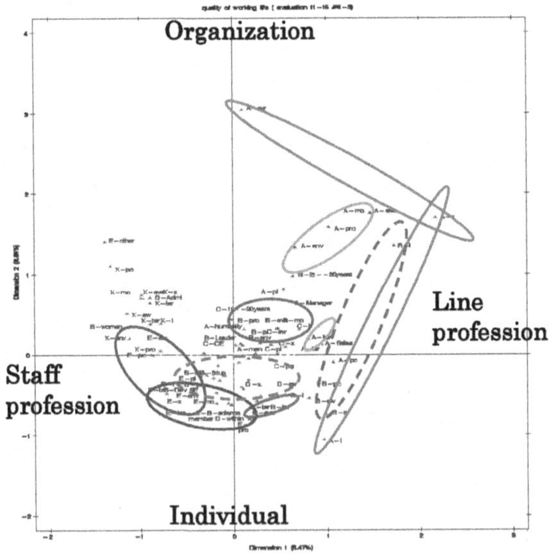

Fig. 9 Scatter graphs of attributes

The skill evaluation of four products has a negative correlation with evaluations in contrast to Axis 1 (+), and the evaluation of four products is negative but E-sw. The employees' satisfaction in a company is proportional to the sufficiency of skills. Even though the scatter graph of variables is negative, changing the management quality of a company strengthens its competitive power.

5.3 Scatter Graph of Employees

(1) Axis 1 in Figure 10

(1.1) Satisfied employees:

Nine out of ten employees who are satisfied or partly satisfied are within the positive region of Axis 1.

(1,2) Unsatisfied employees:

Eighteen out of fifty employees who are unsatisfied are within the positive region of Axis 1.

(1,3) Undecided employees:

Five out of twenty-eight employees are on the positive region of Axis 1.

(2) Axis 2 in Figure 10

(2.1) Satisfied employees:

Ten out of ten employees who are satisfied are within the positive region of Axis 2.

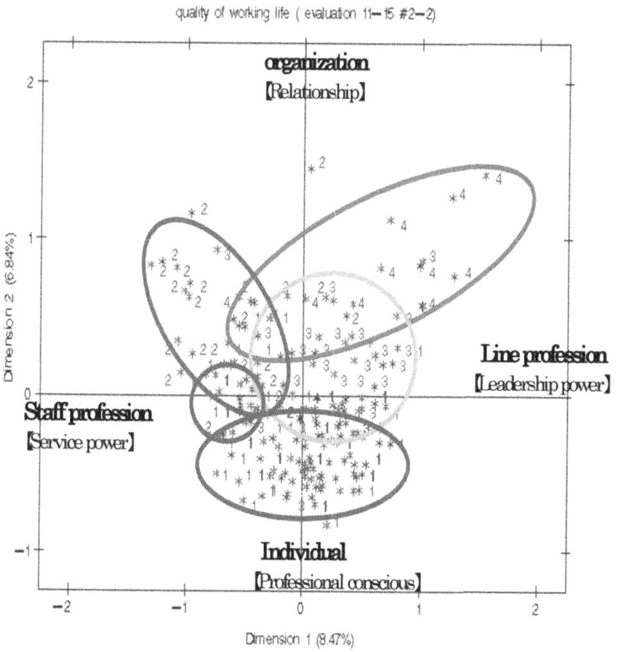

Fig. 10 Scatter graph of total evaluation

(2.2) Partly satisfied employees:
Twenty-two out of twenty-eight employees who are partly satisfied are within the positive region of Axis 2.

(2.3) Unsatisfied employees:
Eleven out of fifty employees who are unsatisfied are within the positive region of Axis 2.

(2.4) Undecided employees:
Twenty-three out of twenty-eight employees are on the positive region of Axis 2.

The quality of working life, everyday life, future prospects and abilities are areas that employees seek to improve.

In the scatter graph, even though many employee responses pull Axes 1 and 2 in a positive direction, there are still many employees in the negative region. This suggests that many employees are disgruntled.

5.4 Scatter Graph of Corporation Awards/Reward System

(1) Axis 1 in Figure 11

(1.1) Satisfied employees:
Six out of nine employees who are satisfied or partly satisfied are within the positive region of Axis 1.

(1.2) Unsatisfied employees:
Twenty out of thirty-nine employees who are unsatisfied are within the positive region of Axis 1. (Totally 29 samples)

(1.3) Undecided employees:
Three out of twenty employees are on the positive region of Axis 1.

(2) Axis 2 in Figure 11

(2.1) Satisfied employees:
Eight out of nine employees who are satisfied are within the positive region of Axis 2.

(2.2) Partly Unsatisfied employees:
Twenty-nine out of thirty-nine employees who are partly satisfied are within the positive region of Axis 2.

(2.3)Unsatisfied employees:
Twenty out of forty-six employees who are unsatisfied are within the positive region of Axis 2.

(2.4) Undecided employees:
Eleven out of twenty employees are on the positive region of Axis 2. The factors that completely satisfy employees are located on Axis 1 and partly on Axis 2. In contrast, employees who are not satisfied are located on the negative regions of Axes 1 and 2.

Thirty-eight employees who are dissatisfied are scattered over the positive region.

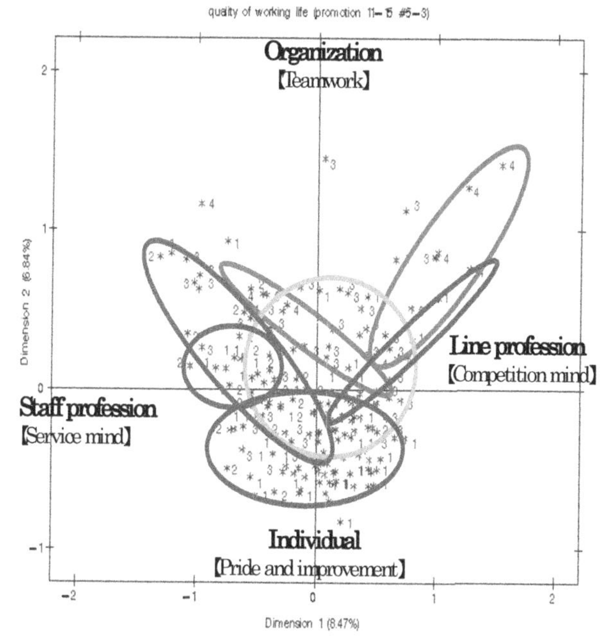

Fig. 11 Scatter graph of corporation awards/reward system

This is an important issue for many companies. Managers must treat and evaluate employees fairly and impartially according to their fiscal plan. Even employees who are in the negative region can be changed into contributing marketing staff if they are treated properly and nurtured. The education and training system in a company should be positioned as one of its business strategies.

6 An Empirical Study on Human Resource Development

In this research, we obtained responses to 47 questions from 167 employees of IT companies. These questions described attributes that characterize employees. Employees with good customer relations help companies earn larger profits and produce brand loyalty. However, it is not easy to determine which employees will have good customer relations. Therefore, we focus on the answer to the question, "Within and outside of the company, do you have good human relations with your customers?"

Let us denote the answer of employee i $(i = 1,2,\cdots,167)$ to the above question as A_i. The objective of this research is to clarify what pattern of the other answers on the questionnaire results in some value of A_i, that is, "yes, I have a good relationship with customers" or "no, I do not have a good relationship with them." We investigated the answers from 167 employees of IT companies and identified the latent structure of the answers as suggesting that managers and companies can provide effective functions to motivate their staff.

We used ROSE (Predki et al. 1998, Predki et al. 1999), developed by the Laboratory of Intelligent Decision Support Systems, to handle rough set analysis problems.

Table 2 Lower and upper approximations

Class number	Number of objects	Lower approximation	Upper approximation	accuracy
1	146	146	146	1.000
2	21	21	21	1.000

Table 2 shows the lower and upper approximations obtained by a rough set analysis. This obtained accuracy is 1.000. We can say the target set is definable on an attribute set (Pawlak et al. 1988). Attribute A_i $(i = 1,2,\cdots,167)$, "within and outside of a company, do you have good human relations with your customers?" is considered to be a decision attribute. The values that A_i $(i = 1,2,\cdots,167)$ takes the values of 1 (yes) or 2 (no). There were 146 yes responses and 21 no responses.

Table 2 shows that the upper and lower approximations are equivalent. Therefore, there is no uncertainty in the classification between classes $D = 1$ and $D = 2$. When decision rules are obtained, the decision rules will help the DM (decision maker) obtain more information about human resources.

6.1 Decision Rules

Decision tables show the covering rate and the elements of rules. This helps to identify these features more clearly. Each rule has its own elements, and these elements are the features of these rules. We want to determine the typical rules for most of employees to help decision makers identify the ideal employee behavior. In this way, a manager can differentiate these features into multiple groups, each with its own policy.

We found 16 such rules, as shown in Appendix A. There are nine decision-related rules that indicate a good relationship with customers and seven decision-related rules that indicate a bad relationship with customers. The total coverage rate is 98.8%. This decision table represents the IT corporation employees' behaviors.

Appendix A indicates that Rule 1 covers 59.59% of the employees. More than half of the people in these IT companies demonstrate these features when they have good relationships with customers. Rule 10 covers 28.57% of the employees. The behavior of these employees towards their customers may not be satisfactory. Rule 10 only covers 28.57% people because out of 167 employees, only 21 employees do not have not good interactions with consumers, but this 28.57% is still meaningful for our results.

7 Results and Discussion

Considering rules 1 and 4 we can see that when employees in IT corporations have good relationships with customers they also demonstrate positive thinking and behavior. They understand the entire system, organization and contacts of their customers and company, they also understand the financial statements. They think it is useful to attended business protocol trainings, they understand the board members and internal operating rules of the company, and they have good relationships with coworkers. These features show that they are very active in their company, so managers must focus on encouraging them to continue to contribute and increase their motivation to benefit the company

Alluding to rules 10 and 12, we see that employees with poor relationships with their customers do not understand the relevant contracts or the system and organization of their company. They do not participate in scheduled trainings, lectures or seminars for business behaviors and career development, they have no contact with board members and no knowledge of customer databases, and they have no solutions for problems with manufacturers. These negative features mean that employees do not care about their jobs or achievements. The managers of these employees must identify the reasons for these negative features and motivate them to work.

These positive and negative decision rules help identify what features and behaviors can bridge the customers and the company.

Maslow's hierarchy of needs is often depicted as a pyramid consisting of five levels: psychological needs, safety, love/belonging, esteem, and self-actualization. Employees who have good relationships with customers are at the level of love/belonging, so they will want to seek the next level, esteem. People need to be engaged to gain recognition with an activity that provides a sense of contribution. People try to improve, so they work hard to prove their ability. These people always want to learn new things and want more information about the company because they see themselves as part of a team and want to perform correctly at their jobs.

Employees who have poor relationships with customers are still at the safety level. They only care about their own employment and their own safety. They only think about themselves, so they do not want to work hard because it does not affect their personal safety. They do not see themselves as part of the company, so they have no interest in doing anything outside of their duties. Managers have to assure the security of their employment; the employees must know that their jobs are protected and that the manager does not fire people easily. They should be moved into the love/belonging level so that the manager can have an effective staff.

Managers must ensure that employees have good relationships with customers, help them perform their work correctly, and occasionally give them challenges and rewards to increase their motivation to improve. This will result in many benefits for the corporation. Whether employees have good or bad relationships with customers is part of the human resources of the corporation. Therefore, managers can give these two groups different functions.

8 Concluding Remarks

This research demonstrates that human resource management and customer relationship management are important for every corporation and suggests ways to deal with this trade-off. Using rough set theory, we reduced superfluous factors to reveal important elements in our decision table. We successfully used rough set theory to address human resources and customer relationship management, and we determined logical ways to manage employees and customers. This information can be adopted in IT corporations to help these companies provide the right service at the right time to satisfy their customers without sacrificing employees' rights. In other words, corporations can increase the satisfaction of employees and customers at the same time.

In this paper, the questionnaire responses were obtained from 167 employees of IT companies. These data were analyzed using a pattern classification model. The following conclusion can be highlihted: to improve the quality of a company, it is most important to ensure employee satisfaction.

References

[Azibi, Vanderpooten 2002] Azibi, R., Vanderpooten, D.: Construction of rule-based assignment models. European Journal of Operational Research 138(2), 274–293 (2002)

[Beynon, Peel 2001] Beynon, M.J., Peel, M.J.: Variable precision rough set theory and data discretisation: an application to corporate failure prediction. Omega 29(6), 561–576 (2001)

[Cascio 1992] Cascio, W.C.: Managing Human Resources: Productivity, Quality of Work Life, Profits. McGraw-Hill, New York (1992)

[Goh, Law 2003] Goh, C., Law, R.: Incorporation the rough sets theory. Chemometrics and Intelligent Laboratory Systems 47(1), 1–16 (2003)

[Greco et al. 2001] Greco, S., Matarazzo, B., Slowinski, R.: Rough sets theory for multicriteria decision analysis. European Journal of Operational Research 129(1), 1–47 (2001)

[Hayashi 1972] Hayashi, C.: Quantification Method,Toyo KeizaiShinpo-sha (in Japanese) ;see also Hayashi, C.,Higuchi, I.,Komazawa,T.(eds.) Statistics in Information Processing, Sangyo Tosho (1972) (in Japanese)

[Jackson et al. 1996] Jackson, A.G., Leclair, S.R., Ohmer, M.C., Ziarko, W., Al-kamhwi, H.: Rough Sets applied to materials data. ACTA Mater 44(11), 4475–4484 (1996)

[JPC-SED 2003] JPC-SED, What is Japan Quality Award 2003, p.4, p.5, p.7, p.153, p.154, p.156.JPC-SED Publishing (2003)

[Khan, Shahabuddin 2003] Khan, M.S.: Reading promotion: Perspective Bangladesh. CDN LAO newsletter 48 (2003)

[Labart et al. 1984] Labart, L., Morineau, A., Warwick, K.M.: Multivariate, Descriptive Statistical Analysis. John Wiley & Sons, New York (1984)

[Li, Wang 2004] Li, R., Wang, Z.O.: Mining classification rules using rough set and neural networks. European Journal of Operational Research 157(2), 439–448 (2004)

[Mirabile 1997] Mirabile, R.J.: Everything you wanted to know about competency modeling. Training & Development 8, 73–77 (1997)

[Mitra et al. 2002] Mitra, S., Pal, S.K., Mitra, P.: Data mining in soft computing framework: a survey. IEEE transactions on Neural Networks 13(1) (2002)

[Okamoto 2005] Okamoto, M.: Introduction management quality. JPC-SED Publishing (2005)

[Pawlak 1982] Pawlak, Z.: Rough sets. International Journal of Computer and Information Science 11(5), 341–356 (1982)

[Pawlak 1984] Pawlak, Z.: Rough classification. International Journal of Man–Machine Studies 20(5), 469–483 (1984)

[Pawlak 1991] Pawlak, Z.: Rough sets.Kluwer Academic Publishers, The Netherlands (1991)

[Pawlak 19996] Pawlak, Z.: Rough set and data analysis. In: Proceedings of the Asian Fuzzy Systems Symposium, Soft Computing in Intelligent Systems and Information, December 11-14,pp.1–6 (1996)

[Pawlak 1999] Pawlak, Z.: Rough classification. Int. J. Human-Computer Studies 51(15), 369–383 (1999)

[Pawlak 2004] Pawlak, Z.: Decision networks. In: Tsumoto, S., Słowiński, R., Komorowski, J., Grzymała-Busse, J.W. (eds.) RSCTC 2004. LNCS (LNAI), vol. 3066, pp. 1–7. Springer, Heidelberg (2004)

[Pawlak 2005] Pawlak, Z.: Rough sets and flow graphs. In: Ślęzak, D., Wang, G., Szczuka, M.S., Düntsch, I., Yao, Y. (eds.) RSFDGrC 2005. LNCS (LNAI), vol. 3641, pp. 1–11. Springer, Heidelberg (2005)

[Pawlak et al. 1988] Pawlak, Z., Wong, S.K.M., Ziarko, W.: Rough sets:Probabilistic versus deterministic approach. International Journal of Man-machine Studies 29, 81–95 (1988)

[Pfeffer 1994] Pfeffer, J.: Competitive advantsgc through people. Harvard University Press, Boston (1994)

[Pfeffer, Veiga 1999] Pfeffer, J., Veiga, J.F.: Putting people first for organizational success. The Academy of Management Executive 13(2), 37–38 (1999)

[Polkowski 2003] Polkowski, L.: Rough Mereology: A Rough Set Paradigm for Unifying Rough Set Theory and Fuzzy Set Theory. Fundamenta Informaticae 54, 67–88 (2003)

[Predki, Wilk 1999] Predki, B., Wilk, S.: Rough Set Based Data Exploration Using ROSE System. In: Raś, Z.W., Skowron, A. (eds.) ISMIS 1999. LNCS (LNAI), vol. 1609, pp. 172–180. Springer, Heidelberg (1999)

[Predki et al. 1998] Predki, B., Slowinski, R., Stefanowski, J., Susmaga, R., Wilk, S.z.: ROSE - Software Implementation of the Rough Set Theory. In: Polkowski, L., Skowron, A. (eds.) Rough Sets and Current Trends in Computing. LNCS (LNAI), vol. 1424, pp. 605–608. Springer, Berlin (1998)

[Quafafou 2000] Quafafou, M.: α-RST: a generalization of rough set theory. Information Sciences 124(4), 301–316 (2000)

[ROIP (2005)] Research Organization Innovation Projects, Guide of Organization innovation practice.p.10,p.35.The SNNO Institute of Management Publishing, (2005)

[Shyng 2007] Shyng, J.Y., Tzeng, G.H., Wang, F.K.: Rough set Theory in Analyzing the Attributes of Combination Values for insurance market. Expert System with Applications 32(1) (2007)

[Tenenhaus, Young 1985] Tenenhaus, M., Young, F.W.: Analysis and Synthesis of Multiple Correspondence Analysis, Optional Scanning, Homogeneity Analysis and Other Methods for Quantifying Categorical Multivariate Data. Psychometrica 50(1), 91–119 (1985)

[Truss 2001] Truss, C.: Complexities and controversies in linking HRM with organizational outcomes. Journal of Management Studies 38(8), 1122–1149 (2001)

[Walczak, Massart 1999] Walczak, B., Massart, D.L.: Rough set theory. Chemometrics and Intelligent Laboratory 47(1), 1–16 (1999)

[Watada 1992] Watada, J.: Fuzzy Quantification Theory. In: Terano, T., Asai, K., Sugeno, M. (eds.) Fuzzy Systems Theory and Its Applications,ch.6, pp. 101–124. Academic Press, London (1992)

[Chai 1981] Watada, J., Tanaka, H., Asai, K.: Analysis of Purchasing Factors Using Fussy Quantification Theory Type III. Journal of the Japan Industrial Management Association 32, 51–65 (1981) (in Japanese)

[Watada J, Tanaka H, Asai K (1983)] Watada, J., Tanaka, H., Asai, K.: Fussy Quantification Type III. The Japanese Journal of Behavior metrics Society of Japan 9(2), 24–32 (1983) (in Japanese)

[Watada et al. 1998] Watada, J., Tanaka, T., Arredondo: Analysis of safety from macro ergonomic point view. The Japanese Journal of Ergonomic 134(6) (1998) (in Japanese)

[Watada et al. 2004] Watada, J., Yabuuchi, Y., Li, M.J.: The Cause and Influence of Accidents in Corporations. In: Proceedings, PICMET 2004, Portland International Conference on Management of Engineering and Technology, pp. 1–8 (2004)

[World Bank 1997] World Bank, World development report: The changing role of the state. Oxford University Press, Oxford (1997)

[Yan Ai-min, Liu Yuan, Liu Zhi-cheng, Chen Zheng] Ai-min, Y., Yuan, L., Zhi-cheng, L., Zheng, C.: Study on Human Resource Niche Concept and Evaluation Indexes, Management Science and Engineering. In:International Conference on ICMSE 2006, pp.1290–1295 (2006)

[Zhai, Shore 2002] Zhai, L.Y., Khoo, L.P., Fok, S.C.: Feature exraction using rough set theory and generic algorithms an application for the simplification of product quality evaluation. Computers & Industrial Engineering 43(4), 661–676 (2002)

APPENDIX A

Rule number	Decision attribute	The minimal covering rule	Covering rate
Rule(1) **59.59%**	1	**Q29** Do you recognize and understand the whole system and organization of your customer's company? Yes **Q30** Do you understand the board members? Yes **Q33** Do you like attending training or seminars to acquire knowledge of financial statements? Yes **Q36** Do you have good human relations with your bosses, senior members, and colleagues within the company? Yes **Q40** Do you understand the internal operating rules of the company? Yes	0.520
Rule(2) **45.21%**	1	**Q29** Do you recognize and understand the whole system and organization of your customer's company? Yes **Q39** Do you understand the contract with the company where you work? Yes **Q2** Is it useful to your business that you attended business protocol training? Yes	0.640
Rule(3) **23.97%**	1	**Q36** Do you have good human relations with your bosses, senior members, and colleagues within the company? Yes **Q4** Did you attend scheduled training, lectures or seminars for business protocol and career development after joining this company? None **Q46** With whom do you consult when you experience trouble with collaboration? The boss **Q48** Are you male or female? Male **Q52** How many books do you read in a month? About 5 books	0.700
Rule(4) **10.27%**	1	**Q18** Regarding the method for learning skills for software products, what is your general product knowledge level of software products? Low **Q38** Are you satisfied with the promotion and reward programs within the company? Yes **Q42** Regarding the contracted rules and collaboration with computer manufacturers, do you understand the yearly measures and policies of the manufacturer? No	0.724
Rule(10) **28.57%**	2	**Q39** Do you understand the contract with the company where you work? No **Q4** Did you attend scheduled training, lectures or seminars for business protocol and career development after joining this company? None **Q31** Can you contact the board members of your company? No **Q50** What is your job title? General level	0.880
Rule(11) **23.81%**	2	**Q29** Regarding the knowledge of the company system and organization, do you recognize and understand the whole system and organization of your customer's company? No **Q33** Do you like to attend training or seminars to acquire knowledge of financial statements if you have the chance? Yes **Q9** Do you have a database of your customers? Outside of my duty **Q45** Did you find any solutions for troubles in collaboration with a manufacturer? No	0.910
Rule(12) **19.05%**	2	**Q45** Did you find any solutions to problems in collaboration with a manufacturer? No **Q5**. How do you find the business manners of other employees around you? Bad **Q28** Do you follow and backup your proposals with specifications and estimations after giving them to customers? I propose an estimation. **Q43** Do you understand dealing rules (price rate of products) with a manufacturer and the sales promotion program? No **Q53** Do you prefer to play a sport? Yes	0.934
Rule(13) **14.29%**	2	**Q30** Do you understand the board members? No **Q33** Do you like to attend any training or seminars to acquire knowledge of financial statements if you have the chance? Yes **Q43** Do you understand dealing rules (price rate of products) with a manufacturer and the sales promotion program? No **Q12** What is your general product knowledge level of the I series? Low **Q32** Regarding knowledge of financial statements, did you learn how to read and analyze financial statements? Yes	0.976
			Total 0.988

Environmental Applications of Granular Computing and Intelligent Systems

Wang-Kun Chen

Abstract. This paper presents the environmental applications of granular computing. First, the relevance of information granulation in the description of environmental phenomena is discussed. A granular prediction model of time series of a dust storm concentration is described. This example is used to explain the technique of information granulation of an environmental phenomenon. Then the issue of environmental management is discussed. Granular computing helps us establish the pattern recognition technique which is also very helpful in environmental management. In addition, this study presents an approach to extract interpretable rules of natural hazards from available data. Finally, the multi-objective design of a granular hierarchy model is presented to determine the optimal management strategy of air quality. The environmental application experiments show that granular computing comes as a promising vehicle for solving social problems related to protection of the environment.

Keywords: Fuzzy theory, Granular computing, Time series, Forecasting, Decision making, Environmental management, Economic evaluation.

1 Introduction

The last two decades have seen a dramatic increase in the research of natural phenomena simulated by granular computing and intelligent systems. This has mainly been due to the introduction of the fuzzy theory which allows complicated environmental features to be characterized. Granular computing is an efficient paradigm for simulating environmental phenomena and for solving the interaction problem between the natural and the social systems. This paper discusses some environmental applications of this new field.

Fuzzy logic was introduced in 1965 by Lotfi A. Zadeh, professor of computer science at the University of California in Berkeley. The main feature of the fuzzy set theory is associated with the concept of membership function (Zadeh 1965), (Zadeh 1975), (Pedrycz and Gomide 2007), (Klir and Yuan 2005). In the present

Wang-Kun Chen
Department of Environment and Property Management, Jinwen University of Science and Technology, Taipei, Taiwan

W. Pedrycz and S.-M. Chen (Eds.): Granular Computing and Intell. Sys., ISRL 13, pp. 275–301.
springerlink.com

paper, the environmental phenomenon is first discussed in a granular model for the prediction of the time series of a dust storm concentration. In the sequel, granular computing and pattern recognition are proposed for environmental management. The main focus here is on two important topics, pattern recognition in the environment, and an environmental design of higher order.

The variable granulation of environmental impact assessment is then explained, including the methodology of fuzzy filtering and image processing in the emission inventory. Next we discuss the system granulation of an environmental economic evaluation, which is divided into two main parts. The first part is concerned with the fuzzy set in risk evaluation of an environmental disaster, and the second part is the multi-objective design of a granular hierarchy model for the management strategy of the air quality.

2 Value Granulation of an Environmental Phenomenon

The environment plays an important role in our daily life. The air we breathe, the water we drink, and the soils we use for growing the plant we eat are all in the natural environment. Thus any phenomenon in the environment will affect our quality of living. For example, earthquakes, the tsunami, and typhoons are all natural events that cause damage to society. Having a better understanding of such phenomena will make it easier to protect ourselves against the consequences of these natural hazards.

To clearly describe events that occur in the environment, we use "time series" to record and describe the event. For example, a typhoon can be described by the position of its center and the wind speed along its trajectory; the seriousness of an earthquake is indicated by the intensity of its shockwaves; and the concentration variation in time domain tells us the variations inside a dust storm. If we are able to predict the value of this time series, then it will also be easier to avoid the loss of human life as well as economic losses.

In the present study we used an example to explain the technique of value granulation of an environmental phenomenon. The example used here is a granular model for predicting the time series of a dust storm concentration. Because the dynamic behavior of a dust storm is too complicated it is almost impossible for the classical solution by diffusion equation to have a precise solution on a global scale. Therefore, it is appropriate to consider the use of granular computing to realize this task.

2.1 A Granular Model for Predicting the Time Series of a Dust Storm Concentration

The granular model for predicting the time series of a dust storm concentration is described as follows. A dust storm is a major environmental phenomenon causing much human suffering and financial losses. A dust storm can be represented by its particle concentration. For example, PM_{10} or $PM_{2.5}$, refer to the concentration of particles with a diameter smaller than 10 μm or 2.5 μm, respectively. Predicting the particle concentration of a dust storm is important, but it is very difficult because there are many factors that can affect the concentration of PM_{10} or $PM_{2.5}$. These factors include wind speed, wind direction, temperature and topography, etc. It is very hard to predict the concentration of a dust storm using a physical model. Therefore, the application of a fuzzy time series is a suitable choice for the prediction of a dust storm concentration.

We used the definition of a fuzzy time series as proposed by Chen based on Song and Chisson's research on the fuzzy set theory. (Song and Chisson 1993), (Song and Chisson 1994), (Chen 1996), (Chen 2002), (Chen and Hwang 2000), (Chen and Hsu 2004). If we let U be the universe of discourse, U = $\{u_1, u_2, u_3, \ldots, u_n\}$, then a fuzzy set A in the universe of discourse U is defined as $A_i = fA_1(u_1)/u_1 + fA_2(u_2)/u_2 + \ldots + fA_n(u_n)/u_n$, where f_A is the membership function of the fuzzy set A, f_A: U → [0, 1], $f_A(u_i)$ denotes the grade of membership of u_i in the fuzzy set A, $f_A(u_i) \in [0, 1]$, and $1 \leqq i \leqq n$.

Let X (t) (t = 0, 1, 2...) be the universe of discourse in X (t), and let F (t) be the collection of f_i (t) (i = 1, 2...), then F (t) is a fuzzy time series of X (t) (t = ..., 0, 1, 2 ...)

If we assume that there is a fuzzy relationship R(t, t-1) such that F(t) = F(t-1)$_\circ$ R (t, t-1), where R (t, t-1) denotes the fuzzy relationship of the first order fuzzy time series between F (t-1) and F (t), and if F (t-1) =A_i, and F (t) =A_j are fuzzy sets, and then the fuzzy relationship can be represented as $A_i \rightarrow A_j$.

If F(t) is caused by F(t-1), F(t-2), F(t-3),...and F(t-n), then the fuzzy relationship is represented by F(t-n),...F(t-2),F(t-1) →F(t). This is the high-order model as proposed by Chen.

In this section, we use the multi-step fuzzy time series presented by Chen as the high-order fuzzy time series for forecasting the PM_{10} concentration (Chen et al. 1998), (Hsu and Chen 2003), (Chen 2010). This method determines the trend of the data by adjusting the length of each interval in the universe of discourse. The proposed method is explained as follows:

Defining the Fuzzy Interval Set for a Dust Storm

Let U be the universe of discourse, U = $[D_{min} - D_1, D_{max} + D_2]$, where D_{min} and D_{max} denote the minimum and maximum PM_{10} concentration, as shown in Table 1, respectively, where D_1 and D_2 are two suitable values, so that U can be divided into several intervals of equal length $u_1, u_2, u_3, \ldots\ldots u_n$. From Table 1, we can see that D_{min} = 34.44 and that D_{max} = 62.20. Then, if we let D_1 = 4.44 and D_2 = 0.3, such that U = $[u_1, u_2, u_3, u_4, u_5, u_6, u_7]$, and let the length of each interval be 5, then the universe of discourse U can be divided into $u_1, u_2, u_3, u_4, u_5, u_6, u_7$, where u_1 =[30,35], u_2 =[35,40], u_3 =[40,45], u_4 =[45,50], u_5 =[50,55], u_6 =[55,60], and u_7 = [60,65].

Table 1 The monthly averaged PM_{10} concentration. [2007.1~2008~12](From the monitoring result of Taiwan EPA)

time	PM_{10} concentration	time	PM_{10} concentration
2007.1	46.63	2008.1	38.76
2007.2	58.65	2008.2	42.88
2007.3	62.20 D_{max}	2008.3	55.54
2007.4	61.57	2008.4	48.57
2007.5	59.72	2008.5	42.06
2007.6	37.78	2008.6	38.67
2007.7	43.99	2008.7	40.86
2007.8	37.17	2008.8	44.27
2007.9	44.18	2008.9	47.54
2007.10	34.44 D_{min}	2008.10	44.84
2007.11	35.28	2008.11	42.89
2007.12	5.0.00	2008.12	47.40

Calculation of the Statistical Distribution of the PM_{10} Concentration in Each Interval

We sort the intervals based on the number of PM_{10} concentration data falling into each interval in descending order. The number of data, $N(u_n)$, in each interval is $N(u_1) = 1$, $N(u_2) = 5$, $N(u_3) = 8$, $N(u_4) = 4$, $N(u_5) = 1$, $N(u_6) = 3$, $N(u_7) = 2$.

The intervals with no distributed data were discarded, and those with a large amount of data were re-divided into additional sub-intervals. The idea being that intervals containing a high number of historical PM_{10} concentration data get divided into more sub-intervals to improve the accuracy of the prediction. Finally, the universe of discourse $[u_1, u_2, u_3, u_4, u_5, u_6, u_7]$ is re-divided into intervals u_1, u_2, u_3, u_4, u_5, u_6, u_7, u_8, u_9, u_{10}, u_{11}, u_{12}, u_{13}, u_{14}, u_{15}, u_{16}, u_{17}, as shown in the following example.

We define the following fuzzy intervals: $u_1 = [32.5,35]$; $u_2 = [35, 36]$; $u_3 = [37,38]$; $u_4 = [38,39]$; $u_5 = [40,41]$; $u_6 = [42,42.5]$; $u_7 = [42.5,43]$; $u_8 = [43,44]$; $u_9 = [44,44.5]$; $u_{10} = [44.5,45]$; $u_{11} = [46,47]$; $u_{12} = [47,48]$; $u_{13} = [48,49]$; $u_{14} = [50,55]$; $u_{15} = [55,57.5]$; $u_{16} = [57.5,60]$; $u_{17} = [60,62.5]$.

Defining Each Fuzzy Set A_i Based on the Re-Divided Interval u_i

We define each fuzzy set A_i based on the re-divided interval u_i as $A_i = fA_1(u_1)/u_1 + fA_2(u_2)/u_2 +\ldots+ fA_{17}(u_{17})/u_{17}$, and fuzzify the historical PM_{10} concentration in Table 3, where fuzzy set A_i denotes a linguistic value of the PM_{10} concentration represented by fuzzy set. $1 \leqq i$.

Table 2 Fuzzy set expressing PM_{10} concentration

A_i	Fuzzy relationship
A_1	$A_1 = 1/u_1+0.5/u_2+\ldots+0/u_{15}+0/u_{16}+0/u_{17}$
A_2	$A_2=0.5/u_1+1/u_2+0.5/u_3+\ldots+0/u_{15}+0/u_{16}+0/u_{17}$
A_3	$A_3=0/u_1+0.5/u_2+1/u_3+0.5/u_4\ldots+0/u_{15}+0/u_{16}+0/u_{17}$
A_4	$A_4=0/u_1+\ldots+0.5/u_3+1/u_4+0.5/u_5+\ldots+0/u_{15}+0/u_{16}+0/u_{17}$
A_5	$A_5=0/u_1+\ldots+0.5/u_4+1/u_5+0.5/u_6+\ldots+0/u_{15}+0/u_{16}+0/u_{17}$
A_6	$A_6=0/u_1+\ldots+0.5/u_5+1/u_6+0.5/u_7+\ldots+0/u_{15}+0/u_{16}+0/u_{17}$
A_7	$A_7=0/u_1+\ldots+0.5/u_6+1/u_7+0.5/u_8+\ldots+0/u_{15}+0/u_{16}+0/u_{17}$
A_8	$A_8=0/u_1+\ldots+0.5/u_7+1/u_8+0.5/u_9+\ldots+0/u_{15}+0/u_{16}+0/u_{17}$
A_9	$A_9=0/u_1+\ldots+0.5/u_8+1/u_9+0.5/u_{10}+\ldots+0/u_{15}+0/u_{16}+0/u_{17}$
A_{10}	$A_0=0/u_1+\ldots+0.5/u_9+1/u_{10}+0.5/u_{11}+\ldots+0/u_{15}+0/u_{16}+0/u_{17}$
A_{11}	$A_{11}=0/u_1+\ldots+0.5/u_{10}+1/u_{11}+0.5/u_{12}+\ldots+0/u_{15}+0/u_{16}+0/u_{17}$
A_{12}	$A_{12}=0/u_1+\ldots+0.5/u_{11}+1/u_{12}+0.5/u_{13}+\ldots+0/u_{15}+0/u_{16}+0/u_{17}$
A_{13}	$A_{13}=0/u_1+\ldots+0.5/u_{12}+1/u_{13}+0.5/u_{14}+0/u_{15}+0/u_{16}+0/u_{17}$
A_{14}	$A_{14}=0/u_1+\ldots+0.5/u_{13}+1/u_{14}+0.5/u_{15}+0/u_{16}+0/u_{17}$
A_{15}	$A_{15}=0/u_1+\ldots+0.5/u_{14}+1/u_{15}+0.5/u_{16}+0/u_{17}$
A_{16}	$A_{16}=0/u_1+\ldots+0.5/u_{15}+1/u_{16}+0.5/u_{17}$
A_{17}	$A_{17}=0/u_1+\ldots+0/u_{15}+0.5/u_{16}+1/u_{17}$

The fuzzified historical PM_{10} concentration is shown in Table 3, which is based on the re-divided intervals derived from the above.

Table 3 PM_{10} Concentration and fuzzy PM_{10} concentration

time	PM_{10} concentration	concentration intervals	Fuzzy PM_{10} Concentration	time	PM_{10} concentration	concentration intervals	Fuzzy PM_{10} Concentration
2007.1	46.63	u_{11} = [46,47]	47.00	2008.1	38.76	u_4 = [38,39]	39.00
2007.2	58.65	u_{16}= [57.5, 60].	57.50	2008.2	42.88	u_7 = [42.5,43]	43.00
2007.3	62.20	u_{17}= [60,62.5]	62.50	2008.3	55.54	u_{15} = [55,57.5]	55.00
2007.4	61.57	u_{17}= [60,62.5]	62.50	2008.4	48.57	u_{13} = [48,49]	49.00
2007.5	59.72	u_{16}= [57.5,60]	60.00	2008.5	42.06	u_7 = [42,42.5]	42.00
2007.6	37.78	u_3 = [37,38]	38.00	2008.6	38.67	u_4 = [38,39]	39.00
2007.7	43.99	u_8 = [43,44]	44.00	2008.7	40.86	u_5 = [40,41]	41.00
2007.8	37.17	u_3 = [37,38]	37.00	2008.8	44.27	u_9 = [44,44.5]	44.50
2007.9	44.18	u_9 = [44,44.5]	44.00	2008.9	47.54	u_{12} = [47,48]	48.00
2007.10	34.44	u_1 = [32.5,35]	35.00	2008.10	44.84	U_{10} = [44.5,45]	45.00
2007.11	35.28	u_2 = [35.36]	35.00	2008.11	42.89	u_7 = [42.5,43]	43.00
2007.12	50.00	u_{14} = [50,55]	50.00	2008.12	47.40	u_{12} = [47,48]	47.00

Establishing Fuzzy Logical Relationships Based on the Fuzzified PM_{10} Concentration.

If the fuzzified PM_{10} concentration for months i and i+1 are given as A_j and A_k, respectively, then we can construct the relationship "$A_j \rightarrow A_k$", where A_j and A_k are called the current state and the next state, respectively, of the PM_{10} concentration. For example, let's consider the fifth-order "A_{11}, A_{16}, A_{17}, A_{17}, $A_{16} \rightarrow A_3$", where the relationship denotes the fuzzified PM_{10} concentration, as shown in Table 3.

If the fuzzified PM_{10} concentration of month i is A_j and the fuzzy representation is shown as "$A_j \rightarrow A_{k1(x1)}, A_{k2(x2)} \ldots A_{kp(xp)}$", then the estimated PM_{10} concentration of month i is determined as

$$C(i) = \frac{X_1 \times m_{k1} + X_2 \times m_{k2} + \cdots + X_p \times m_{kp}}{X_1 + X_2 + \cdots + X_p}, \tag{1}$$

where X_i denotes the number of fuzzy logical relationships "$A_j \rightarrow A_k$" in the fuzzy logical relationship group, $1 \leqq i \leqq p$, m_{k1}, m_{k2},......where m_{kp} is the mid point of the interval u_{k1},u_{k2},......and u_{kp} respectively, and where the maximum membership values of A_{k1}, A_{k2},......and A_{kp} occur at intervals u_{k1},u_{k2},...... and u_{kp}, respectively.

Use of the High-Order Difference to Determine the Upward or Downward Trend

Chen derived a rule to determine the upward and downward trend of the time series. The second order difference between any two neighboring months of the historical PM_{10} concentration can be used for forecasting the trend (Chen, 2008).

Table 4 Fuzzy logical relationship

Two-step fuzzy logical relationship reported in different groups	Five steps fuzzy logical relationship
Group 1 $A_1 \rightarrow A_2$	$A_{11}, A_{16}, A_{17}, A_{17}, A_{16} \rightarrow A_3$
Group 2 $A_2 \rightarrow A_{14}$	$A_{16}, A_{17}, A_{17}, A_{16}, A_3 \rightarrow A_8$
Group 3 $A_3 \rightarrow A_8, A_9$	$A_{17}, A_{17}, A_{16}, A_3, A_8 \rightarrow A_3$
Group 4 $A_4 \rightarrow A_5, A_7$	$A_{17}, A_{16}, A_3, A_8, A_3 \rightarrow A_9$
Group 5 $A_5 \rightarrow A_9$	$A_{16}, A_3, A_8, A_3, A_9 \rightarrow A_1$
Group 6 $A_6 \rightarrow A_4$	$A_3, A_8, A_3, A_9, A_1 \rightarrow A_2$
Group 7 $A_7 \rightarrow A_{12}, A_{15}$	$A_8, A_3, A_9, A_1, A_2 \rightarrow A_{14}$
Group 8 $A_8 \rightarrow A_3$	$A_3, A_9, A_1, A_2, A_{14} \rightarrow A_4$
Group 9 $A_9 \rightarrow A_1, A_7, A_{12}$	$A_9, A_1, A_2, A_{14}, A_4 \rightarrow A_7$
Group 10 $A_{11} \rightarrow A_{16}$	$A_1, A_2, A_{14}, A_4, A_7 \rightarrow A_{15}$
Group 11 $A_{12} \rightarrow A_9$	$A_2, A_{14}, A_4, A_7, A_{15} \rightarrow A_{13}$
Group 12 $A_{13} \rightarrow A_6$	$A_{14}, A_4, A_7, A_{15}, A_{13} \rightarrow A_7$
Group 13 $A_{14} \rightarrow A_4$	$A_4, A_7, A_{15}, A_{13}, A_7 \rightarrow A_4$
Group 14 $A_{15} \rightarrow A_{13}$	$A_7, A_{15}, A_{13}, A_7, A_4 \rightarrow A_5$
Group 15 $A_{16} \rightarrow A_3, A_{17}$	$A_{15}, A_{13}, A_7, A_4, A_5 \rightarrow A_9$
Group 16 $A_{17} \rightarrow A_{16}, A_{17}$	$A_{13}, A_7, A_4, A_5, A_9 \rightarrow A_{12}$
	$A_7, A_4, A_5, A_9, A_{12} \rightarrow A_{10}$
	$A_4, A_5, A_9, A_{12}, A_{10} \rightarrow A_7$
	$A_5, A_9, A_{12}, A_{10}, A_7 \rightarrow A_{12}$

The second order difference is calculated as follows $Y_n = Y_{n-1} - Y_{n-2}$,

We also subdivided each interval into four intervals of equal length, with the 0.25 point and 0.75 point of each interval being used as the upward and downward forecasting points. From the fuzzy logical relationship described above, we are then able to determine the PM_{10} concentration.

Determine the Value of the α-Cut and the High-Order Difference

The α-cut value determines the fuzzified PM_{10} concentration of the interval. It is common to use the triangle function and choose the value of the α-cut equal to 0.5 for estimation. Another important factor is the value of the high-order difference since it dominates the trend of the variation in PM_{10} concentration. Considering that the effect of these factors may improve the accuracy of the forecasting substantially, it is better to include them in the more advanced multi-step fuzzy time series (MFT) model in the next stage.

Since the characteristics of PM_{10} are influenced by many factors, another suitable function, α-cut value, could be used for PM_{10} forecasting. Chen compared the results of the estimated particulate concentration by MFT and the neural network fuzzy time series (NFT) which provided a good result. The PM_{10} concentration predicted by the above method is shown in Chen's work. (Chen 2010)

The mean square error of the results using the neural network fuzzy model with the minimum value indicates the NFT model with the best predicting capability for PM_{10} concentration. The second best model is the multi-step fuzzy model compared with the mean square error value. The proposed method is better than the traditional fuzzy model with 8 and 16 equal intervals. The results predicted by the traditional auto-regressive model, and the linear regression model were all inferior to the MFT and NFT models. This is because that the traditional method of statistics and pattern recognition are parametric or non-parametric models. However, the high-order fuzzy time series and the neural network fuzzy time series recognize the pattern by another method. Pattern recognition of concentration is important for forecasting the particulate concentration, and therefore it is necessary to use a more advanced tool in predicting the PM_{10} concentration.

The "patterned" and "un-patterned" data were taken as examples in this study. Both the MFT and NFT method can identify the pattern in the concentration variation in the atmosphere. The mean square error of the forecasted results of these two models is better than the linear regression model and the autoregressive model. Since the prediction of PM_{10} also includes the hourly and the daily concentrations, it is suggested that both methods are applied for a more detailed analysis of PM_{10} concentration for both space and time under different resolutions.

By comparing the results, it becomes evident that the proposed NFT method has the smallest mean square error among the seven forecasting methods. The MFT model has the second smallest mean square error. In other words these two methods have the highest accuracy among the traditional fuzzy time series, linear regressive model, and the auto-regressive model. It is therefore recommended to apply both these methods in the prediction.

3 Granular Computing for Environmental Management

The environmental phenomena are very complex and not easy to realize. The first step is to determine the pattern of its variation. Granular computing helps us to establish the pattern recognition technique which is important in environmental management. (Duda et al. 2001), (Schuermann 1996)

In this section, we present an example for in-depth understanding of granular computing.

3.1 Pattern Recognition of Environmental Risks

Chen et al. investigated environmental risk management by multivariable analysis of pattern structure. Their study introduced a new framework for the management methodology of environmental risk by using the concept of pattern recognition with multivariable analysis (PRMA). They then developed an ideal pattern structure relationship for environmental risk management by multivariable analysis. The PRMA uses the pattern structure with the characteristic matrix and estimates the relationship by the multivariable analysis method. The candidate indexes of each pattern can be selected based on previous research. (Chen et al. 2010)

Based on the proposed approach, the link between the events in the environment, the change in the ecosystem, the economic losses, and the response of the measurement system can be evaluated. This system will improve the management of risk response and upgrade the quality of the management system. In addition, the methodology can also be used in the future as a basis for the development of an environmental risk response system.

3.2 Environmental Risk Pattern Structures and How to Manage Them

Environmental disasters also interact with the economic system. For example, an earthquake may result in loss of human life and major economic losses by destroying homes, personal property and the infrastructure in a community. Predicting the environmental damage is a challenge for scientists. As shown in Figure 1, at least four aspects must be considered to precisely describe the risk management of a natural hazard. The first is the characteristic of the natural hazard itself. The second is the change that will occur in the ecosystem as a result of the hazard. The third is the economic loss as a result of the hazard, and finally, the response measures for the natural hazard.

Fig. 1 shows the ideal structure of **PRMA**. In this figure, we consider the subject under the following categories: (1) the natural disaster itself; (2) the change in the ecosystem; (3) the economic loss, (4) the response for risk management. Using this framework, the engineers can better the risks of their project under various levels of risk. The arrows in the figure represent the direction of influence. For example, the pattern of ecological change influences the pattern of economic loss. Also, the pattern of the characteristics of the environmental event will influence the pattern of the economic losses.

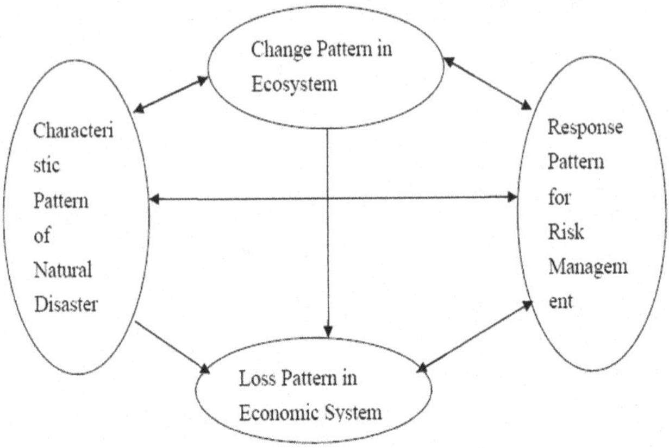

Fig. 1 Pattern structures of environmental risk and its management

Pattern of Characteristic in an Environmental Event

An environmental event is any observable occurrence or extraordinary occurrence of environmental phenomena, including air pollution, water pollution, typhoons, earthquakes, floods, etc. There are many variables in environmental events. For example, the nature of an earthquake could be described by the following factors: strength, location, frequency, time, duration, etc.

Pattern of Change in the Ecosystem

An environmental event may cause a change in the ecosystem. In turn, this change may result in an environmental risk. There are a variety of risk types due to changes in the environment. Other researchers listed the following factors: the use of land and property, social and economic factors, noise and vibration, visual amenities, urban design, traffic, soil and geology, surface water and ground water, flora and fauna, air quality, aboriginal heritage, non-indigenous heritage, hazards, natural resources and waste, etc.

Pattern of Economic Losses

Both the environmental event itself and the resulting change of the ecosystem may cause economic losses. This loss pattern in the economic system includes the following factors: property damage, personal injury and death, agricultural loss, damage to the infra-structure, indirect economic losses, post-disaster reconstruction costs, etc. (Patrick et al. 2007)

Pattern of Response for Risk Management

Response measures are the steps we take in the event of an environmental event. In this study, a proper response pattern for risk management is formed by combining the response measures. In a risk management system the response pattern includes the following factors: reinsurance compensation, super fund, major disaster securities market, social public disclosure, education and training, emergency response, human resources, etc.

3.3 Variable Granulation of the Environmental Impact Assessment

Variable granulation describes a technique for reducing dimensionality, wordiness, and redundancy. It is often used to abstract the data and derive the knowledge from the information. An environmental event may have certain impacts on society. These impacts can be assessed in different ways, using different variables. This procedure is called an" environmental impact assessment". The impact of an environmental event can be represented by X = {xi}, where, xi is a different variable that represents the impact of this event. X is the total impact of this event.

Variable granulation processes the complex information from these variables. In other words, it derives knowledge from the original information. Variable granulation collects entities which originated from environmental factors. We will use the example of fuzzy filtering and image processing for an emission inventory to explain the environmental application of granular computing.

A Fuzzy Representation of Environmental Quality

Let's assume that the environmental quality set is $U = \{ u_1, u_2,\ldots, u_m\}$, where, the terms, $u_1,u_2,\ldots, u_m$ are the value of the environmental factors. The standards for environmental quality assessment are set as $V = \{v_1, v_2\ldots v_m\}$, where the terms, $v_1,v_2\ldots v_m$ are the corresponding set of u_i.

In research on environmental quality, U is a fuzzy vector, and V is a two-dimensional matrix. Given U and V, the fuzzy relationship between the *f*actor domain (pollution factor) and the assessment domain (assessment criteria) can be expressed by the fuzzy relationship matrix R :

$$R = \begin{bmatrix} r_{11} & r_{12} & \cdots & r_{1n} \\ r_{21} & r_{22} & \cdots & r_{2n} \\ \vdots & \cdots & \cdots & \vdots \\ r_{m1} & r_{m2} & \cdots & r_{mn} \end{bmatrix} \quad (2)$$

By definition of the fuzzy relationship, r_{ij} represents the environmental quality of the ith pollutant, and it is also the membership equation of the jth environmental quality. Therefore, the fuzzy relationship matrix R in the first row is $R_i = (r_{i1}, r_{i2} \dots r_{in})$, i = 1, 2... n. This equation is the membership on the subordinate level of the environmental quality standards.

The jth column of the fuzzy relationships represents the membership of each pollutant to the environmental quality standard. It can be written as $R_j = (r_{j1}, r_{j2} \dots r_{jn})$, j= 1, 2... n.

If the fuzzy set on the factors domain U is :

$$A = \frac{a_1}{u_1} + \frac{a_2}{u_2} + \frac{a_3}{u_3} + \cdots + \frac{a_m}{u_m} \tag{3}$$

where, a_i represents the measured influence scale of a single factor u_i, and i_t can be regarded as the weight coefficient of the ith pollutant to the environmental quality, then the fuzzy subset in the assessment domain can be expressed as :

$$B = \frac{b_1}{v_1} + \frac{b_2}{v_2} + \frac{b_3}{v_3} + \cdots + \frac{b_m}{v_m} \tag{4}$$

where the term *bj* represents the membership degree of v_j to the fuzzy subset of the comprehensive assessment. This indicates the degree of membership of the jth grade index of the environmental quality standards in the integrated environmental classification index.

If the fuzzy vector A and the fuzzy relation matrix R are known, then the fuzzy sets of the comprehensive assessment can be expressed as $B = A.\ R$.

Typical Model for the Calculation of the Environmental Impact

Based on the computational methods of the fuzzy sets, there are four models which can be used to calculate the value of b_j.

Model 1 is expressed as $M_1(\cap, \cup)$. as along with an underlying detailed formula

$$\begin{aligned} b_j &= \bigcup_{i=1}^{m} (a_i \wedge r_{ij}) \\ &= Max\{Min(a_1, r_{1j}), Min(a_2, r_{2j}), \cdots, Min(a_{m1}, r_{mj})\} \\ & \qquad j = 1,2,\cdots,n \end{aligned} \tag{5}$$

In the above formula, $\cup$ and $\cap$ are the intersection and the union of the fuzzy set, respectively. The $M_1(\cap, \cup)$ model is calculated according to max-min rule.

Model 2 is $M_2(\bullet, \cup)$. as along with a detailed description

$$\begin{aligned} b_j &= \bigcup_{i=1}^{m} (a_i \bullet r_{ij}) \\ &= Max\{a_1 \bullet r_{1j}, a_2 \bullet r_{2j}, \cdots, a_{m1} \bullet r_{mj}\} \\ & \qquad j = 1,2,\cdots,n \end{aligned} \tag{6}$$

In the above formula, "·" represents an algebraic product. Models M_1 and M_2 highlight the two main types of factors. Therefore, in the present study we chose the maximum membership value of the jth environmental standard among all the factors.

Model 3 is $M_3(\bullet, \oplus)$. and comes with the formula

$$b_j = \sum_{i=1}^{m} a_i \bullet r_{ij}$$

$$= Min\left\{1, \sum_{i=1}^{m} a_i \bullet r_{ij}\right\} \quad (7)$$

$$j = 1,2,\cdots,n$$

This is the weighted model, where a_i can be sought as the weighting factor of certain parameter, its maximum membership value is 1.

Model 4 of the form $M_4(\cap, \oplus)$. is governed by the expression

$$b_j = \sum_{i=1}^{m} a_i \cap r_{ij}$$

$$= Min\left\{1, \sum_{i=1}^{m} (a_i, r_{ij})\right\} \quad (8)$$

$$j = 1,2,\cdots,n$$

Since model 4 follows the min-min principle, it seldom applies to environmental quality assessment.

3.4 Fuzzy Filtering and Image Processing of the Pollution Inventory

The classification of natural hazards is important in managing disaster risks. However, the complexity of natural phenomena makes it hard to identify the pattern of the hazard. This section presents an approach to extract the interpretable rules of a natural hazard from the information obtained. First, we generate the rule for data clustering according to Bayes' theorem. Then the pattern was determined. Then we calculated the optimized category for the fuzzy system and transferred it into the neural network for refining the knowledge obtained. The optimized fuzzy system then extracted the understandable knowledge from the measured results of a natural disaster in for example the ocean, such as a typhoon or a tsunami, etc. Different neural network methods can be used in the algorithm.

The data clustering can be classified using the Bayes' classifier. This classifier is based on Bayes' theorem to determine which pattern should be identified to obtain the desired experimental results. Any error can be minimized through statistical probability analysis.

Pattern C has a characteristic value x, with P(x) is being the probability of this value. P(c) is the prior probability value of pattern C calculated by the random number generator. Thus, based on the conditional probability theory, Bayes' theorem can be written as

$$P(c/x) = \frac{P(c \cap x)}{P(x)} = \frac{P(c)P(x/c)}{P(x)} \tag{9}$$

where P(c/x) denotes the probability of the characteristic value x falling into pattern C. P(x/c) denotes the probability of x among the data set from pattern C.

If we assume that there are k patterns in the experimental results $\{c_1, c_2 \ldots c_k\}$, and that all these data are independent, then we get the following equation.

$$P(c/x) = P(x \cap c_1) + P(x \cap c_2) + \cdots + P(x \cap c_n) = P(c_1)P(x/c_1) + P(c_2)P(x/c_2) + \cdots + P(c_n)P(x/c_n) \tag{10}$$

Figure 2 shows the relation of x and c in the experimental results.

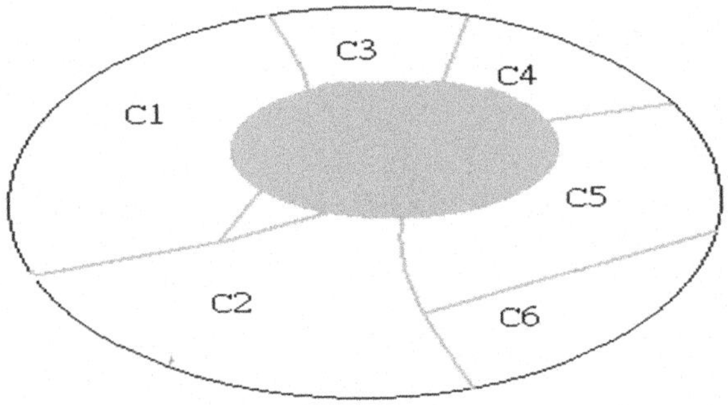

Fig. 2 The data clustering and pattern recognition by Bayes' theorem.

The above equation can be written as

$$P(c/x) = \frac{P(c)P(x/c)}{P(x)} \tag{11}$$

The following equation can be used for k patterns:

$$P(c/x) = \frac{P(c)P(x/c)}{P(c_1)P(x/c_1) + P(c_2)P(x/c_2) + \cdots + P(c_n)P(x/c_n)} \quad (12)$$

In determining which pattern the characteristic value x belong to, we simply calculate the likelihood ratio R between C_i and C_j

$$\begin{aligned} R &= \frac{P(c_i/x)}{P(c_j/x)} \\ &= \frac{P(c_i)P(x/c_i)}{P(c_j)P(x/c_j)} \end{aligned} \quad (13)$$

If R>1, the x values tends to classify to c_i, otherwise, if R<1, c represents the value x is in c_j.

In the real data calculated from the experimental results, P (c_i) is the probability for the ith pattern, and $P(x/c_i)$ is the probability density function obtained from the ith pattern.

If there is more than one characteristic value, say $(x_1,x_2 \ldots x_k)$, then the conditional probability can be represented by

$$\begin{aligned} &P(x_1, x_2, \cdots x_k / c_i) \\ &= \frac{P(c_i)P(x_1/c_i)P(x_2/c_i)\cdots P(x_d/c_i)}{P(x_1/c_i)P(x_2/c_i)\cdots P(x_k/c_i)} \end{aligned} \quad (14)$$

Bayes' theorem with pattern k and characteristic value reads as follows

$$\begin{aligned} &P(c_i / x_1, x_2, \cdots x_d) \\ &= \frac{P(c_i)P(x_1/c_i)P(x_2/c_i)\cdots P(x_d/c_i)}{\sum_{i=1}^{k} P(c_i)P(x_1/c_i)P(x_2/c_i)\cdots P(x_d/c_i)} \end{aligned} \quad (15)$$

3.5 Choosing the Best Pattern by Fuzzy Optimization

Experimental data within the pattern are categorized using the optimization theory for the fuzzy category model. A fuzzy set is a set containing the element that has various degrees of membership in the set. The notation for a fuzzy set when the universe of discourse, X, is finite and can be written as:

$$\begin{aligned} &A \\ &= \mu_A(x_i)/x_i \\ &= \mu_A(x_1)/x_1 + \mu_A(x_2)/x_2 + \cdots + \mu_A(x_n)/x_n \end{aligned} \quad (16)$$

Where μ_A denotes the membership function of fuzzy set A, μ_A: X → [0, 1]. $\mu_A(x_i)$ denotes the grade of membership of x_i belonging to fuzzy set A, and $\mu_A(x_i) \in [0, 1]$.

The standard fuzzy set operations include union, intersection, and complement, and are defined as $\mu_{A\cup B}$, $\mu_{A\cap B}$ and $\mu_{\bar{A}}(x)$ on X, respectively. The simplified fuzzy number arithmetic operation between the triangular fuzzy number A_1 and A_2 are: addition, subtraction, multiplication, and division.

We will track the knowledge of the natural phenomenon, such as typhoon, tsunami, or earthquakes, etc. The characteristic of this phenomenon can be extracted by the best fit pattern and can be determined by the following optimization theory.

If there are n observed values in the observed domain $\{C_i\}$, with i = 1, 2, 3…n, and a calculated value, $\{ \beta_i \}$, i = 1, 2, 3… n, then there is an error △E which can be defined as

$$\Delta E = \sum_{i=1}^{n} \left(C_i - \beta_i\right)^2 \tag{17}$$

Where, C_i is the measured value in i^{th} point, and β_i is the calculated value in the ith point. The optimum solution for the pattern matrix is obtained when the value of △E is a minimum, that is, △E = 0. The above procedure will ensure that the "best fit pattern" has the minimum error for the real environment.

The procedure to determine the best fit pattern is as follows.

- Define the domain area.
- Define each pattern as pattern (i, j…, z).
- Assign each pattern a value. from the lowest to the highest.
- Calculate the value at each point caused by each assumption.
- Calculate the value of △E for different pattern sets.
- Compare the value of △E and choose the pattern value.
- The pattern with the minimum △E value is the best solution.

4 System Granulation of an Environmental and Economic Evaluation

A system refers to a group of related parts that work together. The economy is a system. The environment is too. Determining the relationship between an environmental event and the economic system is called an "environmental and economic evaluation".

In this paper we used two examples to explain the system granulation of an environmental and economic evaluation. The first is a fuzzy set for the risk evaluation in case of an environmental disaster. The second is the multi-objective design of a granular hierarchy model for developing a strategy of air quality management.

4.1 Fuzzy Set for the Risk Evaluation in Case of an Environmental Disaster

Owing to the present climate change and global warming, managing environmental risk has become a matter of concern. The environmental risk as a result of a natural hazard is very complex, making it hard to express in a simple formula. The other issue is that the vagueness and the lack of precise data regarding the disaster makes evaluating these hazards very difficult. This makes it difficult for the manager to make a decisive judgment. (Coppendale 1995), (Cooper 2003), (Choi et al. 2010)

The development of granular computing helps us to solve this problem. In the present study we use an example to investigate natural hazard risk management using the fuzzy multi-criteria evaluation method for pattern recognition. The fuzzy representation for the evaluation of the risk due to natural hazards in different areas was used to convert the observed data into a fuzzy number. (Carr and Tah 2001), (Chapman and Ward 2004), (Ahmad et al. 2010)

Fuzzy Representation of a Natural Hazard

The damage caused by a natural hazard can be represented by a fuzzy set as $X = \{x_1, x_2 \ldots x_n\}$, where, x_i represents the domain area for the evaluation. $x_i = x_1, x_2 \ldots x_n$, and there are j (j = 1, 2... n) areas to be evaluated. The damage of natural hazards includes many different kinds of loss including injury and loss of life, damaged homes and damaged infra-structure, etc. Thus, we defined a subset to represent all the different kinds of losses.

The value of the loss is a characteristic value that represents the risk level of the natural hazard. Let's assume that there is 1th item of loss to be calculated in the jth area, and that the subset can be written as follows.

$$X_j = \{x_{1j}, x_{2j}, \cdots, x_{nj}\}^T \tag{18}$$

The above equation gives us the matrix, X_{ln}, which represents the loss in the jth area.

$$X_{ln} = \begin{bmatrix} x_{11} & x_{12} & \cdots & \cdots & x_{1n} \\ x_{21} & x_{22} & \cdots & \cdots & x_{2n} \\ \cdot & & \cdots & \cdots & \cdot \\ \cdot & \cdot & \cdots & \cdots & \cdot \\ x_{l1} & x_{l2} & \cdots & \cdots & x_{ln} \end{bmatrix} \tag{19}$$

Where, x_{ij} (i = 1, 2…l ; j = 1, 2…n) represents the ith item of loss in the jth area for evaluation.

Triangular Fuzzy Set for Environmental Risk Items

A natural hazard is a very complicated phenomenon. Owing to the limit of the present measurement technology, the lack of information and the subjective judgment by the observer, there are many uncertainties in the observed data. Therefore, we have to use fuzzy numbers to describe the real phenomenon.

For the ith damage of the jth area, the fuzzy equation can be written as $F_{ij} = ({}^{L}F_{ij}, {}^{M}F_{ij}, {}^{R}F_{ij})$, where F_{ij} is the characteristic value of X_{ij}, ${}^{L}F_{ij}$ is the lowest value of F_{ij}, ${}^{R}F_{ij}$ is the maximum value of F_{ij}, ${}^{M}F_{ij}$ is the most probable value of F_{ij}, and ${}^{L}F_{ij} < {}^{M}F_{ij} < {}^{R}F_{ij}$.

The value of ${}^{L}F_{ij}$, ${}^{M}F_{ij}$, ${}^{R}F_{ij}$ is explained by Figure 3.

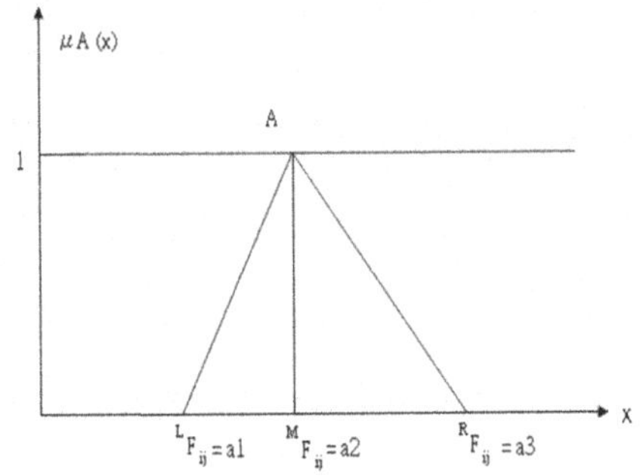

Fig. 3 Membership function of the triangular fuzzy numbers (TFN)

If X is the set of all the objects x, and if there is a fuzzy subset A for X represented by (x, A(x)), x ∈ X, and the range of X is within [0, 1], then set A can be written as: $A = \{(x, \mu_A(x))x \in X\}$, where A(x) is the membership function of A, and its value is in the range of [0, 1]. When the value approaches 1, it means that the membership is higher, and when the value is near to zero, it means the membership is low.

We used a triangular fuzzy number (TFN) to represent the membership function $\mu_A(x)$ represented by (a_1, a_2, a_3). Thus the characteristic number of the ith damage of the jth area is $a_1 = {}^{L}F_{ij}$, $a_2 = {}^{M}F_{ij}$, $a_3 = {}^{R}F_{ij}$. The membership function of TFN then becomes

$$\mu_A(x) = \begin{cases} 0 \quad , & x \langle a_1 \\ \dfrac{x - a_1}{a_2 - a_1} \quad , & a_1 \le x \le a_2 \\ \dfrac{a_3 - x}{a_3 - a_2} \quad , & a_2 \le x \le a_3 \\ 0 \quad , & x \rangle a_3 \end{cases} \tag{20}$$

There are many different kinds of observed data for a natural hazard. This data can be a definite value from an instrument, or it may be a vague and imprecise linguistic description. In this study we used different methods to calculate the observed data, as described below.

If the original results are data within an interval(or regarded as an interval), then we can define the right hand side of this interval as ${}^{R}F_{ij}$, and the left hand side of this interval as ${}^{L}F_{ij}$. The ${}^{M}F_{ij}$ value is calculated by the following equation, ${}^{M}F_{ij} = ({}^{L}F_{ij} + {}^{R}F_{ij})/2$.

Linguistic Variable in Description of the Degree of Damage from a Natural Hazard

In Table 5, type A is the linguistic variable describing the degree of damage from a natural hazard, for example a typhoon or an earthquake. Type B is the qualitative description measured by remote sensing. In this paper we defined its degree of into five different grades.

Table 5 Linguistic and qualitative description of the observations of the natural hazard

-	Type A , Linguistic variable	Type B , Qualitative description
1	very safe	very clear
2	safe	clear
3	medium	medium
4	serious	unclear
5	very serious	very unclear
note	Damage by a typhoon	The results of remote sensors

For the data shown in Table 5, we defined the linguistic fuzzy number as: ${}^{linguistic}F_{ij} = (q\text{-}2, q\text{-}1, q)$, where the linguistic F_{ij} is the fuzzy number with r grades of degree. q-2 is the lowest value of the observation, q-1 is the most probable value of the observation, and q is the highest value of the observation.

Method for Validating the Results of an Environmental Disaster

For different kinds of damage we must normalize the value into the interval of [0, 1]. The process for normalization is realized in the form

$$normalized\,F_{ij} = \left\{ \frac{\left({}^{L}F_{ij} - {}^{L}F_{imin}\right)}{\left({}^{R}F_{imax} - {}^{L}F_{imin}\right)}, \frac{\left({}^{M}F_{ij} - {}^{M}F_{imin}\right)}{\left({}^{R}F_{imax} - {}^{L}F_{imin}\right)}, \frac{\left({}^{R}F_{ij} - {}^{R}F_{imin}\right)}{\left({}^{R}F_{imax} - {}^{L}F_{imin}\right)}, \right\} \forall i,j. \quad (21)$$

Where ${}^{R}F_{imax}$ and ${}^{L}F_{imin}$ represent the maximum value of the right hand side and the minimum value of the left hand side for all i^{th} items. This can be represented by the following two relationships.

$$^{R}F_{imax} = MAX\{^{R}F_{ij}\}, \forall i \quad (22)$$

$$^{L}F_{imax} = MIN\{^{L}F_{ij}\}, \forall i \quad (23)$$

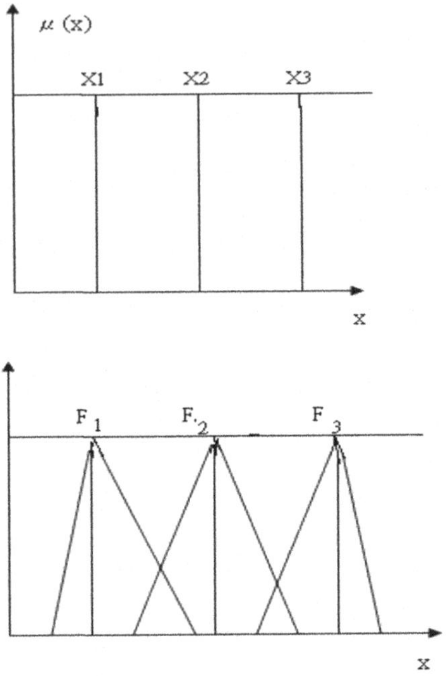

Fig. 4 Converting the original results into a normalized fuzzy number

The weighting factor of each item can be evaluated by the expert opinion method. Thus, we have weighting factor W_i for different items. The overall fuzzy number for the natural hazard is estimated as follows.

$$^{aggregate}F_{ij} = \sum\left(W_{ij} \otimes X_{ij}\right), \qquad \forall j \quad (24)$$

Where $^{aggregate}F_{ij}$ is the summation of the fuzzy number, and $\otimes$ denotes the operation of the fuzzy number. The operation on the two fuzzy numbers $A_1 = (a_{11}, a_{12}, a_{13})$ and $A_2 = (a_{21}, a_{22}, a_{23})$ is defined as:

$$B_1 = A_1 \oplus A_2 = [(a_{11}+a_{21}),\ (a_{12}+a_{22})\ (a_{13}+a_{23})\ ] \quad (25)$$

$$B_2 = A_1 \otimes A_2 = [(a_{11}\times a_{21}),\ (a_{12}\times a_{22})\ (a_{13}\times a_{23})\ ] \quad (26)$$

In determining the membership function, the tendency of the decision maker must be considered. There are two kinds of decision makers, one is the optimist and the other one is the pessimist. The decision F_{ij} can be regarded as ${}^{decision}F_{ij} = ({}^{L}F_{ij}, {}^{M}F_{ij}, {}^{R}F_{ij})$, where, ${}^{L}FS_j$ is the left hand limit of FS_j, ${}^{R}FS_j$ is the right hand side limit of FS_j, and ${}^{M}FS_j$ is the most probable value of FS_j.

FS_j can be represented by a fuzzy set T with membership function $\mu_T(x_j)$, where $\mu_T(x)$ is represented by the following fuzzy set, $\mu_T(x_j) = \lambda\mu_T(x_j) + (1-\lambda)\mu_T(x_j)$, and where λ is the weighting factor for the decision maker. The value of λ ranges from 0 to 1. The decision maker with $\lambda = 1$ is the most optimistic, and $\lambda = 0$ is the most pessimistic.

The fuzzy set for the most optimistic and the most pessimistic decision maker can be defined as

$$^{optimic}F_{ij} = \{x_j | \mu_T(x_j)\}, \forall j \tag{27}$$

$$\mu_T(x_j) = highest\{^{decision}F_{ij} \cap H_{\min}\} \tag{28}$$

Where, $\mu_T(x_j)$ represents the membership of the maximum damage for the most optimistic decision maker.

For the most pessimistic decision maker, we have the fuzzy set

$$^{perssimic}F_{ij} = \{x_j | \mu_T(x_j)\}, \forall j \tag{29}$$

$$\mu_T(x_j) = 1 - highest\{^{decision}F_{ij} \cap H_{\min}\} \tag{30}$$

The relationships between each pattern are described by the fuzzy membership function. The aggregate risks of a natural hazard, such as a typhoon or earthquake is then calculated by the fuzzy multiple or additional operation. In this framework, the degree of risk of a natural hazard can be estimated more comprehensively, and the response measures can be chosen more appropriately.

4.2 A Multi-objective Design for a Granular Hierarchy Model for an Air Quality Management Strategy

Air quality management is important in modern society. Previous management strategy focused on achieving air quality standards. However, the phenomenon of air pollution is extremely complicated, and is affected by many external factors at the same time. To achieve the required air quality standards, these external factors must be taken into account. Therefore, there is a need to develop a tool that can consider the various external factors influencing air quality. (Li et al. 2008), (Byun et al. 2003) Therefore this study presents a knowledge-based air quality management system based on the fuzzy logic principle. (Cai and Chen 2009)

The external environmental costs caused by air pollution were studied using the fuzzy theory. The so-called "fuzzy decision index (FDI)" was derived and applied. An integrated score of multiple assessments was derived by fuzzy logic. The knowledge database established in this method include: emission sources, meteorological data, topographical data, and population density distribution. The external costs of the air pollutants calculated by this method can provide the government with a good reference on the air pollution decision making, such as determining the air pollution control fee.

The first step is setting up the Fuzzy set. The so-called factors set are a collection of various factors for evaluation. This can be represented as $U = [u_1, u_2 \ldots u_n]$, which represents the set of the factors to be considered in the external costs estimation of air pollution.

The second step is to establish the factor's weighting set. Because various factors possess a different degree of importance, it is essential that each factor u_i is given a different weight a_i according to its degree of importance.

Determine the Weighting Factor of the Fuzzy Strategy Sets

The first step in determining the weighting factors of the fuzzy strategy sets is to identify the aim for assessing the factors for set U. The second step is to determine the matrix.

Let A represents the target, and u_i represents assessment factors, $u_i \in U_i$, i = 1, 2, 3…n, where u_{ij} expresses the relative importance of the numerical u_i to u_j, and j = 1, 2, 3 ... n. The value of u_{ij} is according to the rules as follows.

According to the description above, we obtain matrix P, which is known as the determinant matrix of A ~ U.

$$P = \begin{array}{c} \begin{matrix} u_1 & u_2 & \cdots & u_n \end{matrix} \\ \begin{bmatrix} u_{11} & u_{12} & \cdots & u_{1n} \\ u_{21} & u_{22} & \cdots & u_{2n} \\ \cdot & \cdot & \cdot & \cdot \\ \cdot & \cdot & \cdot & \cdot \\ \cdot & \cdot & \cdot & \cdot \\ u_{n1} & u_{n2} & \cdots & u_{nn} \end{bmatrix} \end{array} \begin{matrix} u_1 \\ u_2 \\ \cdot \\ \cdot \\ u_n \end{matrix} \tag{31}$$

From matrix A~U we calculate the maximum corresponding characteristic vector. This vector is the sequence of importance of each assessment factor, or the distribution of the weighting factors.

The determinant matrix A ~ U can be calculated by the square root method. Product W_i of each element in the determinant matrix is calculated by the following equation.

$$\text{Mi} = \prod_{j=1}^{n} u_{ij}, (i, j = 1, 2, \cdots), n) \tag{32}$$

Then we calculate M_i's n^{th} power root, W_i

$$W_i = \sqrt[n]{M_i} \tag{33}$$

For vector W, W = [$w_1, w_2 \ldots w_n$]^T. After normalization, its weighting become

$$W_i = W_i / (\sum_{j=1}^{n} W_i) \tag{34}$$

Then, W = [$w_1, w_2 \ldots w_n$]^T , where W is the characteristic vector. The maximum characteristic root of the matrix is then calculated by the following equation

$$\begin{aligned} \lambda_{\max} &= \sum_{i=1}^{n} \frac{(pw)_i}{nW_i} \\ &= \frac{1}{n} \sum_{i=1}^{n} \frac{(pw)_i}{W_i} \end{aligned} \tag{35}$$

$$PW = \begin{matrix} (pw)_1 \\ (pw)_2 \\ \cdot \\ \cdot \\ \cdot \\ (pw)_n \end{matrix} = \begin{bmatrix} u_{11} & u_{12} & \cdot\ \cdot\ \cdot\ \cdot\ \cdot & u_{1n} \\ u_{21} & u_{22} & \cdot & u_{2n} \\ \cdot & \cdot & \cdot & \\ u_{n1} & u_{n2} & \cdot & u_{nn} \end{bmatrix} \bullet \begin{bmatrix} w_1 \\ w_2 \\ \cdot \\ w_n \end{bmatrix} \tag{36}$$

where $(pw)_i$ represents the ith component of vector pw.

Parameter Optimization and Validation of Results

The characteristic vector is the weighting factor. In order to minimize any errors when selecting the appropriate parameters it is necessary to reexamine the consistency of the judge matrix. The equation for the consistency test is CR = CI /RI, where CR is the random consistency ratio of the determinant matrix. CI is the consistency index of the determinant matrix, and can be obtained by the following equation.

$$CR = \frac{1}{n-1} (\lambda_{\max} - n), \tag{37}$$

When CR has good consistency, and the weighting factor is reasonable, then the results are acceptable. The weighting factor set Ã is a subset of set U, which can be represented as Ã = {a1, a_2, $a_3 \ldots a_i \ldots a_n$}, and

$$\sum_{i=1}^{n} a_i = 1 \tag{38}$$

The Membership Function and the Triangular Fuzzy Number Representation

The formula of the S-function can be used to represent the relationship of the variables.

$$U_{Ri}(X_j, Y_k) = \check{S}(a;b,c,) = \begin{cases} 0, & x \le a \\ 2\left[\dfrac{x-a}{c-a}\right]^2 & a \le x \le b \\ 1-2\left[\dfrac{x-c}{c-a}\right]^2 & b \le x \le c \\ 1, & x \ge c \end{cases}$$

$$U_{Ri}(x_i, y_i) \quad 1 \leqq I \leqq n \tag{39}$$

$U_{Ri}(X_j, Y_k)$ represents the membership degree of (x_i, y_i), and

$$R_i = \int_{XxY} U_R[\ (x_i, y_i)/(x_i, y_i)\ ] \tag{40}$$

The relationship matrix of X to Y reads as follows.

$$Ri = \begin{array}{c} \quad y_1 \quad y_2 \quad y_3 \quad y_4 \quad y_5 \quad y_6 \quad y_7 \quad y_8 \quad y_9 \\ \begin{bmatrix} x_1\ U_R(x_1,y_1) & \cdots\cdots\cdots\cdots\cdots\cdots\cdots & U_R(x_1,y_9) \\ x_2\ U_R(x_2,y_2) & & U_R(x_2,y_9) \\ x_3\ U_R(x_3,y_3) & \ddots & U_R(x_3,y_9) \\ x_4\ U_R(x_4,y_4) & & U_R(x_4,y_9) \\ x_5\ U_R(x_5,y_5) & \cdots\cdots\cdots\cdots\cdots\cdots\cdots & U_R(x_5,y_9) \end{bmatrix} \end{array} \tag{41}$$

In the above equations, R_i represents the parameters for the different time intervals. Based on the above formula $Y_k = \sum_{k=(1\sim9)}\sum_{j=(1\sim5)} U_R(X_j, Y_k)$, $1 \leq k \leq 9$.

Processing the α-cut is important in the fuzzy operation. Let $R = (r_{ij})_{m \times n}$,

$$R_\alpha = \{(x_j, y_k) | (x_j, y_k) \geq \alpha, \forall x \in X..and..y \in Y\} \tag{42}$$

$$R_\alpha = (r_{ij})_{m \times n},$$

$$= \begin{cases} 1, & r_{jk} \geq \alpha \\ 0, & r_{jk} \langle \alpha \end{cases} \tag{43}$$

We take the α-cut as {0.2, 0.4} and {0.6, 0.8}, and make the divided point according to the original membership degree. We then substitute the original formula. From the R_i fuzzy related matrix, we take the four values to calculate the (α-cut) calculation.

According to the above equation, this then becomes $R_\alpha = \{0.2\ R_{0.2} \cup 0.4R_{0.4} \cup 0.6R_{0.6} \cup 0.8R_{0.8}\}$, and, $U_\alpha(X_j, Y_k)$ the related matrix is represented as follows.

$$R_\alpha = \int_{XxY} [U_\alpha(X_j, Y_k)/(X_j, Y_k)] \tag{44}$$

$U_{Ri}(X_j,Y_k)$ represents the membership degree of (X_j,Y_k). The X to Y relationship matrix is as follows.

$$
\begin{array}{c} \\ \\ R_\alpha = \begin{array}{c} x_1 \\ x_2 \\ x_3 \\ x_4 \\ x_5 \end{array} \end{array}
\begin{array}{c} y_1 \quad y_2 \quad y_3 \quad y_4 \quad y_5 \quad y_6 \quad y_7 \quad y_8 \quad y_9 \\ \\
\begin{bmatrix} U_\alpha(x_1,y_1) & \cdots\cdots\cdots\cdots\cdots\cdots\cdots & U_\alpha(x_1,y_9) \\ U_\alpha(x_2,y_2) & & U_\alpha(x_2,y_9) \\ U_\alpha(x_3,y_3) & \ddots & U_\alpha(x_3,y_9) \\ U_\beta(x_4,y_4) & & U_\alpha(x_4,y_9) \\ U_\alpha(x_5,y_5) & \cdots\cdots\cdots\cdots\cdots\cdots\cdots & U_\alpha(x_5,y_9) \end{bmatrix} \end{array} \tag{45}
$$

If we calculate $Y_{k\alpha} = \sum_{k=(1\sim9)}\sum_{j=(1\sim5)} U_{R\alpha}(X_j, Y_k)$, $1 \le k\alpha \le 9$, using the above formula, then the values$(Y_{1_\alpha} \sim Y_{9_\alpha})$ can be obtained under the constant Ri_α.

5 Conclusions

This paper reported a comprehensive approach based on granular computing to describe the soft computing technique applied in the environmental field. In particular, three main areas were described, value granulation of the environmental phenomenon, granular computing for environmental management, and system granulation for environmental economic evaluation. These descriptions bring a rough estimation, but provide very useful information for environmental management. For example, to establish an emergency response system in case of an environmental disaster, to select the best strategy for environmental quality management, and to predict the concentration of a dust storm, among many others.

The results obtained by granular computing are useful in situations where limited information is available to the decision maker. Therefore, the computing technique can provide the decision makers with a systematic thinking and precise solutions for their strategy analysis. Of course the environmental applications of granular computing are not limited to the fields mentioned above. Additional research on this topic will be forthcoming soon.

References

Ahmad, M., Mohammad, S.M., Meysam, M.S.: Project risk identification and analysis based on group decision making methodology in a fuzzy environment. International Journal of Management Science and Engineering Management 5(2), 108–118 (2010)

Byun, D.W., Kim, S.T., Cheng, Kim, F., Cuclis, S., Moon, N.: Information infrastructure for air quality modeling and analysis: Application to the Houston-Galveston ozone non-attainment area. J. Environ. Info. 2(2), 38–57 (2003)

Cai, D.L., Chen, W.K.: Knowledge-based air quality management study by fuzzy logic principle. In: Proceedings of the 8th International Conference on Machine Learning and Cybernetics, Baoding, China, pp. 3064–3069 (2009)

Carr, V., Tah, J.: A fuzzy approach to construction project risk assessment and analysis. Advances in Engineering Software 32, 847–857 (2001)

Chapman, C., Ward, S.: Project risk management: processes, techniques and insights, 2nd edn. John Wiley and Sons Ltd, Chichester (2004)

Chen, S.M.: Forecasting enrollments based on fuzzy time series. Fuzzy Sets and Systems 81(3), 311–319 (1996)

Chen, S.M.: Forecasting enrollments based on high-order fuzzy time series. An International Journal Cybernetics and System 33(1), 1–16 (2002)

Chen, S.M., Hsu, C.C.: A new approach for handling forecasting problem using high-order fuzzy time series. Intelligent Automation and Soft Computing 14(1), 29–43 (2008)

Chen, S.M., Huarng, J.R., Lee, C.H.: Handling forecasting problem using fuzzy time series. Fuzzy Sets and Systems 100(2), 217–229 (1998)

Chen, S.M., Hwang, J.R.: Temperature predicting using fuzzy time series. IEEE Transaction on System, Man, and Cybernetics- Part B: Cybernetics 30(2), 263–275 (2000)

Chen, S., Hsu, C.C.: A new method to forecast enrollments using fuzzy time series. International Journal of Applied Science and Engineering 2(3), 234–244 (2004)

Chen, W.K.: An approach to pattern recognition by fuzzy category and neural network simulation. In: Proceedings of the 9th International Conference on Machine Learning and Cybernetic, Cingdao, China, pp. 3042–3048 (2010)

Chen, W.K., Juang, Y.R., Cai, D.L.: Courseware design and assessment methodology by fuzzy theory - A case study of energy saving course. In: Proceedings of the 8th International Conference on Machine Learning and Cybernetic, Baoding, China, pp. 3042–3048 (2009)

Chen, W.K., Sui, G.J., Tang, D.L., Cai, D.L., Wang, J.S.: Study of environmental risk management by multivariable analysis of pattern structure. In: Proceedings of The Fourth International Conference on Management Science and Engineering Management, Chungli, Taiwan, pp. 144–148 (2010)

Choi, H.G., Ahn, J.O., Jeung, H.S., Kim, J.S.: A framework for managing risks on concurrent engineering basis. International Journal of Management Science and Engineering Management 5(1), 44–52 (2010)

Cooper, L.: A research agenda to reduce risk in new product development through knowledge management: a practitioner perspective. Journal of Engineering and Technology Management 20(1-2), 117–140 (2003)

Coppendale, J.: Manage risk in product and process development and avoid unpleasant surprises. Engineering Management Journal 5(1), 35–38 (1995)

Duda, P.O., Hart, P.E., Stork, D.G.: Pattern classification, 2nd edn. Wiley, New York (2001)

Hsu, C.C., Chen, S.M.: A new method for forecasting enrollments based on high-order fuzzy time series. In: Proceeding of the 2003 Jont Conference on AI Fuzzy System and Grey system, Taipei, Taiwan (2003)

Klir, G., Yuan, B.: Fuzzy sets and fuzzy logic: Theory and Applications. Pearson Education Taiwan Ltd, London (2005)

Li, H.L., Huang, G.H., Zou, Y.: An integrated fuzzy-stochastic modeling approach for assessing health-impact risk from air pollution. Stochastic Environmental Research and Risk Assessment 22(6), 789–803 (2008)

Murray, T.J., Pipino, L.L., Gigch, J.P.: A pilot study of fuzzy set. modification of Delphi. Human Systems Management 5, 76–80 (1985)

Patrick, X., et al.: Understanding the key risks in construction projects in China. International Journal of Project Management 25, 601–614 (2007)

Pedrcyz, W., Gomide, F.: Fuzzy systems engineering, pp. 101–135. John Wiley & Sons, USA (2007)

Saaty, T.L.: The analytic hierarchy process: planning, priority setting, resource allocation. McGraw-Hill, New York (1980)

Schuermann, J.: Pattern recognition. Wiley & Sons, Chichester (1996)

Song, Q., Chissom, B.S.: Fuzzy time series and its models. Fuzzy Sets and Systems 54(3), 269–277 (1993)

Song, Q., Chissom, B.S.: Forecasting enrollments with fuzzy time series – Part I. Fuzzy Sets and Systems 54(1), 1–9 (1993)

Song, Q., Chissom, B.S.: Forecasting enrollments with fuzzy time series Part II. Fuzzy Sets and Systems 62(1), 1–8 (1994)

Zadeh, L.A.: Fuzzy sets. Information and Control (8), 338–353 (1965)

Zadeh, L.A.: Fuzzy logic and approximate reasoning. Synthese (30), 407–428 (1975)

Index

Author Index